Advances in

MAGNETIC RESONANCE

VOLUME 7

Contributors to This Volume

BERNARD R. APPLEMAN
JOHN L. BJORKSTAM
G. L. CLOSS
BENJAMIN P. DAILEY
DANIEL KIVELSON
KENNETH OGAN

Advances in MAGNETIC RESONANCE

EDITED BY

JOHN S. WAUGH

DEPARTMENT OF CHEMISTRY
MASSACHUSETTS INSTITUTE OF TECHNOLOGY
CAMBRIDGE, MASSACHUSETTS

VOLUME 7

1974

ACADEMIC PRESS New York and London

A Subsidiary of Harcourt Brace Jovanovich, Publishers

ACADEMIC PRESS, INC.
111 Fifth Avenue, New York, New York 10003

United Kingdom Edition published by
ACADEMIC PRESS, INC. (LONDON) LTD.
24/28 Oval Road, London NW1

LIBRARY OF CONGRESS CATALOG CARD NUMBER: 65-26774

ISBN 0-12-025507-3

PRINTED IN THE UNITED STATES OF AMERICA

Contents

NMR Studies of Collective Atomic Motion Near Ferroelectric Phase Transitions

John L. Bjorkstam

Spin Relaxation Theory in Terms of Mori's Formalism

Daniel Kivelson and Kenneth Ogan

Chemically Induced Dynamic Nuclear Polarization

G. L. Closs

Magnetic Shielding and Susceptibility Anisotropies

Bernard R. Appleman and Benjamin P. Dailey

Contributors

Numbers in parentheses indicate the pages on which the authors' contributions begin.

BERNARD R. APPLEMAN (231), Department of Chemistry, Columbia University, New York, New York

JOHN L. BJORKSTAM (1), University of Washington, Seattle, Washington

G. L. CLOSS (157), Department of Chemistry, The University of Chicago, Chicago, Illinois

BENJAMIN P. DAILEY (231), Department of Chemistry, Columbia University, New York, New York

DANIEL KIVELSON (71), Department of Chemistry, University of California, Los Angeles, California

KENNETH OGAN* (71), Department of Chemistry, University of California, Los Angeles, California

* Present address: Department of Physiology, Boston University School of Medicine, Boston, and Department of Physics, M.I.T., Cambridge, Massachusetts.

Preface

The present volume continues the eclectic tradition of *Advances in Magnetic Resonance* in combining the outlooks of the physicist and the chemist, which are jointly valuable to the essentially interdisciplinary subject matter of magnetic resonance.

John L. Bjorkstam, a certified physicist, writes from a physicist's point of view on what I hope he will forgive me for calling the essentially chemical subject of collective atomic motions in crystals as studied by NMR. Daniel Kivelson and Kenneth Ogan, who are by conventional definition chemists, expound in a more formally physical–theoretical manner on Mori's approach to irreversible processes as applied to the theory of spin relaxation. The article on chemically induced dynamic nuclear polarization is written by G. L. Closs, who is, at one and the same time, a pure organic chemist of considerable repute and one of the chief innovators in the interpretation of CIDNP.

Finally, Bernard R. Appleman and Benjamin P. Dailey discuss in great and authoritative detail the important anisotropic behavior of nuclear magnetic shielding and electronic magnetic susceptibility in molecules.

JOHN S. WAUGH

Contents of Previous Volumes

Volume 1

Volume 2

Volume 3

Volume 4

Volume 5

Volume 6

NMR Studies of Collective Atomic Motion Near Ferroelectric Phase Transitions

JOHN L. BJORKSTAM

UNIVERSITY OF WASHINGTON, SEATTLE, WASHINGTON

I. Introduction

The purpose of this presentation is to illustrate the utility of NMR as a tool for studying collective atomic motion near phase transitions. Rather than attempt a comprehensive survey, attention will focus upon a particular class of materials, i.e. the KH_2PO_4-type ferroelectrics (and antiferroelectrics). This rather narrow focus is not meant to suggest a unique importance for these materials. Nevertheless, it is true that no other phase transition has been so extensively investigated by NMR. This family of crystals is rich in the variety of phenomena which lend themselves to such studies. The KH_2PO_4 family has been described in terms of several theoretical models so that a substantial theoretical framework is available for comparison with experiment. Finally, the fact that this group has been so extensively investigated by other methods allows a demonstration of the way in which NMR experiments benefit from, and provide a complement to, the other standard tools.

Neutron and light scattering continue to be primary tools for studying wavelength and frequency dependence of critical fluctuations near phase transitions. However, there is a range of large correlation lengths, κ^{-1}, and wavelengths, $\lambda = (2\pi/q)$, near the origin of the (κ, q) plane, inaccessible to such experiments. The limiting factor is instrument resolution. Those fluctuations characterized by correlation times (or reciprocal frequencies) and wavelengths too long to cause a measurable change in the scattered beam can provide a very effective mechanism for changing spin-lattice relaxation time, linewidth, and spectrum at NMR frequencies.

It is customary to refer to that region of the (κ, q) plane for which $q \gg \kappa$ (i.e. wavelength small compared with correlation length) as the critical region, while $q \ll \kappa$ designates the hydrodynamic range. Such a restriction will not be followed in this paper. The term "critical fluctuation" will be used in a rather loose sense to designate any fluctuations associated with the phase transition.

In order to provide a self-contained discussion for NMR specialists unfamiliar with phase transitions, as well as ferroelectricity experts with limited exposure to the NMR method, some sections will seem oversimplified to both groups. In particular the reader reasonably familiar with KH_2PO_4-type ferroelectrics may skip Section II and Appendix B. Sections III,B and III,C will probably be superfluous reading for the NMR specialist. However, the remainder of Section III does provide a summary of the earlier NMR work. It is a necessary background for the more recent results of Sections IV and V.

In the concluding comments brief mention is made concerning similar studies of other phase transitions. No attempt is made to be comprehensive. The intention is simply to broaden slightly the horizon with respect to NMR as a tool for studying the dynamics of phase transitions.

II. KH_2PO_4-Type Crystals: Structure and Properties

KH_2PO_4 (KDP)-type crystals are of a common MH_2RO_4 form, where M can be either ^{39}K, ^{87}Rb, ^{133}Cs, or NH_4, R will be ^{31}P or ^{75}As, and H is either hydrogen or its deuterium isotope. Aside from the designation KDP for KH_2PO_4, KDA is commonly used for KH_2AsO_4, ADP for $NH_4H_2PO_4$, and ADA for $NH_4H_2AsO_4$. We will extend this usage to the ^{87}Rb and ^{133}Cs isomorphs, writing RDP, CDA, etc., and refer to the deuterium isomorphs of all the above as KD*P, KD*A, and so on. An important property of these materials is that the ^{39}K, ^{87}Rb, and ^{133}Cs isomorphs are ferroelectric, while the NH_4 isomorphs are antiferroelectric.

Ferroelectric crystals are a subgroup of the polar crystals. The latter have a structure such that their unit cell has a nonvanishing electric dipole moment, $\int \rho(\mathbf{r})\mathbf{r}\,dV \neq 0$, where $\rho(r)$ is the charge density (both nuclear and electric) at the point r. In ferroelectric crystals the direction of that spontaneous polarization can be reversed by the application of an electric field. In most ferroelectrics (including the KDP group) the polarization itself disappears at a critical (or Curie) temperature, T_C, above which the crystals become nonpolar. Antiferroelectric crystals also have a critical point at which structural changes are observed, but no net spontaneous polarization appears. Instead, they have intertwined sublattices with equal but opposite polarization.

A full discussion of the various aspects of ferro- and antiferroelectricity in KDP-type crystals may be found in the books on ferroelectricity by Jona and Shirane[1] and Fatuzzo and Merz[2] or in the book-length review by Känzig.[3] Results relevant to the NMR work will be summarized in this chapter.

A. The Ferroelectric Crystals

1. *Structure*

A complete structure determination, both below and above the Curie temperature, has so far been carried out only for KDP.[4,5] For KDP, $T_C = 122°K$. A c-axis projection of the structure in Fig. 1 shows a network of PO_4 tetrahedra linked by hydrogen bonds. The c coordinates give ^{31}P locations along the c-axis in fractions of a unit cell. The PO_4 groups are symmetric with respect to the c-axis, with two oxygens positioned $\simeq 1/8$ above the phosphorus and the other two $\simeq 1/8$ below. Hydrogen bonds link "lower" oxygens of one PO_4 group with "upper" oxygens of an adjacent group and are consequently roughly parallel to the plane of projection. The ^{39}K atoms (not shown in Fig. 1) are midway between PO_4 groups along the c direction.

In the paraelectric phase (i.e. for $T > T_C$) the unit cell is tetragonal, and PO_4 groups are nearly regular tetrahedra. In neutron diffraction experiments, hydrogens show as an elongated distribution, centered in the bond.[5] The ferroelectric phase ($T < T_C$) unit cell is orthorhombic with

[1] F. Jona and G. Shirane, "Ferroelectric Crystals." Pergamon, Oxford, 1962.

[2] E. Fatuzzo and W. J. Merz, "Ferroelectricity." North-Holland Publ., Amsterdam, 1967.

[3] W. Känzig, *Solid State Phys.* **4**, 1–197 (1957).

[4] B. Frazer and R. Pepinsky, *Acta Crystallogr.* **6**, 273 (1953).

[5] G. Bacon and R. Pease, *Proc. Roy. Soc., Ser. A* **220**, 397 (1953); **230**, 359 (1955).

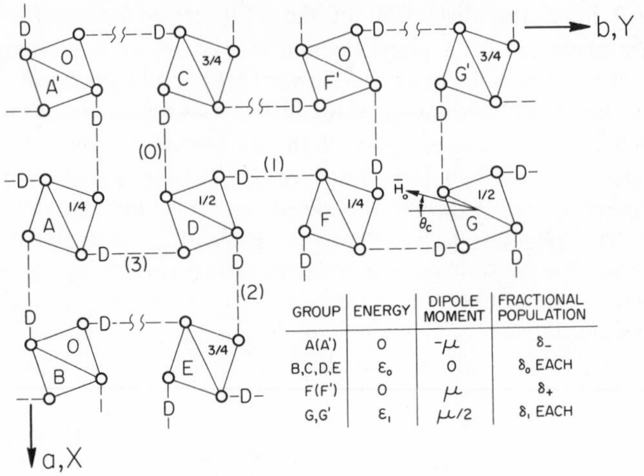

GROUP	ENERGY	DIPOLE MOMENT	FRACTIONAL POPULATION
A(A')	0	$-\mu$	δ_-
B,C,D,E	ε_0	0	δ_0 EACH
F(F')	0	μ	δ_+
G,G'	ε_1	$\mu/2$	δ_1 EACH

FIG. 1. *c*-Axis projection of KDP, showing the six Slater configurations (A–F), as well as the Takagi groups (G, G′).

distorted PO_4 tetrahedra, and hydrogen peaks are shifted to one side of the bond.[5] Hydrogens may be moved across the bond by polarization reversal. Hydrogen arrangement is such that the crystal is divided into regions consisting of H_2PO_4 groups with hydrogens close to either the upper or lower O–O bonds (such as groups A, A′ or F, F′ of Fig. 1). These regions constitute ferroelectric domains with, respectively, negative and positive *c*-axis polarization.[1] H_2PO_4 configurations such as A and F will therefore be referred to as "polar groups." The configurations labeled B through E in Fig. 1 are called "nonpolar groups" since they have no dipole moment along the polarization axis. The orthorhombic unit cell and the most simple tetragonal unit cell (the so-called West cell[6]) have their *c*-axes in common. The orthorhombic *a*- and *b*-axes correspond to the base diagonals of the tetragonal cell. A base diagonal which in various regions of the crystal becomes the orthorhombic *a*-axis will in other regions become the orthorhombic *b*-axis. The result is a twinning of the crystal below T_C, in addition to the ferroelectric domain structure. It is, nevertheless, standard practice to relate experimental data, below as well as above T_C, to the tetragonal axes. The lattice parameters are then[7]: $a_0 = 7.426$ and $c_0 = 6.930$ right above T_C, and $a_0 = 7.430$, $c_0 = 6.934$ for the "pseudo-tetragonal" cell just below T_C. (Values are in Å.)

[6] J. West, *Z. Kristallogr., Kristallgeometrie, Kristallphys., Kristallchem.* **74**, 306 (1930); Ref. 3, p. 149.

[7] W. Cook, *J. Appl. Phys.* **38**, 1637 (1967).

Figure 2 illustrates distortion of the PO_4 group below T_C. The PO_4 groups are projected on a plane defined by the crystalline c-axis and the direction which bisects the upper and lower O–O bond directions, a direction which will henceforth be designated as the O–O bisector. The notation O(H) is used for the oxygen atoms with nearby hydrogens. In addition to the distortion, there is a 1.1° rotation of the PO_4 group about the c-axis. Interatomic distances in the PO_4 tetrahedron above and below T_C are given in Fig. 2. The largest relative change is in position of the ^{31}P atom, as evidenced by the large change in P–O bond lengths. Both the ^{31}P atom

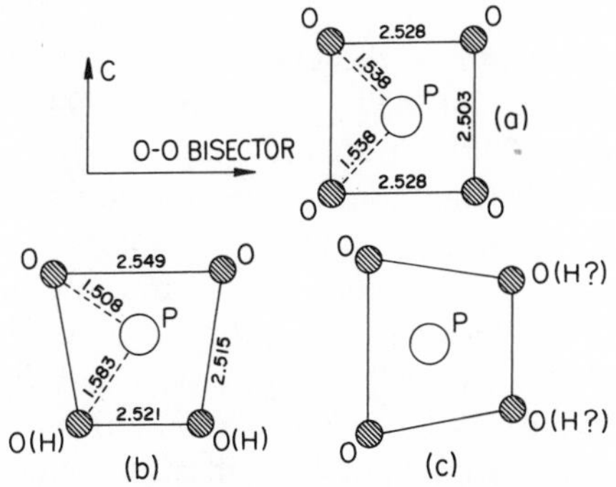

FIG. 2. Structure of the PO_4 tetrahedron: (a) KDP and ADP above T_C, (b) KDP below T_C, (c) ADP below T_C. Atomic distances are in Å.

and ^{39}K across the (H)O–O(H) bond from ^{31}P move away from the (H)O–O(H) bond upon going through the transition. The measured polarization can be accounted for on the basis of a P^{5+}, O^{2-}, K^+ ionic system and the observed displacements.[1] Below T_C, the equilibrium positions of the hydrogen are at 1.05 Å from either end of the hydrogen bond, while at the transition the length of the bond changes from 2.475 Å (132°K) to 2.486 Å (77°K).

As already mentioned, such detailed information is available for KDP only. For KDA, the positions of heavy atoms (i.e. all but the hydrogens) are known in the paraelectric phase,[8] while for the other compounds only the unit cell dimensions are known. In paraelectric KDA, the AsO_4 structure is symmetric with respect to the ^{75}As atom, but not as regular a tetrahedron as is the PO_4 group in KDP. A comparison of the relevant, room temperature, structural information in KDA and KDP is presented in Table I. The ^{75}As or ^{31}P atom is placed at (0, 0, 0), and the O positions are: $(\pm x, \pm y, z)$, $(\mp y, \pm x, -z)$. The ratio of the upper (or lower) O–O

[8] R. Wyckoff, "Crystal Structures," 2nd Ed., Vol. 3, pp. 160–165. Wiley (Interscience), New York, 1965.

TABLE I

ROOM TEMPERATURE STRUCTURAL DATA

	Cell dimensions (Å)		Oxygen parameters		
	a_0	c_0	x	y	z
KDP	7.448	6.977	0.0828	0.1486	0.1261
KDA	7.630	7.163	0.085	0.1575	0.150
ADP	7.510	7.564	0.085	0.146	0.115
ADA	7.699	7.729	0.086	0.162	0.136

bond length to the length of the side O–O bonds is 2.528/2.503 = 1.01 for KDP, but only 2.726/2.887 = 0.945 for KDA. From the data of Table I, orientations of the tetrahedra may also be obtained. The angle between the upper O–O bond and the nearest *a*-axis is 29.1° in KDP and 28.3° in KDA.

Differences in unit cell dimensions, similar to those shown in Table I, are also found for the ^{87}Rb isomorphs, with the arsenate again having the largest cell.[7] Size of the unit cell also increases with increasing size of the monovalent ion. Of the ^{133}Cs compounds, only the arsenates have a KDP structure; CDP (and presumably CD*P) has an orthorhombic unit cell in the paraelectric phase.[8] Curie temperatures for the various compounds are given in Table II.

TABLE II

CURIE TEMPERATURES (°K) OF KDP-TYPE CRYSTALS

	H_2PO_4	D_2PO_4	H_2AsO_4	D_2AsO_4
K	123	222	96	168
Rb	147	228	110	185
Cs			143	213
NH_4	148		216	
ND_4		245		304

2. *Models for the Phase Transition*

The first theory for the ferroelectric transition in KDP was given by Slater in 1941.[9] Now, thirty years later, and with the actual nature of the transition being strongly debated, the Slater theory is still very much alive for describing equilibrium properties. Slater proposed that (aside from the K^+ ions which would play no role) the crystal be considered as an

[9] J. Slater, *J. Chem. Phys.* **9**, 16 (1941).

assembly of (H_2PO_4) dipoles with "defect" (or "Takagi") groups, such as $(HPO_4)^{2-}$ and (H_3PO_4), of sufficiently high energy for their presence to be negligible. The six possible orientations of the $(H_2PO_4)^-$ dipole are shown in Fig. 1. They are the groups labeled A through F. (The "defect" groups are designated G, G′.) Slater further assigned to the two dipole orientations along the ferroelectric polarization axis (groups A and F of Fig. 1) a lower energy than to the remaining ones, with ε_0 representing that energy difference. The fractional number of dipoles perpendicular, parallel, and antiparallel to the c-axis is designated δ_0, δ_+, and δ_-, respectively. The reduced polarization of the crystal is $p \equiv \delta_+ - \delta_-$. An approximate statistical analysis of the system then leads to a minimum free energy for

$$\delta_0 = \{4 - 2[p^2(4 - \exp(2\varepsilon_0/kT)) + \exp(2\varepsilon_0/kT)]^{1/2}\}/[4 - \exp(2\varepsilon_0/kT)]. \quad (1)$$

In the paraelectric phase, $p = 0$; one obtains from Eq. (1) and the condition $\delta_0 + \delta_+ + \delta_- = 1$ the result

$$\delta_+ = \delta_- = (\delta_0/4)\exp(\varepsilon_0/kT). \quad (2)$$

This is a Boltzmann distribution over the energy parameter ε_0 (a result that will be used later). The Slater model predicts a first-order transition from a disordered paraelectric state to the completely ordered ferroelectric state. A somewhat high value for the transition entropy change is obtained. The theory predicts a Curie law dependence for the dielectric susceptibility in the paraelectric phase:

$$\chi = C/(T - T_C). \quad (3)$$

The Curie constant, C, is in reasonable agreement with experiment.

Refinements of the Slater model have subsequently been made by Takagi,[10] who included the "defect" groups in his analysis, as well as by Senko,[11] and Silsbee, Uehling, and Schmidt,[12] who added the effect of long range forces. A review of these developments has been given by Uehling.[13] Where the Slater model predicted a discontinuous change in polarization at T_C, the modifications made it possible to account for an experimentally observed, more gradual increase in polarization.

Over the last decade several alternatives to the Slater approach have been developed. While their full implication is still being explored, comprehensive reviews of their basic aspects have been given.[14,15] They

[10] Y. Takagi, *J. Phys. Soc. Jap.* **3**, 271, 273 (1948).
[11] M. Senko, *Phys. Rev.* **121**, 1599 (1961).
[12] H. Silsbee, E. Uehling, and V. Schmidt, *Phys. Rev. A* **133**, 165 (1964).
[13] E. Uehling, *Lect. Theor. Phys.* **5**, 138–217 (1963).
[14] R. Blinc, "Theory of Condensed Matter," p. 395. IAEA, Vienna, 1968.
[15] M. Tokunaga, *Ferroelectrics* **1**, 195 (1970).

generally emphasize the dynamic and microscopic aspects of the phase transition, rather than the overall statistical ones. Blinc[16] focused attention on the dynamics of the proton disorder by characterizing it as a tunneling motion in a double well potential along the hydrogen bond. When the energy barrier between the two wells of a double well potential is larger than the ground state energy of the proton in a single well potential (but not large with respect to the energy of excited proton states), two closely spaced "ground state" energy levels result from the mixing of the "single-well" ground states.[17] It is shown that a particle will then move between the two wells at a rate $\Delta E/2h$, where ΔE is the energy separation of the two mixed states. ΔE decreases exponentially with increasing mass of the tunneling particle. Blinc proposed the double well model to account for the large isotope effect observed on replacing KDP protons by deuterons. These isotope effects include a shift in Curie temperature (see Table II) from 123°K (KDP) to 222°K (KD*P), and an approximately 5 orders of magnitude larger domain wall mobility in KDP than KD*P.[18] It must be noted, however, that individual proton motion does not occur in the double minimum potential well required for tunneling.[19] A long range collective motion of surrounding hydrogens can cause the potential minimum of individual hydrogens to shift from one side of the bond to the other. It should be noted that in KDP there are no closed loops of hydrogen bonds in which such collective motion can occur with the involvement of only a few hydrogens.

Since it is primarily the K-PO_4 ionic displacement, rather than the proton ordering, which accounts for the observed polarization, proton–lattice coupling was investigated in relation to the proposed proton-tunneling. No connection with the *c*-axis polarization was established however. Tokunaga and Matsubara[20] noted this deficiency and emphasized the need for a model which includes both protonic and ionic motion. One such model had been proposed already by Cochran,[21] but it failed to account for the order–disorder nature of the transition. It was an attempt by Cochran to extend to KDP his earlier[22] lattice dynamics theory, which

[16] R. Blinc, *J. Phys. Chem. Solids* **13**, 204 (1960); R. Blinc and M. Ribaric, *Phys. Rev.* **130**, 1816 (1963); R. Blinc and S. Svetina, *Phys. Rev.* **147**, 423, 430 (1966).

[17] E. Merzbacher, "Quantum Mechanics." Wiley, New York, 1961; R. Blinc and D. Hadzi, *Mol. Phys.* **1**, 391 (1958).

[18] J. Bjorkstam and R. Oettel, *Proc. Int. Meet. Ferroelec. Prague* **2**, 91 (1966).

[19] C. Reid, *J. Chem. Phys.* **30**, 182 (1959).

[20] M. Tokunaga and T. Matsubara, *Progr. Theor. Phys.* **35**, 581 (1966); M. Tokunaga, *Progr. Theor. Phys.* **36**, 857 (1966). The parameter ε_1 as used by these authors corresponds to $(\varepsilon_1 - \varepsilon_0)$ of SUS.12

[21] W. Cochran, *Advan. Phys.* **10**, 401 (1961).

[22] W. Cochran, *Advan. Phys.* **9**, 387 (1960).

successfully accounts for displacive ferroelectric phase transitions. The theory is based on the fact that in ionic crystals optical mode lattice vibrations correspond to polarization oscillations of equal frequency. Long range Coulomb interaction between ions taking part in such a mode can oppose the short range forces. If for a particular vibrational mode of the crystal these long range forces cancel the short range forces, the crystal becomes unstable and a phase transition is observed. The unstable or "soft" mode was identified as a degenerate optical branch of the vibration spectrum which, for zero wavenumber, has a frequency dependence

$$\omega^2(0) = \gamma(T - T_C). \tag{4}$$

No such optical mode is clearly evident in KD*P.

In recent years, numerous reports claiming the observation of (variously described) ferroelectric, proton-tunneling, condensing, collective, soft, and low-frequency modes (or any combination of them), have been published.[23–25] No generally accepted interpretation of all those observations has so far been given. Taken together, they may be seen as a substantiation of the essential features of a dynamic theory proposed by Kobayashi.[26] The Kobayashi model assumes a coupling between the proton-tunneling mode and the c-axis optical mode vibration of the K-PO_4 complex. Two coupled modes are obtained, one (the in-phase mode, ω_-) where the protons and heavy atoms move jointly closer to, or away from, their ferroelectric positions (see Fig. 3), and the other (ω_+) where protons and nearby K and P atoms move toward, or away from, the same point in space. The phase transition occurs when, due to increased order in the proton motion as $T \to T_C$, the ω_- mode becomes unstable. The temperature dependence of the ω_- mode is of the form given in Eq. (4). While incorporating elements of a lattice dynamics nature, the Kobayashi model remains an order–disorder description of the transition. That this is required of any successful model for KDP is demonstrated by the experimental observations discussed in the following sections. It also is indicated by an observation made by Jona and Shirane[1]: The Curie constants, C, are of

[23] I. Aref'ev, P. Bazhulin, and T. Mikhal'tseva, *Sov. Phys.—Solid State* **7**, 1948 (1966); Y. Imry, I. Pelah, E. Wiener, and H. Zafrir, *Solid State Commun.* **5**, 41 (1967); R. Blinc and S. Zumer, *Phys. Rev. Lett.* **21**, 1004 (1968); T. Plesser and H. Stiller, *Solid State Commun.* **7**, 323 (1969); F. Sugawara and T. Nakamura, *J. Phys. Soc. Jap.* **28**, 158 (1970); J. Skalyo, B. Frazer, and G. Shirane, *Phys. Rev. B* **1**, 278 (1970); B. Lavrencic, I. Levstek, B. Zeks, R. Blinc, and D. Hadzi, *Chem. Phys. Lett.* **5**, 441 (1970); M. Brunstein, J. Grinberg, I. Pelah, and E. Weiner, *Solid State Commun.* **8**, 1212 (1970).

[24] I. Kaminow and T. Damen, *Phys. Rev. Lett.* **20**, 1105 (1968).

[25] K. White, W. Taylor, R. Katiyar, and S. Kay, *Phys. Lett. A* **33**, 175 (1970).

[26] K. Kobayashi, *J. Phys. Soc. Jap.* **24**, 497 (1968).

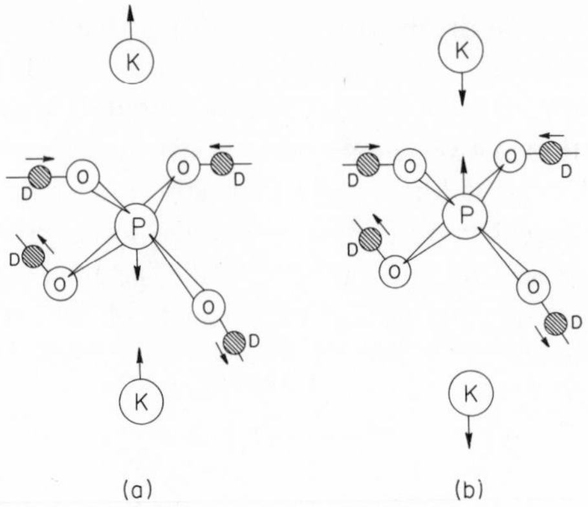

FIG. 3. Coupled hydrogen-lattice modes of the Kobayashi model: (a) the in-phase mode, ω_-, (b) the out-of-phase mode, ω_+.

the order of 10^5 °K for displacive ferroelectrics, and 10^3 °K for order–disorder ferroelectrics. For KDP[27] and KD*P,[28] $C \simeq 3 \times 10^3$ °K.

To end this survey of theories for the ferroelectric transition in KDP, one more approach should be mentioned. It is a new formalism, rather than a new model, but an extremely useful one. De Gennes[29] assigned to each proton a fictitious spin 1/2, with "quantum" numbers $S_z = 1/2$ when the proton is at the "correct" equilibrium site along the hydrogen bond, and $S_z = -1/2$ when it occupies the "wrong" one. The order–disorder problem for the protons is then formally analogous to the Ising problem of ferromagnetism, and "quasi-spin wave" excitations corresponding to the proton tunneling motion may be obtained. The formalism was used by Kobayashi to obtain the above described dynamic theory. It was also used to obtain a computer simulation for the order–disorder phenomena in a KDP-type system.[30] The results of that simulation will be discussed in Section IV,A,2,b.

B. The Antiferroelectric Crystals

1. *Structure*

No complete structure determination is available for any of the antiferroelectric KDP isomorphs. For ADP, the heavy atom positions were

[27] G. Busch, *Helv. Phys. Acta* **11**, 269 (1938).

[28] R. Mayer and J. Bjorkstam, *J. Phys. Chem. Solids* **23**, 619 (1962); T. Sliker and S. Burlage, *J. Appl. Phys.* **34**, 1837 (1963).

[29] P. De Gennes, *Solid State Commun.* **9**, 132 (1963).

[30] N. Ogita, A. Veda, T. Matsubara, H. Matsuda, and F. Yonezawa, *J. Phys. Soc. Jap.* **26**, Suppl. 145 (1969).

obtained, both above and below T_C, by means of X-ray diffraction.[31] The ADP hydrogen positions were obtained by neutron diffraction[32] for the paraelectric structure only. The antiferroelectric KDP-type crystals shatter upon passing through the phase transition, making a successful neutron diffraction analysis below T_C impossible. For ADA, only the positions of heavy atoms above T_C are available.[33] The lattice parameters for ADP and ADA have been measured over a wide temperature range.[7] Whereas the ferroelectric crystals show a fairly isotropic and small ($\sim 0.05\%$) increase in unit cell size at the Curie temperature, changes in the antiferroelectrics are anisotropic and more drastic. The change of a_0 is $\simeq +0.5\%$, while for c_0 it is $\simeq -0.7\%$. This large and nonuniform change of the unit cell is held responsible for shattering of the crystals at the transition. The Curie temperatures of the antiferroelectrics have been included in Table II.

The paraelectric phase structure of the antiferroelectric crystals is tetragonal and completely isomorphous with the KDP structure. Table I lists room temperature cell dimensions and oxygen parameters of ADP and ADA. The ratios of upper and lower O–O bond lengths to the remaining ones are: 2.470/2.460 = 1.003 for ADP, and 2.824/2.899 = 0.975 for ADA. Comparison with similar bond lengths obtained for KDP and KDA reveals that the PO_4 groups in ADP and KDP have essentially the same size. The same observation holds for AsO_4 groups in ADA and KDA. The NH_4 tetrahedra are oriented with two of their edges parallel to the (a, a) plane. The ratio of upper to side edges of the tetrahedron is 1.680/1.622 = 1.035. The upper H–H direction is rotated 23° about the c-axis (counterclockwise) from the upper O–O direction of the PO_4 groups above and below the NH_4 group. N–H···O hydrogen bonds are established with the other four surrounding PO_4 groups.

Below T_C, ADP has an orthorhombic structure with the tetragonal a, a, c-axes becoming the orthorhombic a, b, c-axes. As was the case for KDP, a twinning of the crystal is observed, with a given tetragonal a-axis turning into the orthorhombic a-or b-axis in various parts of the crystal. This twinning leads to X-ray ($hk0$) reflections which are difficult to resolve and therefore to a sizeable (but unspecified) uncertainty in the x-parameters of the atoms below T_C.[31] Positions of the N–PO_4 complex are well enough established, nevertheless, to reveal the PO_4 group distortion shown in Fig. 2c, and the emergence of two sublattices with opposite polarization along the orthorhombic b-axis. The hydrogen assignments of Fig. 2c were made on the basis of known hydrogen locations in KDP for entirely

[31] R. Keeling and R. Pepinsky, *Z. Krystallogr., Kristallgeometrie, Kristallphys., Kristallchem.* **106**, 236 (1955).
[32] L. Tenzer, B. Frazer, and R. Pepinsky, *Acta Crystallogr.* **11**, 505 (1958).
[33] C. Delain, *C. R. Acad. Sci.* **247**, 1451 (1958).

analogous displacements of the heavy atoms. They also correspond to the assignments made on the basis of deuteron magnetic resonance in AD*P.[34]

2. *Theory of the Transition*

The two approaches to the transition theory for KDP, statistical (Slater) and vibrational (Cochran), have their counterparts in the theory for antiferroelectrics. Not as much attention has been paid to the antiferroelectric case however, primarily because it promises to be more difficult to resolve. The "duality" noted in the ferroelectrics, namely that while the proton system undergoes the ordering at T_C, other atoms account for the polarization, is also present in the antiferroelectrics. In the antiferroelectrics the NH_4^+ ion also plays a distinct role in the transition. The fact that the ammonium isomorphs are antiferroelectric and not ferroelectric is in itself convincing evidence of this. It was also demonstrated by Matthias *et al.*[35] who found that partial substitution of Tl^+ for NH_4^+ changes the transition temperature, until, for 33% or more Tl, the material behaves as a normal dielectric.

Nagamiya[36] showed that the dielectric, piezoelectric, and elastic properties of paraelectric ADP could be accounted for on the basis of Slater's theory for KDP, if allowance was made for a negative value of the energy parameter ε_0. In other words, the dipole orientations perpendicular to the *c*-axis would be energetically favored. For the antiferroelectric phase Nagamiya proposed the dipole arrangement which was later verified by the X-ray results. The calculated entropy change, due to H_2PO_4 dipole ordering at the transition, was $\simeq 20\%$ too low. An increase, at T_C, in the activation energy for NH_4 reorientation has since been shown[37] to account for an entropy increase on the order of the needed 20%.

The other, lattice-dynamics, viewpoint is again due to Cochran.[21] He proposed that antiferroelectricity in ADP be considered the result of a lattice instability against a mode with wavevector $\mathbf{q} = (0, 0, 2\pi/c)$, i.e. at the edge of the Brillouin zone. The "ferroelectric" mode would still be present but the "antiferroelectric" mode frequency would go to zero first. Kobayashi[26] points out that his model for KDP predicts such a mode if a negative value is assigned to the proton dipole–dipole interaction. Raman,[38] infrared,[39] and far infrared[40] studies have so far been unable to find any

[34] E. Uehling and J. Soest, *Bull. Amer. Phys. Soc.* **12**, 290 (1967); J. Soest, Ph.D. Thesis, Univ. of Washington, Seattle, 1967.
[35] B. Matthias, W. Merz, and P. Scherrer, *Helv. Phys. Acta* **20**, 273 (1947).
[36] T. Nagamiya, *Progr. Theor. Phys.* **7**, 275 (1952).
[37] D. Genin, D. O'Reilly, and T. Tsang, *Phys. Rev.* **167**, 445 (1968).
[38] I. Aref'ev and P. Bazhulin, *Sov. Phys.—Solid State* **7**, 326 (1965).
[39] E. Wiener, S. Levin, and I. Pelah, *J. Chem. Phys.* **52**, 2891 (1970).
[40] I. Aref'ev, P. Bazhulin, and I. Zheludev, *Sov. Phys.—Solid State* **7**, 2290 (1966).

significant difference in the KDP and ADP vibration spectra. On the other hand, temperature-dependent quasi-elastic neutron scattering has been observed at the Brillouin zone edge in AD*P.[41] It was interpreted in terms of a "condensing, highly damped, optic-phonon mode which initiates the transition to the antiferroelectric state." The ambiguity involved in such interpretations is perhaps best demonstrated by the fact that completely analogous scattering (near $\mathbf{q} = 0$) in KD*P has been discounted as evidence for a lattice-mode model.[42]

No model has been proposed so far which explicitly takes the NH_4^+ ions into account. Continuing comparative studies between the ferroelectrics and antiferroelectrics will be useful to establish more clearly the similarities and differences.

III. Introduction to Deuteron NMR Studies

The focus of this section is upon those aspects of the deuteron and proton NMR studies not associated with critical fluctuations near the ferroelectric phase transition temperature, T_C. Particular attention is given to the spectrum above T_C, and those motional contributions to spin-lattice relaxation, T_1, which must be separated off to leave only the contribution from critical effects. Since there can be many contributions to T_1, such a separation is always necessary. Unless specified otherwise, T_1 represents the time constant for return of the spin energy levels to an equilibrium population, i.e. the Zeeman spin-lattice relaxation time.

In order to provide a self-contained discussion some early results from a review article by Blinc[43] will be summarized. For the nonspecialist in NMR an elementary discussion of the connection between transition probabilities and fluctuations in the Hamiltonian is given in Section III,B.

A. The DNMR Spectrum and Deuteron Motion

Under usual experimental conditions, interaction of the deuteron electric quadrupole moment with the electric field gradient (EFG), established by the surrounding charge distribution at the deuteron site, causes a small perturbation of the nuclear Zeeman levels. In a 10 kG field, for example, the $\Delta m = \pm 1$ transition frequency of the $I = 1$ deuteron (6.54 MHz) is perturbed by $\lesssim 300$ kHz. Techniques for extracting the EFG tensor at the

[41] H. Meister, J. Skalyo, B. Frazer, and G. Shirane, *Phys. Rev.* **184**, 550 (1969).
[42] G. Paul, W. Cochran, W. Buyers, and R. Cowley, *Phys. Rev. B* **2**, 4603 (1970).
[43] R. Blinc, *Advan. Magn. Resonance* **3**, 141 (1968).

nuclear site have been extensively discussed in review articles on quadrupole effects.[44] The most straightforward is that of Volkoff.[45] Defining a set of axes X, Y, Z with respect to the crystal, the frequency splitting $2\Delta\nu$ between the $(m=1)\leftrightarrow(m=0)$ and $(m=0)\leftrightarrow(m=-1)$ transitions of a spin $I=1$ nucleus, of quadrupole moment Q, is given by

$$(2\Delta\nu)_i = (3eQ/4h)\{-V_{ii} + (V_{jj} - V_{kk})\cos 2\theta_i - 2V_{jk}\sin 2\theta_i\}, \tag{5}$$

where i, j, k are cyclic indices, and V_{ij} the ijth element of the EFG tensor at the nuclear site. The angle θ_i between the magnetic field $\bar{H}_0$ and j-axis is measured clockwise when viewing along the i-axis, which is taken normal to $\bar{H}_0$. Since the V_{ii} satisfy Laplace's equation, $\sum_{i=1}^{3} V_{ii} = 0$, the EFG tensor has five independent components which can be determined by applying Eq. (5) to rotation data about the three orthogonal axes, X, Y, Z. The tensor so determined may be diagonalized to yield components $|V_{zz}| \geqslant |V_{yy}| \geqslant |V_{xx}|$ and their direction cosines with respect to the crystal X, Y, Z system. Convenient measures of the quadrupole interaction are the electric quadrupole coupling constant $|eQV_{zz}/h|$, and asymmetry factor, $\eta \equiv (V_{xx} - V_{yy})/V_{zz}$.

Initial DNMR experiments on KD*P gave a room temperature EFG tensor of nearly cylindrical symmetry ($\eta = 0.049$) in each bond, with principal axis V_{zz} essentially along the O–D···O bond direction.[46] The coupling constant, $|eQV_{zz}/h| = 119.7\,\text{kHz}$, is essentially temperature independent above T_C.

1. *Evidence for Interbond Motion*

The spectra of X and Y bond deuterons are easily resolved if $\bar{H}_0$ is chosen $\perp c(Z)$-axis and makes an angle other than 45° with respect to either bond direction. The quadrupole perturbed Zeeman energy levels of the X and Y bonds are then as illustrated in Fig. 4. To first order, the quadrupole perturbation leaves the $m=0$ level undisplaced. The transition frequencies satisfy $\nu_{X+} > \nu_{Y-} > \nu_0 > \nu_{Y+} > \nu_{X-}$, with ν_0 the unperturbed Zeeman frequency. With the spectrum swept in the order $\nu_{X-}, \nu_{Y+}, \nu_{Y-}$, ν_{X+}, at rf levels sufficient to cause partial saturation, resonance ν_{Y+} is enhanced in comparison with ν_{Y-}. Sweeping resonance ν_{X+} first enhances the relative amplitude of ν_{Y-}. This can be understood by noting that sweeping resonance ν_{X-} tends to equalize the populations of the X bond $m=0$ and $m=+1$ levels, thus increasing the population difference between

[44] M. Cohen and F. Reif, *Solid State Phys.* **5**, 321 (1957); W. Gretschischkin and N. Ajnoinder, *Usp. Fiz. Nauk* **80**, 597 (1963).

[45] G. Volkoff, *Can. J. Phys.* **31**, 820 (1953).

[46] J. Bjorkstam and E. Uehling, *Phys. Rev.* **114**, 961 (1959); *Bull. Amer. Phys. Soc.* **5**, 345 (1960).

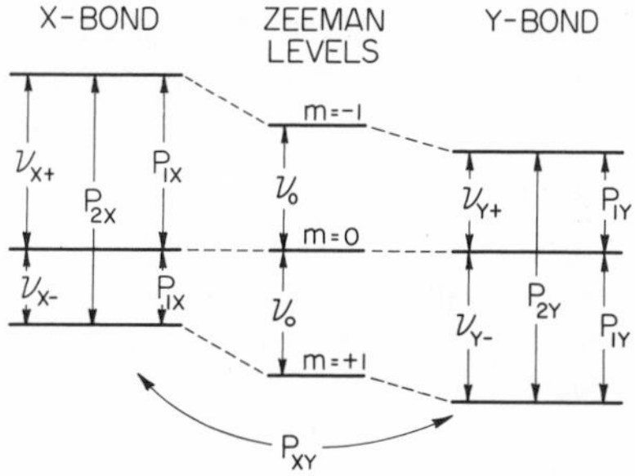

FIG. 4. Quadrupole-perturbed, Zeeman energy levels of X and Y bond deuterons in KD*P.

the $m = 0$ and $m = -1$ levels over the equilibrium values. This enhances the $m = -1 \leftrightarrow m = 0$ transition. Now if X and Y bond deuterons exchange locations during the time required to sweep through ν_{X-}, the $m = +1 \leftrightarrow m = 0$ resonance of both X and Y bond deuterons is partially saturated. The population difference between the $m = 0$ and $m = -1$ deuteron levels in both X and Y bonds is thus increased, and resonance ν_{Y+} is enhanced.

An extensive pulsed NMR investigation of the above phenomenon was carried out by Schmidt and Uehling.[47] It was thus possible to saturate the ν_{X-} transition in a time either short or long compared with the correlation time τ_{XY} for exchange of X and Y bond deuterons. Such experiments allow unambiguous determination of τ_{XY}. The above authors found $\tau_{XY} \simeq 0.5$ sec at room temperature, with an activation energy of 0.58 eV. Their measurements on electrical conductivity gave the same activation energy, thus demonstrating the dominant role played by deuteron jumping in the conduction process. Calculation of the interbound jumping contribution to the deuteron spin-lattice relaxation rate, $(T_1)_{XY}^{-1}$, will be included in Section III,C,1.

2. *Evidence for Intrabond Motion*

Neutron diffraction results above T_C showed an elongated hydrogen distribution centered in the bond.[5] Whether this should be best represented by a large anisotropic vibration about a potential minimum at the bond center, or rapid jumping between the two off-center equilibrium positions observed for $T < T_C$, was not apparent. Evidence from DNMR is consistent with the latter point of view.[48]

Upon cooling through T_C the reduction from tetragonal to orthorhombic symmetry, discussed in Section II,A,1, is accompanied by a small shear

[47] V. Schmidt and E. Uehling, *Phys. Rev.* **126**, 447 (1962).
[48] J. Bjorkstam, *Phys. Rev.* **153**, 599 (1967).

($X_y \simeq 1/2°$) in the X, Y-plane. When the crystal is cooled without a polarizing electric field it breaks into an equal number of antiparallel, oppositely polarized domains. The opposite shear in antiparallel domains causes a small splitting of the DNMR spectrum in Z-rotation.[46] For rotation about either the X or Y crystal axis, taken normal to $\overline{H}_0$, 180° symmetry of the quadrupole interaction rules out any such domain splittings.

The X and Y bonds of Fig. 1 are approximately 0.5° out of the X, Y-plane. They will be designated as $X_{\pm}$, $Y_{\pm}$ bonds. X_+ is an X-directed bond with

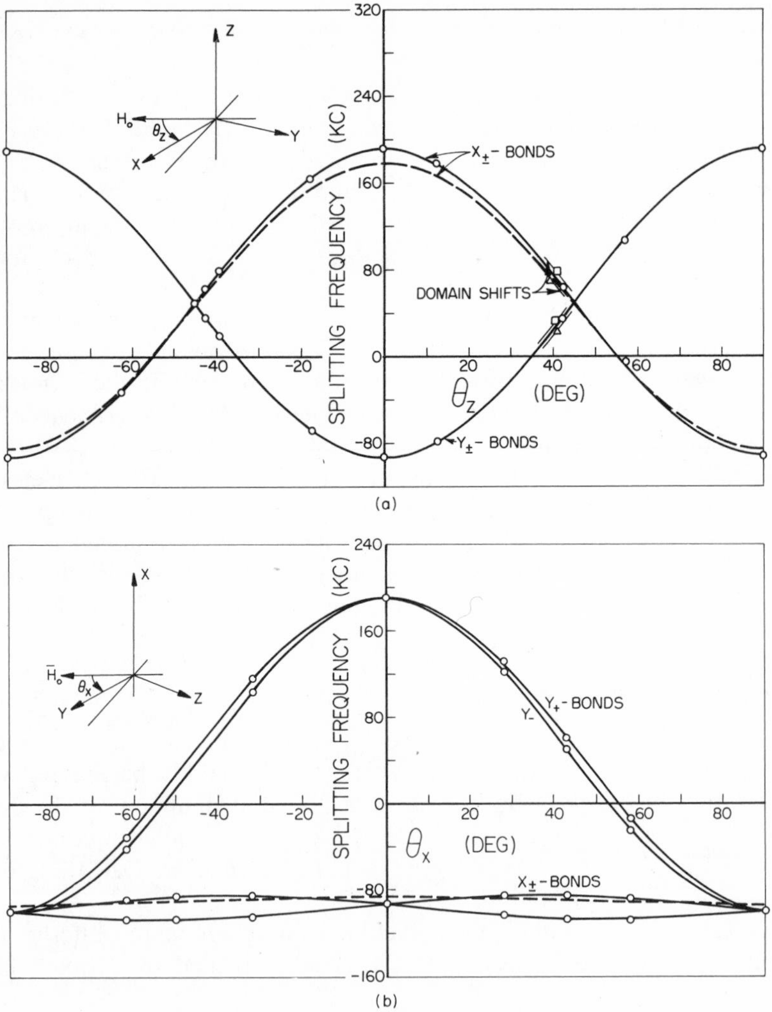

FIG. 5 (a, b). See facing page for legend.

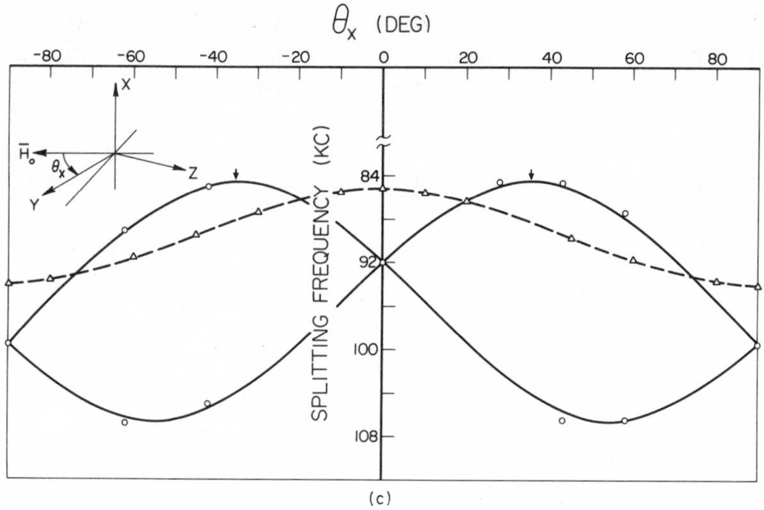

FIG. 5. Deuteron quadrupole splitting in KD*P at $T = -100°C$ ($T < T_C$). (a) Z-rotation, (b) X-rotation, (c) X-rotation (expanded scale for X bonds only). Dashed curves show the X bond results at $T = 30°C$.[48]

"upper" end along the $+Z(c)$-axis, etc. Because of the crystal symmetry, X-rotation data for $Y_{\pm}$ bonds may be utilized to determine V_{XZ} elements for the X bond. Thus the X- and Z-rotation data of Fig. 5 are sufficient for a complete determination of the EFG tensors through the use of Eq. (5).

The significant change in spectrum which occurs upon cooling through T_C is evident from the X-rotation data for $X_{\pm}$ bonds in Fig. 5c. Since in this case the rotation axis is within $\pm 1°$ of the largest component of the diagonalized EFG tensor (V_{zz}) for $X_{\pm}$ bonds, extrema in the data correspond to orientations where the other components of the diagonalized tensor, V_{yy} and V_{xx}, are essentially parallel to $\bar{H}_0$. It is apparent that the high temperature result (dotted curve) is (except for a small constant shift which reflects the change in V_{zz} upon cooling below T_C) just an average of the two low temperature (solid) curves whose extrema are displaced by $\pm 35.5°$ from that of the high temperature data. Eigenvectors of the $X_{\pm}$ bond EFG tensors are summarized in Table III. The low temperature EFG tensor components, evaluated in the crystal system, are included in this table for convenience in later calculations.

Precisely the results described above had been predicted by Chiba.[49] Below T_C the intermediate component V_{yy} of the diagonalized EFG tensor is supposed normal to the plane formed by the deuteron, its nearest oxygen,

[49] T. Chiba, *J. Chem. Phys.* **41**, 1352 (1964).

TABLE III

$X_\pm$ Bond Deuteron EFG Components ($3eQV_{ii}/2h$ in kHz) for KD_2PO_4[48]

$T = 30°C > T_C$		$T = -100°C < T_C$	
Crystal system	Eigenvalues and direction cosines	Crystal system	Eigenvalues and direction cosines
$V_{XX} = 178$	$V_{zz} = 179.2 \begin{cases} \pm 1 \\ 0 \\ 0 \end{cases}$	$V_{XX} = 191.2$	$V_{zz} = 191.2 \begin{cases} \pm 1 \\ 0 \\ 0 \end{cases}$
$V_{YY} = -84.5$		$V_{YY} = -92$	
$V_{ZZ} = -93.5$	$V_{yy} = -94.0 \begin{cases} 0 \\ 0 \\ \pm 1 \end{cases}$	$V_{ZZ} = -99.2$	$V_{yy} = -106.7 \begin{cases} 0 \\ \mp 0.58 \\ +0.814 \end{cases}$
$X_{XZ} = \pm 3.5$		$V_{XZ} = \pm 6.3$	
	$V_{xx} = -85.2 \begin{cases} 0 \\ \pm 1 \\ 0 \end{cases}$		$V_{xx} = -84.5 \begin{cases} 0 \\ \pm 0.81 \\ +0.58 \end{cases}$
$V_{XY} = V_{YZ} = 0$		$V_{YZ} = \pm 10.4$	
		$V_{XY} = 0$	

and the ^{31}P atom of the $(D_2PO_4)^-$ group. From the crystal structure[5] it can be ascertained that normals to the planes so constructed for the two deuterons of a $(D_2PO_4)^-$ group make angles of $\pm 40.5°$ with respect to the $Z(c)$-axis. The 5° deviation between predicted and observed orientation of V_{yy} may be attributed to distortion of the PO_4 tetrahedra below T_C, as well as to contributions from electronic and ionic charge distributions not part of the P–O–D plane. When account is taken of the 7% greater coupling constant below T_C, the $T > T_C$ EFG tensor is within 1% of being a perfect average of that for the two deuteron positions below T_C. This seems to be rather conclusive evidence for the order–disorder nature of the transition, at least insofar as the hydrogen bonds are concerned. Since the high temperature deuteron spectrum shows no line broadening above T_C due to the intrabond motion, it must follow that the intrabond jump time τ_I is much less than the reciprocal of the frequency *difference* between the quadrupole perturbed, Zeeman transitions at either end of the bond below T_C. Thus $\tau_I \lesssim 10^{-5}$ sec at least to within 0.5°C of T_C.

B. Fluctuations and Transition Probabilities

This section contains a brief review which connects the time dependence of a perturbation with its effectiveness in inducing transitions. Consider a system with Hamiltonian

$$\mathscr{H} = \mathscr{H}_0 + \mathscr{H}_1(t). \tag{6}$$

Here $\mathscr{H}_1(t)$ represents a time-dependent perturbation which fluctuates about the average Hamiltonian $\mathscr{H}_0$, with eigenstates $|\alpha\rangle, \ldots, |\beta\rangle, \ldots$, and energy eigenvalues $\hbar\alpha, \ldots, \hbar\beta, \ldots$. A general state of the system may be expressed as

$$|\xi\rangle = \sum_\alpha C_\alpha \{\exp(-i\alpha t)\} |\alpha\rangle,$$

where the time-dependent coefficients obey Schrodinger's equation,

$$i\hbar(dC_\alpha/dt) = \sum_\gamma \langle\alpha|\mathscr{H}_1(t)|\gamma\rangle \{\exp(i\omega_{\alpha\gamma} t)\} C_\gamma,$$

with $\omega_{\alpha\gamma} \equiv (\alpha-\gamma)/\hbar$. If the system is in a definite state $|\beta\rangle$ at $t = 0$, the probability $P_{\alpha\beta}(t)$ that a transition to state $|\alpha\rangle$ has taken place in time t is just $|C_\alpha|^2$. The transition probability per unit time is

$$W_{\alpha\beta} \equiv dP_{\alpha\beta}/dt = C_\alpha(dC_\alpha^*/dt) + \text{c.c.},$$

where c.c. means complex conjugate. With the usual assumption of first-order perturbation theory that the initial state undergoes negligible depletion over the times of interest,

$$C_\alpha(t) = -(i/\hbar) \int_0^t \langle\alpha|\mathscr{H}_1(t')|\beta\rangle \exp(i\omega_{\alpha\beta} t')\, dt'.$$

Denoting the ensemble average by a bar one has

$$\overline{W_{\alpha\beta}(t)} = \hbar^{-2} \int_0^t \overline{\langle\alpha|\mathscr{H}_1(t')|\beta\rangle\langle\beta|\mathscr{H}_1(t'')|\alpha\rangle} \exp[i\omega_{\alpha\beta}(t'-t'')]\, dt' + \text{c.c.} \quad (7)$$

When $\mathscr{H}_1(t)$ represents a stationary, random process, the integrand depends only upon $\tau \equiv t'-t''$, and Eq. (7) becomes

$$\overline{W_{\alpha\beta}(t)} = \int_0^t G_{\alpha\beta}(\tau) \exp(i\omega_{\alpha\beta}\tau)\, d\tau + \text{c.c.}, \quad (8)$$

with the correlation function defined by

$$G_{\alpha\beta}(\tau) \equiv \hbar^{-2}\, \overline{\langle\alpha|\mathscr{H}_1(\tau)|\beta\rangle\langle\beta|\mathscr{H}_1(0)|\alpha\rangle}. \quad (9)$$

When there is negligible depletion of the initial state in a time t, large compared with the correlation time τ, the limits of integration may be taken as $\pm\infty$. Finally, then

$$\overline{W_{\alpha\beta}} = \int_0^\infty G_{\alpha\beta}(\tau) \exp(i\omega_{\alpha\beta}\tau)\, d\tau + \text{c.c.} \quad (10)$$

Thus the relaxation rate which results from the perturbation $\mathscr{H}_1(t)$ will depend upon the spectral density of the autocorrelation function of the matrix elements of $\mathscr{H}_1$, with respect to the eigenstates of $\mathscr{H}_0$.

When considering transitions between nuclear spin states, $\mathscr{H}_1(t)$ may be expanded in terms of products of operators $\mathscr{F}^{(K)}$, which depend upon spatial coordinates of the nuclei, and $\mathscr{A}^{(K)}$ which involve spin operators, i.e.

$$\mathscr{H}_1(t) = \sum_{K=-2}^{+2} \Delta\mathscr{F}_{(t)}^{(K)}\mathscr{A}^{(K)}. \tag{11}$$

It is the $\Delta\mathscr{F}_{(t)}^{(K)}$ which are related to microscopic models for the polarization fluctuations.

The polarization fluctuations may be related to models which consider (1) individual uncorrelated motions, (2) correlated motion of small clusters, or (3) Fourier expansion in terms of collective coordinates. The KDP-type crystals have been used as a model system to illustrate all three methods.

C. Rate Equations for Spin-Lattice Relaxation

Since the quadrupole interaction is a small perturbation on the Zeeman splitting, the transition frequencies of Fig. 4 satisfy the conditions $\nu_{X+} \simeq \nu_{X-} \simeq \nu_0$ and $\nu_{Y+} \simeq \nu_{Y-} \simeq \nu_0$. In addition $kT \gg h\nu_0$ over the entire temperature range of interest. Thus the equilibrium population differences between adjacent pairs of spin states will be essentially equal so that detailed balancing in equilibrium gives for the transition probabilities defined in Fig. 4, $P_{1X_+} \simeq P_{1X_-} \equiv P_{1X}$ and $P_{1Y_+} \simeq P_{1Y_-} \equiv P_{1Y}$. This does not mean that P_{1X} is equal to P_{1Y}, of course. Following Schmidt and Uehling,[47] the variables $x_\pm$ and $y_\pm$ may be used to represent deviations from their equilibrium values of populations in the states $m = \pm 1$ of X and Y bonds, respectively. The rate equation for x_+ is then

$$dx_+/dt = P_{1X}(x_0 - x_+) + P_{2X}(x_- - x_+) + P_{XY}(y_+ - x_+),$$

where $x_0 = -(x_+ + x_-)$ since the sum of the deviations is zero. Three similar equations for the time dependence of x_-, $y_\pm$, together with the condition $y_0 = -(y_+ + y_-)$, completely describe the changes in spin population. Introducing the variables

$$U_\pm \equiv x_+ \pm x_-, \qquad V_\pm \equiv y_+ \pm y_-, \tag{12}$$

the time-dependent equations become

$$dU_+/dt = -(P_{XY} + 3P_{1X})\,U_+ + P_{XY}\,V_+, \tag{13a}$$

$$dV_+/dt = P_{XY}\,U_+ - (P_{XY} + 3P_{1Y})\,V_+, \tag{13b}$$

$$dU_-/dt = -(P_{XY}+P_X)\,U_- + P_{XY}\,V_-, \tag{13c}$$

$$dV_-/dt = P_{XY}\,U_- - (P_{XY}+P_Y)\,V_-, \tag{13d}$$

where

$$P_X \equiv P_{1X} + 2P_{2X} \qquad \text{and} \qquad P_Y \equiv P_{1Y} + 2P_{2Y}. \tag{14}$$

Both U_+ and V_+ return to equilibrium with time constants

$$T_{1\pm}(+) = \{P_{XY}+(3/2)(P_{1X}+P_{1Y})\pm(1/2)\,[9(P_{1X}-P_{1Y})^2+4P_{XY}^2]^{1/2}\}^{-1}, \tag{15}$$

while for U_- and V_- the time constants are

$$T_{1\pm}(-) = \{P_{XY} + (1/2)(P_X+P_Y) \pm (1/2)[(P_X-P_Y)^2 + 4P_{XY}^2]^{1/2}\}^{-1}. \tag{16}$$

It is possible to arrange the experimental circumstances in several ways so that the return to equilibrium is characterized rather well by a single time constant. The separate transition probabilities which make up $T_{1\pm}(\pm)$ can thus be compared with theory.

The final problem remaining is to make a connection between models for the dynamics of the hydrogen motion and the transition probabilities which enter Eqs. (15) and (16). As an example, consider P_{1X}. The contributions to P_{1X} from inter- and intrabond motion, to be designated by $(P_{1X})_{\mathrm{I}}$ and $(P_{1X})_{XY}$, respectively, must be separated. The formalism leading to Eq. (10) assumed a transition time very long compared with the correlation time of the motion responsible for the relaxation process. This assumption allowed the integration limits of Eq. (8) to be taken as $\pm\infty$. Thus, with spin-lattice relaxation due only to XY exchange, for the time scale of interest the X and Y bond spin populations are maintained equal by the XY exchange. In that case, $(P_1)_{XY} \equiv (P_{1X})_{XY} = (P_{1Y})_{XY}$.

The total Hamiltonian (neglecting magnetic dipole–dipole interactions) is

$$\mathscr{H} = \mathscr{H}_{z'} + \mathscr{H}_{\mathrm{Q}}(t), \tag{17}$$

with $\mathscr{H}_{z'}$ the time-independent Zeeman Hamiltonian and $\mathscr{H}_{\mathrm{Q}}(t)$ the quadrupole Hamiltonian which fluctuates with time as a result of the XY exchange. The laboratory system, in which $\mathscr{H}_{z'}$ is diagonal, will be designated by coordinates x', y', z', with the Zeeman magnetic field $\overline{H}_0$ along z'. Equation (17) may be put in the form of Eq. (6) if

$$\mathscr{H}_0 \equiv \mathscr{H}_{z'} + \mathscr{H}_{\mathrm{Q_{avg}}}, \tag{18}$$

$$\mathscr{H}_1(t) \equiv \mathscr{H}_{\mathrm{Q}}(t) - \mathscr{H}_{\mathrm{Q_{avg}}}. \tag{19}$$

Here, $\mathscr{H}_{\mathrm{Q_{avg}}}$ is the quadrupole Hamiltonian average over a time long compared with the correlation time for fluctuations in $\mathscr{H}_1(t)$.

Assuming the average quadrupole interaction to be a small perturbation of $\mathscr{H}_{z'}$, the eigenstates designated previously as $|\alpha\rangle, \ldots, |\beta\rangle, \ldots$, may be taken as $|0\rangle, |\pm 1\rangle, \ldots, |\pm I\rangle$, the eigenstates of $\mathscr{H}_{z'}$ for a nucleus of spin I. Then, for a $\Delta m = \pm 1$ transition, Eqs. (8) and (9) give

$$\begin{aligned}(P_1)_{XY} &= \overline{W}_{1,0} = \overline{W}_{1,0} \\ &= \hbar^{-2}\left\{\int_0^\infty \overline{\langle 1|\mathscr{H}_1(\tau)|0\rangle\langle 0|\mathscr{H}_1(0)|1\rangle} \exp(i\omega\tau)\, d\tau + \text{c.c.}\right\}, \qquad (20)\end{aligned}$$

where $\omega = 2\pi\nu_0$, with ν_0 the Larmor frequency. The only nonvanishing matrix elements of the quadrupole Hamiltonian are[44]

$$\begin{aligned}\langle m|\mathscr{H}_Q|m\rangle &= A[3m^2 - I(I+1)]\, V_0, \\ \langle m\pm 1|\mathscr{H}_Q|m\rangle &= A(2m\pm 1)[(I\pm m+1)(I\mp m)]^{1/2}\, V_{\pm 1}, \qquad (21) \\ \langle m\pm 2|\mathscr{H}_Q|m\rangle &= A[(I\mp m)(I\mp m-1)(I\pm m+1)(I\pm m+2)]^{1/2}\, V_{\mp 2},\end{aligned}$$

where V_0, $V_{\pm 1}$, $V_{\pm 2}$ expressed in the laboratory frame are

$$V_0 = V_{z'z'}, \qquad V_{\pm 1} \equiv V_{x'z'} \pm iV_{y'z'}, \qquad V_{\pm 2} \equiv \tfrac{1}{2}(V_{x'x'} - V_{y'y'}) \pm iV_{x'y'}, \quad (22)$$

and

$$A \equiv eQ/4I(2I-1). \qquad (23)$$

The EFG components V_{ij} are to be evaluated in the coordinate system where $\mathscr{H}_{z'}$ is diagonal.

The matrix elements for the $I = 1$ deuteron then give

$$P_K = 2(A/\hbar)^2[KJ_K(\omega_K)], \qquad K = 1, 2, \qquad (24)$$

where A is defined in Eq. (23), and

$$J_K(\omega_K) \equiv \int_0^\infty \overline{\Delta V_K(\tau)\Delta V_K^*(0)} \exp(i\omega_K\tau)\, d\tau + \text{c.c}, \qquad \omega_K \equiv K\omega. \qquad (25)$$

With $\Delta V_K(\tau)$ the difference between V_K at time τ and its average value, $\overline{\Delta V_K(\tau)\Delta V_{-K}(0)} = \overline{\Delta V_K(\tau)\Delta V_K^*(0)}$, is just the ensemble average of the *fluctuation* in V_K from its average value.

The notation $\Delta V_K^{(i)}$ will be used to designate the difference between V_K evaluated at site i and the properly weighted average over all sites. The correlation function $G(\tau)$ for fluctuations in V_K is defined as

$$G_K(\tau) = \overline{G_K^{(i)}(\tau)} = \overline{\Delta V_K(\tau)\Delta V_K^*(0)}, \qquad (26)$$

where

$$G_K^{(i)}(\tau) = \Delta V_K^{(i)} \sum_{j=1}^{M} P(j, \tau; i)\Delta V_K^{(j)*}, \qquad (27)$$

and

$$\overline{G_K^{(i)}(\tau)} = \sum_{i=1}^{M} \delta_i G_K^{(i)}(\tau). \qquad (28)$$

$P(j, \tau; i)$ is the conditional probability that a spin, initially at site i, is at site j a time τ later, while δ_i is the probability that the spin is initially at site i.

Contributions to T_1 from Interbond Motion

The general results outlined above will now be utilized to evaluate the contribution to the spin-lattice relaxation rate from interbond jumping, i.e. $(T_1)_{XY}^{-1}$. It will be shown that this contribution is unimportant in the temperature range near T_C where critical relaxation effects prove to be the dominant relaxation process.

As pointed out in Section III,A,2, above T_C the EFG tensors at the deuteron sites have nearly cylindrical symmetry with maximum component of the diagonalized tensor, V_{zz}, along the O–D$\cdots$O bond direction. In units of $(2h/3eQ) \times 10^3 \sec^{-1} = (h/6A) \times 10^3 \sec^{-1}$, the only nonvanishing X bond EFG tensor components in the X, Y, Z crystal coordinate system are: $V_{XX}^{(X)} = 178$, $V_{YY}^{(X)} = -84.5$, $V_{ZZ}^{(X)} = -93.5$, and $V_{XZ}^{(X)} = \pm 3.5$. Similarly for Y bonds: $V_{YY}^{(Y)} = 178$, $V_{XX}^{(Y)} = -84.5$, $V_{ZZ}^{(Y)} = -93.5$, and $V_{YZ}^{(Y)} = \pm 3.5$.

With z' along $\bar{H}_0$, a straightforward transformation to the x', y', z' laboratory system gives

$$V_{x'z'} = (V_{XX} - V_{YY}) \sin\theta_Z \cos\theta_Z + V_{XY}(\cos^2\theta_Z - \sin^2\theta_Z),$$

$$V_{y'z'} = V_{XY} \cos\theta_Z - V_{YZ} \sin\theta_Z. \tag{29}$$

With $V_{XY}^{(X)}$, $V_{XY}^{(Y)}$, $V_{XZ}^{(Y)}$, $V_{YZ}^{(X)}$ all zero, and $|V_{XZ}^{(X)}| = |V_{YZ}^{(Y)}|$ negligibly small, one has

$$-V_{x'z'}^{(Y)} = V_{x'z'}^{(X)} = (V_{XX}^{(X)} - V_{YY}^{(X)}) \sin\theta_Z \cos\theta_Z$$

$$= (262.5(10^3)\, h/6A)(\sin 2\theta_Z/4) \tag{30a}$$

$$V_{y'z'}^{(X)} \simeq 0 \simeq V_{y'z'}^{(Y)}. \tag{30b}$$

Since the deuteron spends equal time in both X and Y bonds,

$$-\Delta V_1^{(Y)} = \Delta V_1^{(X)}$$

$$= (1/2)\{(V_{x'z'}^{(X)} + iV_{y'z'}^{(X)}) - (V_{x'z'}^{(Y)} + iV_{y'z'}^{(Y)})\}$$

$$= (263(10^3)\pi\hbar \sin 2\theta_Z)/6A. \tag{31}$$

Equations (26) and (28), specialized to the case of two sites where i can be either X or Y, give

$$\Delta V_K(0)\Delta V_K^*(\tau) \equiv G_K(\tau)$$

$$\equiv \overline{G_K^{(i)}(\tau)}$$

$$= \delta_X G_K^{(X)}(\tau) + \delta_Y G_K^{(Y)}(\tau). \tag{32}$$

With equal occupation probability for both sites, on a time scale short compared with $(T_1)_{XY}$, the weighting factors are $\delta_X = \delta_Y = (1/2)$. Finally, utilizing Eq. (27), Eq. (32) becomes

$$G_K(\tau) = (1/2)\{\Delta V_K^{(X)}[P(X,\tau;X)\Delta V_K^{(X)*} + P(Y,\tau;X)\Delta V_K^{(Y)*}] + \Delta V_K^{(Y)}[P(Y,\tau;Y)\Delta V_K^{(Y)*} + P(X,\tau;Y)\Delta V_K^{(X)*}]\}.$$

For this simple case of jumping between two sites, with equal statistical weight, the time-dependent occupation probabilities, p_X and p_Y, satisfy

$$\dot{p}_X = P_{XY}(p_Y - p_X); \qquad \dot{p}_Y = P_{XY}(p_X - p_Y); \qquad p_X + p_Y = 1. \tag{33}$$

The dot denotes time derivative and, as before, $P_{XY} = P_{YX}$ is the probability per unit time for a deuteron jump between an X and Y bond. Equations (33) have solutions

$$p_X = [1 + A_0 \exp(-2P_{XY}t)]/2, \qquad p_Y = [1 - A_0 \exp(-2P_{XY}t)]/2,$$

with the constant A_0 dependent upon the initial conditions. Thus

$$\begin{aligned} P(X,\tau;X) &= [1+\exp(-2\mathrm{P}_{XY}\tau)]/2 = P(Y,\tau;Y), \\ P(Y,\tau;X) &= [1-\exp(-2P_{XY}\tau)]/2 = P(X,\tau;Y). \end{aligned} \tag{34}$$

With $\delta_X = \delta_Y = 1/2$, Eqs. (27) and (28) substituted in Eq. (32) yield

$$G_K(\tau) = (1/2)\{[1+\exp(-2P_{XY}\tau)][|\Delta V_K^{(X)}|^2 + |\Delta V_K^{(Y)}|^2] + [1-\exp(-2P_{XY}\tau)][\Delta V_K^{(X)}\Delta V_K^{(Y)*} + \Delta V_K^{(Y)}\Delta V_K^{(X)*}]\}. \tag{35}$$

Specializing Eq. (35) to $K = 1$, and substituting from Eq. (31), Eq. (24) becomes

$$\begin{aligned} (P_1^{(Z)})_{XY} &= (263(10^3)\,\pi \sin 2\theta_Z/3)^2 \left\{\int_0^\infty \exp[(-2P_{XY}+i\omega_1)\tau]\,d\tau + \text{c.c.}\right\} \\ &= (263(10^3)\pi \sin 2\theta_Z/3)^2\,\tau_{XY}/(1+\omega_1^2\tau_{XY}^2), \end{aligned} \tag{36}$$

where

$$\tau_{XY} \equiv (4P_{XY})^{-1}.$$

The superscript and subscript on P_1 provide a reminder that this is the contribution for a Z-rotation, due only to the interbond jump mechanism. The experiment discussed in Section III,A,1 is well fitted by the result[47]

$$\tau_{XY} = 1.69 \times 10^{-8}\,T^{-1}\exp(0.58\,ev/kT).$$

With $\nu_1 \equiv \omega_1/2\pi \simeq 10\,\text{MHz}$, $\omega_1^2\tau_{XY}^2 \gg 1$ over the entire temperature range of interest so that Eq. (36) gives

$$(P_1^{(Z)})_{XY} = B\sin^2 2\theta_Z; \qquad B \equiv [263(10^3)/6\nu_1]^2\,\tau_{XY}^{-1}. \tag{37}$$

A similar procedure leads to

$$(P_2^{(Z)})_{XY} = (B/8)\cos^2 2\theta_Z. \tag{38}$$

These transition probabilities are a factor of 4 smaller than given initially.[47] The correction has been pointed out by Schmidt.[50] To complete this portion of the discussion the corresponding X-rotation results are

$$(P_1^{(X)})_{XY} = (B/4)\sin^2 2\theta_X; \qquad (P_2^{(X)})_{XY} = (B/32)(3-\cos 2\theta_X)^2. \tag{39}$$

It follows from the above results that all the spin transition rates *which are a result only of interbond exchange* are $<10^{-4}P_{XY}$. Thus the interbond jumping maintains equal populations in corresponding Zeeman levels of X and Y bond deuterons as they return to a Boltzmann equilibrium after any cw or pulse saturation. The variables U_- and V_- of Eq. (12) are thus equal so that Eqs. (13c) and (13d) become

$$(dU_-/dt) = -P_X U_-, \qquad (dV_-/dt) = -P_Y V_-,$$

with P_X, P_Y given in Eq. (14). The return to a Boltzmann population is then with time constant

$$(T_1)_{XY} = (P_Y)^{-1} = (P_X)^{-1}.$$

Thus in Z-rotation, for example, the rate at which the deuteron spin system returns to a Boltzmann population, due only to interbond exchange, is

$$(T_1^{(Z)})_{XY}^{-1} = (P_1^{(Z)})_{XY} + 2(P_2^{(Z)})_{XY}. \tag{40}$$

While the foregoing results have been treated in more detail than necessary for the NMR specialist, the analysis should provide for the nonspecialist a platform from which to view the central results of the next section with some degree of confidence.

IV. Relaxation Time Studies of Critical Fluctuations Near T_C

Recent nuclear spin relaxation results in KD*P-type crystals constitute an exceptional example of the usefulness of NMR in providing information on the dynamics of collective motion near phase transitions. In this section a review is given of these experiments. After formulating the connection between intrabond deuteron motion and spin-lattice relaxation, the inadequacy of theoretical models, which include only short range correlation of hydrogen motion, is demonstrated by the lack of agreement they give with the critical deuteron T_1 dependence as $T \rightarrow T_C$ from above.

Expanding the deuteron motion in collective coordinates, the Fluctuation

[50] V. Schmidt, *Phys. Rev.* **164**, 749 (1967).

Dissipation Theorem (FDT) may be used to relate time dependence of the deuteron motion to the imaginary part of the dielectric susceptibility. Using the dielectric susceptibility as obtained from a dynamic Ising model, together with inelastic neutron scattering data, the proper choice of a single parameter, the noninteracting deuteron jump time τ_C, leads to excellent agreement between theory and the experimental T_1 behavior. The value of τ_C is found to be consistent with a comparable parameter obtained from dielectric relaxation experiments.

In Section IV,B a comparison between deuteron T_1 results and relaxation results for the heavy atoms demonstrates, at least near T_C, a striking similarity in dynamic behavior. This lends further credibility to the coupled hydrogen-heavy atom mode discussed in Section II,A,2.

A. Deuteron Intrabond Fluctuations

1. *The Connection between Deuteron T_1 and Interbond Motion*

As may be seen from the elementary D_2PO_4 Slater dipoles of Fig. 1, hydrogen motion across the bond reverses the contribution of that particular hydrogen to polarization along the *c*-axis. To completely reverse the orientation of a D_2PO_4 group, with respect to the *c*-axis, actually requires movement across the bond for all four surrounding deuterons of a PO_4 tetrahedron. As was evident from the deuteron quadrupole splitting below T_C, discussed in Section III,A,2, motion across the bond changes orientation of the EFG tensor at the deuteron site. As a result the intrabond motion produces fluctuations in both crystal polarization and the deuteron EFG tensor. The polarization fluctuations are related to the imaginary part of the dielectric susceptibility (χ'') through the FDT, while the EFG fluctuations make a contribution $(T_1)_C^{-1}$ to the deuteron spin-lattice relaxation rate.

It has been well demonstrated that for O–D···O bonds the deuteron quadrupole coupling constant is dependent essentially upon the bond length,[51] and quite independent of the extrabond charge distribution. Thus the change ΔV_K in EFG tensor which results from deuteron motion along the bond is independent of deuteron positions in other bonds. On the other hand, the probability of finding a deuteron on one side of the bond, or the other, does depend upon the surrounding deuteron configuration. As discussed more completely in Appendix A, for the deuteron in bond *m*,

$$\Delta V_K(t) = \tfrac{1}{2}[(1+\sigma_m(t))(V_K(1)-\overline{V_K}) + (1-\sigma_m(t))(V_K(2)-\overline{V_K})]. \quad (41)$$

The variable $\sigma_m(t)$ can take values ± 1. The symbols $V_K(1)$, $V_K(2)$ represent the value of V_K on the two sides of the bond, while $\overline{V_K}$ is the time-averaged value. In the paraelectric phase the deuteron is assumed to

[51] T. Chiba, *J. Chem. Phys.* **41**, 1352 (1964); G. Soda and T. Chiba, *J. Chem. Phys.* **50**, 439 (1969); R. Blinc and D. Hadzi, *Nature* (*London*) **212**, 1307 (1966).

spend equal time on either side of the bond so that

$$\overline{V_K} = [V_K(1) + V_K(2)]/2,$$

and Eq. (41) reduces to

$$\Delta V_K(t) = \sigma_m(t)\Delta V_K/2,$$

with (41a)

$$\Delta V_K \equiv V_K(1) - V_K(2).$$

It should be noted that complete ordering of the deuteron leads to $\sigma_m(t) = 1$, $\overline{V_K} = V_K(1)$, and $\Delta V_K(t) = 0$. Thus, for intrabond motion, Eqs. (24) and (25) give

$$P_K = (A|\Delta V_K|/\hbar)^2 (K/2)\left\{\int_0^{\infty} \overline{\sigma_m(\tau)\sigma_m(0)} \exp(i\omega_K \tau)\, d\tau + \text{c.c.}\right\}. \tag{42}$$

Evaluation of $|\Delta V_K|^2$ is straightforward. It is the proper evaluation of $\overline{\sigma_m(\tau)\sigma_m(0)}$ which is the important ingredient of a successful theory for the anomalous increase in relaxation rate, evident in Fig. 6, as $T \to T_C$.[52]

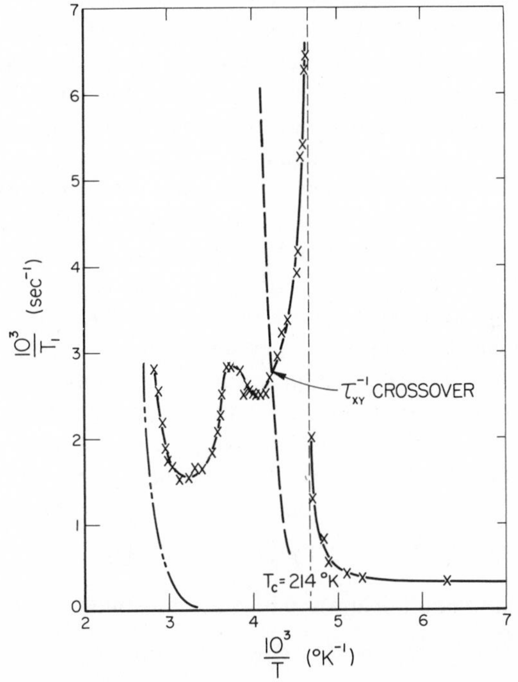

FIG. 6. Temperature dependence of the deuteron spin-lattice relaxation rate in KD*P. The interbond jump rate ($10^3\tau_{XY}^{-1}$), together with the consequent theoretical contribution to relaxation $[10^3(T_1^{-1})_{XY}]$, is included.[52] $\theta_z = 55°$; $\nu_0 = 10.6$ MHz. Solid curve, T_1^{-1}(expt.); dashed curve, τ_{xy}; dot-dash curve, $(T_1^{-1})_{xy}$.

[52] R. Blinc, J. Stepisnik, M. Jamsek-Vilfan, and S. Zumer, *J. Chem. Phys.* **54**, 187 (1971). This paper will be designated as BSJZ.

The calculated contribution of interbond jumping is, as mentioned in the previous section, lower by a factor of 4 than given in BSJZ.[52] The anomalous increase near 270°K has been tentatively attributed to cross-relaxation.[52] An alternative explanation will be given in Section IV,A,3,a.

2. *Small Cluster Models*

a. *The Dynamic SUS Model.* The first proposal for evaluation of $\overline{\sigma_m(\tau)\,\sigma_m(0)}$ was based upon the modified Slater model mentioned briefly in Section II,A,2.[50] In the completely ordered state well below T_C, all D_2PO_4 groups in a domain polarized along the $(+c)$-axis will be of the form F, F′ in Fig. 1. As the crystal is warmed, a pair of defect (Takagi) groups (G, G′) are formed by thermally activated motion of a single deuteron across the bond at a cost of energy $2\varepsilon_1$. With an increase in temperature, further motion causes "chains" to form, consisting of nonpolar groups (B–E, Fig. 1) which terminate in Takagi (T) groups. Keeping in mind that Fig. 1 is a projection only, such chains cannot close on themselves, but have a spiral-like configuration in the general direction of the c-axis. The mean time between intrabond jumps of individual deuterons is designated τ_0, and assumed to be temperature independent. An individual deuteron may undergo many such jumps within a bond before a deuteron in an adjacent bond moves in phase with it, to lengthen the chain and move the T-group to an adjacent location. The mean time for such T-group motion is designated $\tau_{0e} > \tau_0$.

Designating as "correct" that side of the bond at which a deuteron makes a positive contribution to the fractional spontaneous polarization p, in an equilibrium region containing $N\,(D_2PO_4)$-groups there are $N_c = (1+p)(N/V)$ deuterons per unit volume at correct positions, and $(1-p)(N/V)$ at "wrong" positions. The effective jump rate per unit volume is $4(\delta_1+\delta_{-1})(N/V\tau_{0e})$, so that a deuteron in a correct position has a probability per unit time, $2(\delta_1+\delta_{-1})/(1+p)\,\tau_{0e}$, of jumping to the wrong position. The contrary probability has $(1-p)$ in the denominator. If in a region containing $2N$ deuterons N_c fluctuates from its equilibrium value, the return to equilibrium is governed by

$$N_c = [N_c - N(1+p)]\,4(\delta_1+\delta_{-1})/(1-p^2)\,\tau_{0e},$$

so the approach to equilibrium is with time constant

$$\tau_{we} \equiv (1-p^2)\,\tau_{0e}/4(\delta_1+\delta_{-1}).$$

The autocorrelation function then becomes

$$\overline{\sigma_m(\tau)\,\sigma_m(0)} = G(\tau) = [(1-p^2)\exp(-\tau/\tau_{we}) + (1-p)^2]/4, \qquad (43)$$

where τ_{we} and p are now ensemble values. Above T_C, $p=0$, and $\tau_{we} = \tau_{0e}/4(\delta_1+\delta_{-1})$. From the modified Slater theory of SUS,[12] when

$p = 0$, $4(\delta_1 + \delta_{-1}) \simeq \exp(-\varepsilon_1/kT)$.[50] The time-dependent part of $G(\tau)$ then reduces to

$$G(\tau) = [\exp(-\tau/\tau_{we})]/4,$$

with

$$\tau_{we} = \tau_{0e} \exp(\varepsilon_1/kT).$$

Then, as in Eq. (36), Eq. (42) becomes

$$P_K = (A|\Delta V_K|/h)^2 (K/2)[\tau_{we}/2(1 + \omega_K^2 \tau_{we}^2)].$$

Above T_C, dielectric relaxation experiments require $\tau_0 \simeq 10^{-13}$ sec, so that at NMR frequencies, $\omega_K \tau_{we} \ll 1$. The predicted contribution to the deuteron spin-lattice relaxation rate, $(T_1)_C^{-1}$, from intrabond motion is then exponential.

$$(T_1)_C^{-1} = A_0 \exp(\varepsilon_1/kT). \qquad (44)$$

Sufficient error was present in the initial experiments so that this exponential dependence could not be ruled out.[50]

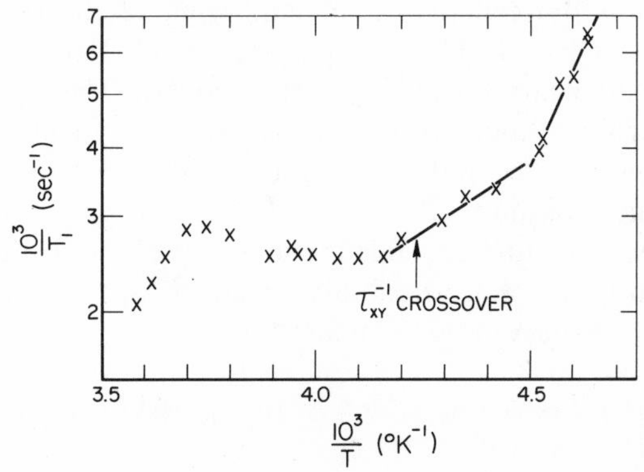

FIG. 7. The data of Fig. 6 replotted in a form suitable for comparison with Eq. (44). $\theta_z = 55°$; $\nu_0 = 10.6$ MHz.

In Fig. 7 the more recent data from Fig. 6 is replotted in a form suitable for comparison with Eq. (44). Discounting the anomalous peak near 270°K $[(10^3/T) \simeq 3.7]$, it is clear that the data is not well fitted by this simple dependence. Over an $\simeq$10°C interval just above T_C, the straight line shown gives $(\varepsilon_1/k) = 3640$°K. This is a factor of 4 larger than the value which gives a good fit to dielectric data.[12,50] The straight line slope at higher temperature gives $(\varepsilon_1/k) = 390$°K.

It should be noted that the theory described above does not take into account the effect upon τ_{0e} which results from the intersection of nonpolar group "chains." It would, of course, be exceedingly difficult to do so. This is a general problem of models which involve individual particle coordinates rather than collective coordinates.

The above model does properly predict a rapid decrease in relaxation rate below T_C where $p \neq 0$, and the Takagi group density (hence "chain" density) becomes small. If it is valid at all above T_C it should be near T_C. Unfortunately the large value for (ε_1/k), given by the straight line just above T_C in Fig. 7, seems to indicate that even in this range the predictions are invalid.

b. *The Matsubara Model.* A somewhat different dynamic cluster treatment of the modified Slater model has been given by Matsubara and co-workers.[30,53] Denoting the two possible positions of the jth deuteron in its bond by Ising spin variables, $S_j(\pm 1/2)$, the effective Hamiltonian for a four-deuteron cluster may be written

$$\mathscr{H} = -U(S_0 S_2 + S_1 S_3) - V(S_0 S_1 + S_1 S_2 + S_2 S_3 + S_3 S_0) - \mu \bar{E}(t)(S_0 + S_1 + S_2 + S_3).$$

Such a cluster is illustrated by the numbering of the bonds around the PO_4 group D in Fig. 1. The two parameters U, V represent J_{ij} for the next nearest and nearest neighbors, respectively. The effective field $\bar{E}(t)$, assumed equal for all deuterons in the cluster, is introduced to represent long range interaction with deuterons outside the cluster. It is evaluated by a self-consistency condition.

It is assumed that time-dependent properties of the deuteron system can be described in terms of a probability density $P(\sigma_1 \cdots \sigma_N; t)$ which satisfies a master equation derived by Glauber,[54]

$$(d/dt)\,P(\sigma_1 \cdots \sigma_N; t) = -\sum_j W_j(\sigma_j; \{\sigma\})\,P(\sigma_1 \cdots \sigma_N; t) + \sum_j W_j(-\sigma_j; \{\sigma\}')\,P(\sigma_1 \ldots, -\sigma_j, \ldots, \sigma_N; t).$$

For convenience the notation $\sigma_j = \pm 1$ is used rather than $S_j \equiv (\sigma_j/2)$. The notation $W_j(\sigma_j; \{\sigma\})$ represents a transition probability per unit time, $\sigma_j \to -\sigma_j$, for the jth deuteron in a configuration $\{\sigma\} = \sigma_1 \cdots \sigma_j \cdots \sigma_N$, while $W_j(-\sigma_j; \{\sigma\}')$ represents the reversed transition, $-\sigma_j \to \sigma_j$, when $\{\sigma\}' = \sigma_1 \cdots -\sigma_j \cdots \sigma_N$. With the imposition of detailed balancing on $W_j(\sigma_j; \{\sigma\})$, moment equations are constructed which take the form

$$-\tau_c(d\langle\{\sigma\}_n\rangle/dt) = \sum_{j=1}^{n} \langle\{\sigma\}_n[1 - \sigma_j \tanh(\beta E_j/2)]\rangle.$$

[53] K. Yoshimitsu and T. Matsubara, *Progr. Theor. Phys., Suppl.* Extra No., p. 109 (1968).
[54] R. Glauber, *J. Math. Phys.* **4**, 294 (1963).

Here $\beta \equiv (kT)^{-1}$, τ_c is a relaxation time, E_j is the local field at deuteron j, and the nth-order moment is defined by

$$\langle\{\sigma\}_n\rangle = \sum_{\sigma=\pm 1}\left(\prod_{j=1}^{n}\sigma_j\right)P(\sigma_1\cdots\sigma_N;t) \quad \sum_{\sigma=\pm 1} P(\sigma_1\cdots\sigma_N;t).$$

Specialization of this formalism to even the four-deuteron cluster leads to very complex algebraic expressions which can only be solved with several approximations. Above T_C the correlation function becomes finally

$$G(\tau) = \text{const.} + A\exp(-\tau/\tau_1) + B\exp(-\tau/\tau_2). \tag{45}$$

The parameters τ_1, τ_2, A, and B are all complicated functions of $(T-T_C)/T_C$ which must be evaluated numerically. Graphs of all parameters as a function of $(T-T_C)/T_C$ were given for the value $(\varepsilon_1/\varepsilon_0) = 5$, which was the best choice for a fit of the model to the static properties of KD*P.[20] For this choice the parameter τ_2 is temperature independent, while $\tau_1 \to \infty$ as $T \to T_C$. Over the entire range of $(T-T_C)/T_C$ for which NMR experiments have been performed in KDP-type crystals, $B \ll A$ and $\tau_2 \ll \tau_1$. While comparison with deuteron relaxation data can only be done numerically, it has been pointed out in BSJZ that the predicted critical anomaly is much sharper than that observed experimentally.

In generalizing their model to larger clusters, Yoshimitsu and Matsubara find a distribution of relaxation times. This might well soften the predicted critical dependence. Again the problem of algebraic complexity makes comparison with experiment virtually impossible.

While the two-cluster models discussed (as well as others[16]) seem quite adequate for describing equilibrium properties, algebraic difficulties preclude their application to dynamic aspects of the critical behavior. However, before introducing a collective coordinate description in the next section, it will be worthwhile to consider a computer simulation of the dynamic Matsubara model. The results will provide a very convenient framework for explaining some unusual features of the ^{75}As spectrum to be presented in Section V,A,2.

The computer version [30] is a 128×128 two-dimensional square lattice of hydrogen bonds, with four hydrogens surrounding each fixed point charge PO_4 group. The Hamiltonian for the ith hydrogen, in the presence of an external field E, may be written as

$$\mathscr{H}_i = -\left(\sum_j J_{ij}\sigma_j + E\right)\sigma_i. \tag{46}$$

Eigenvalues of σ_j are ± 1, with $J_{ij} = V$ for the four nearest hydrogen neighbors, U for the two next nearest neighbors, and 0 for all others. In this version of the two-dimensional Ising model, the dipole moments are normalized to unity.

The potential minima on either side of a constant potential hill of height W depend upon the six neighbor configurations, as well as E. Denoting the entire bracket in Eq. (46) by E_i, the transition probability of a deuteron from the $(+)$ to the $(-)$ side of the bond is expressed as

$$p(\sigma_i \rightarrow -\sigma_i; E_i) = (C/\tau_0)\exp[-(W-\sigma_i E_i)/kT]. \tag{47}$$

An exact solution of the two-dimensional square Ising model with $U = 0$ has been given.[55] From that result a normalized transition temperature, $T_C = 1$, requires $V = 3.526$. The arbitrary constant (C/τ_0) is so chosen that

$$p(\sigma_i \rightarrow -\sigma_i; E_i) = 0.7\exp[-\{(\sigma_i E_i)_{\max} - \sigma_i E_i\}/kT].$$

With transition probabilities calculated in advance for all configurations of nearest and next nearest neighbors, simulation is begun with the completely ordered state (all $\sigma_i = 1$). A bond is chosen by random number selection, and the calculated transition probability compared with another random number between 0 and 1. When the transition probability is larger than this random number, the sign of the spin is changed. After repeating this 2^{15} (i.e. $\simeq 3.3 \times 10^4$) times, one unit of time is considered to have lapsed, and the entire hydrogen configuration is determined. A moving picture was prepared in which a bright spot appears in any bond with its hydrogen on one side, and a dark spot in those bonds with hydrogens on the other side. General features of the moving picture are as follows: (1) At $T = 0$ the entire lattice is ordered and appears dark. (2) As T is increased isolated bright scintillations appear, indicating momentary excursions of isolated hydrogens across the bond. (3) With a further increase of T, extensive bright clusters form within which rapid scintillations may be seen. (4) As $T \rightarrow T_C = 1$, the bright clusters, which are of random size, occupy essentially half the total lattice area. While the cluster boundaries move with time, and small clusters collapse and reform, the time scale for appreciable change in size of large clusters is very large compared with that observed for scintillations within the cluster. (5) Well above T_C, individual hydrogen fluctuations take place almost independently of the configuration of neighbors, and no large clusters form.

It is tempting to associate τ_1 in Eq. (45) with the mean lifetime of an average cluster, with τ_2 characteristic of individual hydrogen motions across the bond. The authors point out the futility of attempting to connect the computer time scale with that of motion in a real crystal. We will return to this model in discussing the ^{75}As NMR results.

[55] L. Onsager, *Phys. Rev.* **65**, 117 (1944); B. Kaufman, *Phys. Rev.* **76**, 1232 (1949).

3. *Collective Coordinates, the Fluctuation Dissipation Theorem* (*FDT*), *and Critical Spin-Lattice Relaxation*

A brief introduction to collective coordinate descriptions of critical fluctuations, and the FDT, is provided in Appendix B. In this section extensive use will be made of methods and results introduced there.

By using a collective coordinate representation of hydrogen bond fluctuations, BSJZ have succeeded in obtaining very satisfactory agreement between theory and the increased deuteron spin-lattice relaxation rate evident near T_c in Fig. 6.[52] The individual deuteron intrabond fluctuation $\sigma_m(t)$, at lattice site $\bar{r}_m$, may be expanded in Fourier components as

$$\sigma_m(t) = N^{-1/2} \sum_{\bar{q}} \sigma_{\bar{q}}(t) \exp(i\bar{q}\cdot\bar{r}_m). \tag{48}$$

Assuming no correlation between the different normal modes $\sigma_{\bar{q}}(t)$ Eq. (42) becomes

$$P_K = (A|\Delta V_K|/\hbar)^2 (K/2N) \left\{ \sum_{\bar{q}} \int_0^\infty \overline{\sigma_{\bar{q}}(\tau)\sigma_{-\bar{q}}(0)} \exp(i\omega_k \tau)\, d\tau + \text{c.c.} \right\}. \tag{49}$$

For the deuterons ΔV_K is essentially independent of the extrabond charge distribution. It is thus unnecessary to expand this parameter in terms of normal modes. Making use of the classical form of the FDT, as given in Eq. (B.7) of Appendix B, this becomes

$$P_K = (A|\Delta V_K|/\hbar)^2 (K/2N) \left\{ (2kT/\omega_K) \sum_{\bar{q}} \chi''_{\bar{q}}(\omega_K) \right\}. \tag{50}$$

In order to evaluate Eq. (50) an appropriate choice must be made for $\chi''_{\bar{q}}(\omega_K)$. As pointed out in Appendix B, NMR experiments are conducted at such low frequencies that distinction between relaxation and resonant behavior in the form of $\chi_{\bar{q}}(\omega) \equiv \chi'_{\bar{q}}(\omega) + i\chi''_{\bar{q}}(\omega)$ has not as yet been demonstrated. A molecular field treatment of the dynamic Ising model gives for the susceptibility, normalized to a unit elementary dipole moment[56] (the development is outlined in Appendix C),

$$\chi_{\bar{q}}(\omega) = \beta/[1 - \beta J_{\bar{q}} - i\omega\tau_c], \tag{51}$$

where

$$\beta = 1/kT; \qquad J_{\bar{q}} = J_{-\bar{q}} = \sum_l J_{lm} \exp(i\bar{q}\cdot(\bar{r}_l - \bar{r}_m)]. \tag{52}$$

The parameter τ_c is described as a noninteracting dipolar relaxation time. It is assumed to be independent of the dipolar interactions, depending

[56] M. Suzuki and R. Kubo, *J. Phys. Soc. Jap.* **24**, 51 (1968).

only upon the temperature of the heat bath with which the dipole system is in contact. Quoting from the reference by Suzuki and Kubo;[56] "Here τ_c is a constant which may in general depend on the temperature and on the spins other than the mth one. The latter dependence will introduce arbitrariness and complication unnecessary for our present purpose so that we assume in the following that τ_c depends only on the temperature. It may be regarded as a constant when the consideration is limited only to the neighborhood of the critical temperature." From this point of view τ_c is the minimum mean time required for a dipolar flip, i.e. for a deuteron to move across the bond.

Equation (51) may be written as

$$\chi_{\bar{q}}(\omega) = \chi_{\bar{q}}(0)/(1-i\omega\tau_{\bar{q}}), \tag{53}$$

where

$$\chi_{\bar{q}}(0) \equiv \beta/(1-\beta J_{\bar{q}}) = [k(T-(J_{\bar{q}}/k))]^{-1} \tag{54}$$

and

$$\tau_{\bar{q}} \equiv \tau_c/(1-\beta J_{\bar{q}}). \tag{55}$$

It follows from Eq. (54) that the Curie temperature $T_{\bar{q}}$ is given by

$$T_{\bar{q}} = J_{\bar{q}}/k. \tag{56}$$

Substituting the result [from Eqs. (53)–(55)],

$$\chi''_{\bar{q}}(\omega) = \beta\omega\tau_c/[(1-\beta J_{\bar{q}})^2 + \omega^2\tau_c^2], \tag{57}$$

together with the definition

$$A_K \equiv K(A|\Delta V_K|/\hbar)^2 \tag{58}$$

in Eq. (50) gives

$$P_K = (A_K\tau_c/N)\sum_{\bar{q}} [(1-\beta J_{\bar{q}})^2 + \omega_K^2\tau_c^2]^{-1}. \tag{59}$$

Replacing the sum by an integration over the Brillouin zone of volume $(2\pi)^3/V$,

$$P_K = A_K\tau_c\mathscr{I}, \tag{60}$$

with

$$\mathscr{I} \equiv [NV/(2\pi)^3] \iiint [(1-\beta J_{\bar{q}})^2 + \omega_K^2\tau_c^2]^{-1}\, d\bar{q}^3. \tag{61}$$

Coherent neutron scattering results on KD*P have been fitted to an Ising model of the form described above.[42] One Ising spin $\sigma_l = \pm 1$ is assigned to each primitive cell with short range interaction extending over eight nearest neighbors plus a Coulomb interaction approximated by its singular form as $q \to 0$. The parameter equivalent to $T_{\bar{q}}$ is written as

$$T_{\bar{q}} = T_0\{1-(8T_\alpha/T_0)[1-\cos(aq_X/2)\cos(aq_Y/2)\cos(cq_Z/2)]-(T_\delta/T_0)\cos^2\theta\}, \tag{62}$$

with a, c the lattice parameters as given in Section II,A,1 and θ the angle between the polarization (c) axis and $\bar{q}$. Two different ways of fitting the experimental data gave the values for T_0, T_α, T_δ in Table IV. The discrepancies give some measure of the uncertainties.

TABLE IV

PARAMETERS T_0, T_α, AND T_δ AS DETERMINED FROM COHERENT NEUTRON SCATTERING[42]

	T_0(°K)	T_α(°K)	T_δ(°K)
Computer fit	228.5 ± 0.7	10.90 ± 0.05	258 ± 1
Graphical analysis	223.5 ± 1	7.3 ± 1.4	180 ± 20

With the approximations

$$a \simeq c \simeq 7.3A°, \qquad \cos(aq_i/2) \simeq 1-(aq_i)^2/8, \tag{63}$$

Eq. (62) becomes

$$T_{\bar{q}} = T_0\{1-\alpha q^2-\delta\cos^2\theta\},$$

a form used previously by Bonera *et al.*[57] Using the "Graphical Analysis" parameters of Table IV which give better agreement with the measured transition temperature and a more reasonable $\delta(<1)$, one has

$$\alpha = 1.75(\text{Å})^2, \qquad \delta = 0.81. \tag{64}$$

In the case of isotropic dipolar interaction ($\delta = 0$), the integral of Eq. (61) has been evaluated.[58] For $T > T_0$ the rather lengthy expression reduces to

$$\mathscr{I}_{\text{iso.}} = B(T/T_0)^2\{\varepsilon^{-1/2}\tan^{-1}u_m - u_m^2/\alpha^{1/2}q_m(1+u_m^2)\}, \tag{65}$$

with

$$B \equiv V/(2\pi)^2\alpha^{3/2}, \qquad u_m \equiv q_m(\alpha/\varepsilon)^{1/2}, \qquad \varepsilon \equiv (T-T_0)/T_0, \tag{66}$$

when

$$\omega\tau_c \ll (T-T_{\bar{q}})/T. \tag{67}$$

The parameter q_m is the radius of an "equivalent" spherical Brillouin zone, of volume equal to that of the actual zone. Thus

$$q_m = (6\pi^2/V)^{1/3} = 0.54(\text{Å})^{-1} \tag{68}$$

for KD*P with unit cell volume $V = 390(\text{Å})^3$.

[57] G. Bonera, F. Borsa, and A. Rigamonti, *Phys. Rev. B* **2**, 2784 (1970).
[58] R. Blinc, S. Zumer, and G. Lahajnar, *Phys. Rev. B* **1**, 4456 (1970).

For large anisotropy ($\delta \simeq 1$), $\mathscr{I}$ has not been given in closed form. The dominant terms when $\varepsilon \to 0$ may be approximated by[57]

$$\mathscr{I}_{\text{aniso.}} \simeq B(T/T_0)^2 \{(\pi/2\delta^{1/2})[\ln \varepsilon^{-1/2} + \ln((\varepsilon+\delta)^{1/2} + \delta^{1/2})] - (\alpha^{1/2} q_m)^{-1}\}. \tag{69}$$

Note that this does reduce to Eq. (65) in the limit $\delta \to 0$, while for $\delta \to 1$ the dominant temperature dependence as $\varepsilon \to 0$ is $(\ln \varepsilon^{-1})$. This may be contrasted with the $\varepsilon^{-1/2}$ dependence of Eq. (65) in the same limit.

In general the integral of Eq. (61) may be expressed as

$$\mathscr{I} = B(T/T_0)^2 \{\varepsilon; \alpha^{1/2} q_m, \delta\}, \tag{70}$$

where the bracket $\{\varepsilon; \alpha^{1/2} q_m, \delta\}$ may be determined by numerical evaluation of the integral as a function of ε, for appropriate values of the parameters $(\alpha^{1/2} q_m)$ and δ. Within the approximation of Eq. (67), "universal" curves of $\{\cdots \text{anis} \cdots\}$ versus $\ln \varepsilon^{-1}$ and $\{\cdots \text{iso} \cdots\}$ versus $\varepsilon^{-1/2}$ are plotted in Fig. 8. In both cases $\alpha^{1/2} q_m = 0.71$, appropriate to KD*P. For $\{\cdots \text{anis} \cdots\}$, $\delta = 0.81$ while for $\{\cdots \text{iso} \cdots\}$, $\delta = 0$.

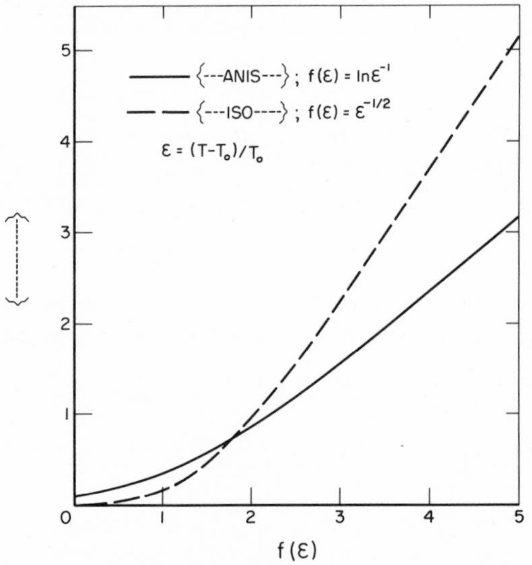

FIG. 8. Theoretical "universal curves" of the bracket in Eq. (70) for the anisotropic ($\alpha^{1/2} q_m = 0.71$, $\delta = 0.81$), as well as the isotropic ($\alpha^{1/2} q_m = 0.71$, $\delta = 0$) case for KD*P.

a. *The Collective Model and T_1 in KD*P.* If the interbond exchange rate P_{XY} is negligible compared with P_X it follows from Eq. (13c) that U_- returns to equilibrium with time constant $T_{1X} = P_{1X} + 2P_{2X}$. Orienting the crystal such that the two transitions for X bond deuterons are at the same frequency, the population deviations satisfy $x_+ - x_0 = x_0 - x_- = (x_+ - x_-)/2$. This condition is realized at $\theta_Z = 55°$, as may be seen from Fig. 5. Under these circumstances the contribution to spin-lattice relaxation from intra-bond motion follows from Eq. (60), i.e.

$$(T_1)_C^{-1} = P_1 + 2P_2 = (A_1 + 2A_2)\tau_c \mathscr{I}. \tag{71}$$

It is apparent from the $T < T_C$ results of Table III that, as an X bond deuteron moves from one end of the bond to the other, only the V_{YZ} component of the EFG tensor fluctuates. Making use of Eqs. (22) a straightforward transformation to the laboratory system gives

$$|\Delta V_1|^2 = 4V_{YZ}^2 \sin^2\theta_Z, \qquad |\Delta V_2|^2 = 4V_{YZ}^2 \cos^2\theta_Z. \tag{72}$$

Substituting from Eqs. (23), (58), and (72) in Eq. (71) gives

$$(T_1)_C^{-1} = (1 + 3\cos^2\theta_Z)(eQV_{YZ}/2\hbar)^2 \tau_c \mathscr{I}, \tag{73}$$

where

$$(eQV_{YZ}/2\hbar) = [2\pi(21kHz)/18].$$

Substituting in Eq. (73) from Eq. (70), together with the appropriate numerical factors for KD*P, gives

$$(T_1)_C^{-1}(T_0/T)^2 = 8.1(10^9)\tau_c\{\ldots\}. \tag{74}$$

From the date of Fig. 6 and the anisotropic "universal curve" of Fig. 8, one may make the plot shown in Fig. 9. The slope of Fig. 9, together with Eq. (74), leads to the value

$$\tau_c = 2.7 \times 10^{-13}\,\text{sec}. \tag{75}$$

The data of Fig. 6 is also plotted in Fig. 9 as a function of $\{\cdots \text{iso} \cdots\}$ to show the distinctly better fit which results from assuming anisotropic dipolar interaction.

The correlation time τ_c may be compared with the reciprocal of a parameter A described by Hill and Ichiki[59] as having "... the dimensions of frequency and might be thought of as a maximum rate at which a single dipolar unit cell could reorient itself. Any fluctuations involving more than one dipole, or a dipole with a large, inhibiting, local field would occur at a slower rate." These authors are able to fit the dielectric relaxation data, over the temperature range $0 < (T - T_C) \lesssim 35°\text{C}$, in KD*P with a temperature-independent value of $A^{-1} = 8(10^{-13})$ sec.[60] This agreement within a factor of $\simeq 3$ must be considered quite satisfactory in view of the diversity of the experiments. It should be kept in mind that four deuterons

[59] R. Hill and S. Ichiki, Phys. Rev. **128**, 1140 (1962).
[60] R. Hill and S. Ichiki, Phys. Rev. **130**, 150 (1963).

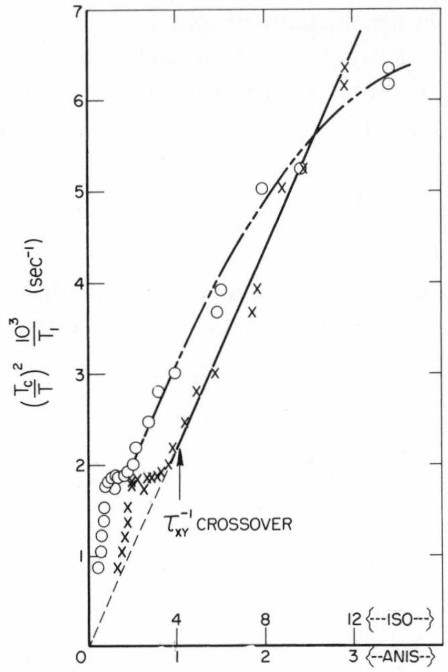

FIG. 9. The deuteron relaxation data of Fig. 6 replotted in a form suitable for comparison with Eq. (74). Dependence of the brackets $\{\cdots \text{anis} \cdots\}$ (solid line) and $\{\cdots \text{iso} \cdots\}$ (dot-dash line) upon $\varepsilon \equiv (T - T_C)/T_C$ is given in Fig. 8. $\theta_z = 55°$; $\nu_0 = 10.6$ MHz.

must move across their bonds to reverse one D_2PO_4 dipole. With modes corresponding to an appreciable range of $\bar{q}$ values contributing to the deuteron motion, the mean time for a single deuteron to move across a bond will be appreciably less than that for all four deuterons surrounding a PO_4 group to change locations in such a way as to reverse the D_2PO_4 dipole.

For experiments in the radio-frequency range it is clear that

$$\omega\tau_c \ll (T - T_0)/T \leqslant (T - T_{\bar{q}})/T$$

is well satisfied for $(T - T_0) \gtrsim 10^{-2}$°C. This validates the assumption of Eq. (67).

In describing the physical significance of τ_c the possibility of a dependence upon the heat bath temperature, with which the dipole system is in contact, has been left open. If one assumes a dependence of the form $\tau_0 \exp(E_a/kT)$, the rather good straight line fit of Fig. 9 implies a very high activation energy, E_a. In making such a fit to the data over an extended temperature range, it is necessary that proper account be taken of other contributions to the relaxation rate which may be temperature dependent. Here again the KD*P results illustrate the subtlety of pitfalls which may occur.

It is quite clear that the peak in relaxation rate near 260°K of Fig. 6 is not described by Eq. (71). It was suggested in BSJZ that the origin might be "cross relaxation between a multiple ^{39}K transition and the deuteron system." A more plausible explanation is apparent if one considers the interbond exchange rate τ_{XY}^{-1}, shown as a dotted curve in Fig. 6. Quite clearly the anomally occurs just above the temperature at which $(\tau_{XY})^{-1} \simeq (T_1)^{-1}$. The T_1 data was obtained using a cw spectrometer to saturate the deuteron transition at high rf field, and then observe the recovery in successive passages through resonance. Presumably the saturation is accomplished in a time short compared with T_1. At temperatures low enough so that $\tau_{XY} \gg T_1$, the X bond deuteron spin system under observation is independent of the Y bond system. However, as the temperature is increased, and τ_{XY} becomes comparable with T_1, appreciable interbond exchange occurs during the relaxation time measurement. With τ_{XY} long compared to the time required for initial spin system saturation, only the X bond system is saturated. However, during the recovery time a mixing occurs between saturated X bond spins and unsaturated Y bond spins. The X bond spin population appears to recover its equilibrium population more quickly, thus shortening the apparent T_1. The relaxation will not be precisely exponential under such circumstances but can probably be fitted to an exponential, within the experimental error.

At even higher temperatures a further complication enters the interpretation. In the range $260°\text{K} \lesssim T \lesssim 330°\text{K}$ the interbond exchange rate is fast enough so that during saturation of the X bond resonance both the X and Y bond spin systems are saturated. However, the previously discussed interbond contribution to the total spin-lattice relaxation rate is still unimportant. The interbond exchange will nevertheless modify the intrabond contribution. In evaluating Eq. (72) one must average $\cos^2 \theta_Z$ over the X and Y bonds, for which θ_Z differs by 90°. This effect leads to an $\simeq 20\%$ increase in relaxation rate near room temperature. Thus, for comparison with the foregoing theory of intrabond relaxation, the data points near room temperature should be lowered by $\simeq 20\%$. Of course nonequality of adjacent level spacings for deuterons during their "dwell time" in Y bonds will introduce a slight perturbation from simple exponential relaxation.

This rather lengthy discussion has been presented to emphasize the importance of having an abundance of good data from a small temperature interval very near T_0. Fortunately, contributions to the spin-lattice relaxation rate from other mechanisms usually have much less temperature dependence near T_0 than those from critical effects.

In plotting the data of Fig. 9 no "background" contribution to the relaxation rate has been subtracted. Subtraction of a constant background has little effect upon the slope. The primary effect is to shift the entire curve downward. Within the experimental accuracy of the data near T_0, any attempt to make such a subtraction seems unwarranted.

The assumption has been made that T_0 is the "free" crystal phase transition temperature. This needs some justification. As has been pointed out by Paul *et al.*,[42] the neutron scattering measurements are at sufficiently large wave vectors that "... frequencies in the spectrum of the ferroelectric mode were significantly lower than the frequency of the acoustic mode of the same wave vector. The acoustic mode can therefore follow the ferroelectric fluctuations, and conditions, extrapolated to zero wave vector, are those of the free crystal. Thus T_0 should be equal to the transition temperature." In contrast Raman scattering is at much smaller wave vectors so it should have singular behavior at the Curie temperature of the clamped crystal.

B. Heavy Atom Fluctuations

1. ^{31}P *Spin-Lattice Relaxation*

While attention has thus far been focused upon the deuteron experiments, similar critical relaxation effects were noted earlier in the spin-lattice relaxation rate of ^{31}P nuclei in KH_2PO_4.[61] The sharp increase in (T_1^{-1}),

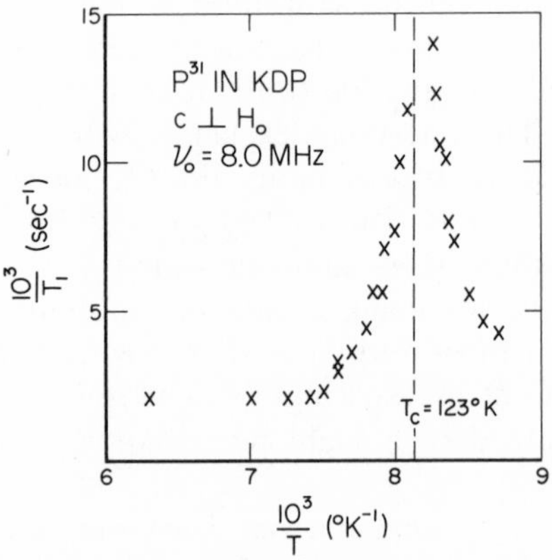

FIG. 10. Temperature dependence of the ^{31}P spin-lattice relaxation rate in KDP.[61] $c \perp H_0$; $\nu_0 = 8.0$ MHz.

[61] R. Blinc and S. Zumer, *Phys. Rev. Lett.* **21**, 1004 (1968).

evident near $T = T_C$ in Fig. 10, results from modulation of the proton–phosphorous magnetic dipolar interaction as a result of hydrogen intra-bond motion.

Neglecting any coupling between ^{31}P nuclei, the relaxation rate for the jth ^{31}P nucleus as a result of its dipolar coupling with surrounding protons is[62]

$$(T_1)^{-1} = A_P^2\left\{\sum_{K=0}^{2} d_K J^{(K)}(\omega_K)\right\}, \tag{76}$$

where

$$A_P^2 \equiv 3(\gamma_H \gamma_P \hbar/2)^2, \qquad d_0 = 1/12, \qquad d_1 = 3/2, \qquad d_2 = 3/4,$$
$$\omega_0 \equiv (\omega_H - \omega_P), \qquad \omega_1 = \omega_P, \qquad \omega_2 = (\omega_H + \omega_P), \tag{77}$$

and

$$J^{(K)}(\omega_K) = \sum_i \int_{-\infty}^{\infty} \overline{\Delta F_{ij}^{(K)}(\tau)\Delta F_{ij}^{(K)}(0)^*} \exp(i\omega_K \tau)\, d\tau. \tag{78}$$

The summation is over all surrounding protons. In writing Eq. (76) it has been assumed that the much shorter proton T_1 maintains the proton spin system in equilibrium when the ^{31}P system is observed. Subscripts H and P refer to protons and ^{31}P, respectively, with γ and ω the gyromagnetic ratios, and Larmor frequencies, of these $I = 1/2$ nuclei.

Instead of Eq. (41a) one now has (above T_c)

$$\Delta F_{ij}^{(K)}(t) = \sigma_i(\Delta F_{ij}^{(K)}/2), \tag{79}$$

where $\Delta F_{ij}^{(K)}$ is the difference in interaction coefficient between the jth ^{31}P nucleus and the ith proton, with the proton at the two ($T < T_c$) equilibrium sites of the bond.

The $F_{ij}^{(K)}$ are given by

$$F_{ij}^{(0)} = (1 - 3\cos^2\theta_{ij})/r_{ij}^3,$$
$$F_{ij}^{(1)} = (\sin\theta_{ij}\cos\theta_{ij})\exp(-i\phi_{ij})/r_{ij}^3, \tag{80}$$
$$F_{ij}^{(2)} = (\sin^2\theta_{ij})\exp(-2i\phi_{ij})/r_{ij}^3,$$

with θ_{ij}, ϕ_{ij} the usual polar angles between the external magnetic field $\bar{H}_0$ and the internuclear vector $\bar{r}_{ij}$ from the ith to the jth spin.

Expanding in normal modes as before, and assuming that the long wavelength ($\bar{q} \to 0$) values of $F_{ij}^{(K)}$ are valid for all $\bar{q}$,

$$J^{(K)}(\omega_K) = (4N)^{-1}\left[\sum_i \overline{|\Delta F_{ij}^{(K)}|^2}\right] \sum_{\bar{q}} \int_{-\infty}^{\infty} \overline{\sigma_{\bar{q}}(\tau)\sigma_{-\bar{q}}(0)} \exp(i\omega_K\tau)\, d\tau. \tag{81}$$

[62] A. Abragam, "The Principles of Nuclear Magnetism." Oxford Univ. Press, London and New York, 1961.

The approximation, while reasonable since primarily low $\bar{q}$ values of susceptibility are strongly enhanced near T_c, will somewhat overestimate the critical relaxation rate. The rigid lattice ($\bar{q}=0$) sums, as determined from the known crystal structure, are[61]

$$\sum_i \overline{|\Delta F_{ij}^{(K)}|^2} = S^{(K)},$$

with

$$S^{(0)} = .0.69(10^{-3})\,\text{Å}^{-6}, \qquad S^{(1)} = 1.65(10^{-3})\,\text{Å}^{-6},$$
$$S^{(2)} = 5.82(10^{-3})\,\text{Å}^{-6}.$$

Using again the classical form of the FDT, Eq. (76) becomes

$$(T_1)^{-1} = (A_{\text{P}}^2/4N)\left[\sum_{K=0}^{2} d_K S^{(K)}\right](2kT/\omega_K)\sum_{\bar{q}} \chi_{\bar{q}}''(\omega_K).$$

With $\chi_{\bar{q}}''(\omega)$ as given in Eq. (57), one then has

$$(T_1)^{-1} = \left[\sum_{K=0}^{2} A'_K\right]\tau_c \mathscr{I}, \qquad A'_K \equiv (A_{\text{P}}^2\,\text{d}_K S^{(K)}/2). \tag{82}$$

The factor which multiplies $\tau_c\mathscr{I}$ has numerical value, $2.4\times 10^8\,\text{sec}^{-2}$. Using the same values for (α,δ) as in KD*P, the final result is given by Eq. (74) with the numerical coefficient (8.1) replaced by (10.1); i.e., for ^{31}P,

$$(T_1)_{\text{C}}^{-1}(T_0/T)^2 = 10.1(10^9)\,\tau_c\{\cdots\}. \tag{74a}$$

The slope of the $(T > T_c)$ ^{31}P data, as plotted against $f(\varepsilon) = \{\cdots \text{anis}\cdots\}$ in Fig. 11, is essentially independent of any background relaxation rate $[10^3/(T_1)_{\text{B}}]$ over the entire range from 0 to $1.9\,\text{sec}^{-1}$, which is the high temperature asymptote of Fig. 10. Comparing Eq. (74a) with the slope of this data as $T\to T_0$ gives $\tau_c = 4.3\times 10^{-13}\,\text{sec}$. Thus the noninteracting deuteron dipolar relaxation time just above T_0 in KD*P, as determined from the deuteron relaxation data, is $\simeq 40\%$ less than the corresponding parameter obtained from the ^{31}P results in KDP. In view of the experimental errors, and approximations, it would be presumptuous to consider this difference significant. Rather, in view of the considerable difference in experiments, the fact that the values are so nearly the same does give credibility to the numerical value.

The ^{31}P data has been plotted in Fig. 11 against $\{\cdots \text{iso}\cdots\}$ to demonstrate the somewhat poorer fit to a straight line obtained under the assumption of isotropic electric dipolar interaction. It should be pointed out that a straight line through the appropriate data just above T_0 should project back through the origin in the absence of some other background

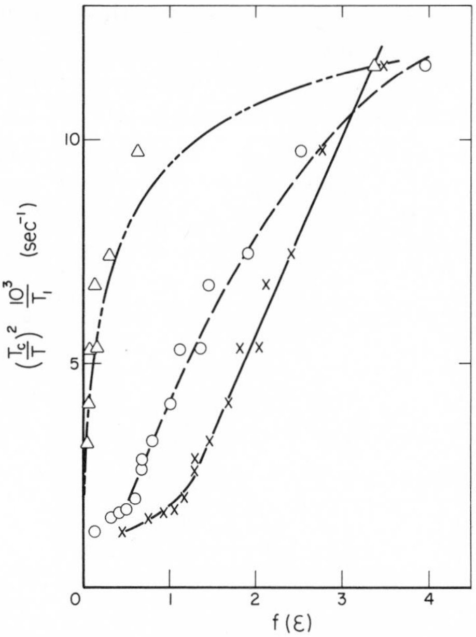

FIG. 11. The ^{31}P data of Fig. 10, plotted in a form suitable for comparison with Eqs. (83a) and (83b). Solid line, $f(\varepsilon) = \{\cdots \text{anis} \cdots\}$; dashed line, $f(\varepsilon) = \{\cdots \text{iso} \cdots\}$; dot-dash line, $f(\varepsilon) = 10^{-2}/\varepsilon^2 T$.

relaxation processes. If such a background is present, the projection should exhibit a positive ordinate intercept as $f(\varepsilon) \to 0$. The fact that this is not so, assuming the same values of (α, δ) as for KD*P, may reflect some difference in these parameters between KDP and KD*P. An equally probable reason is experimental error since a precise location for the measured transition temperature was not given in the original data.[61]

Finally, the data was originally analyzed in terms of a single lattice mode model, with $\chi(\omega) = \omega_0^2\,\chi(0)/(\omega_0^2 - \omega^2 + i\omega 2\Gamma)$.[61] This led to the result

$$(T_1)^{-1}(T_0/T)^2 = A'(\Gamma/T)[T_0/T - T_0)]^2 \equiv (A'\Gamma/\varepsilon^2 T), \qquad (83)$$

with A' a constant. As may be seen from Fig. 11, and as pointed out in the original reference, this model requires a temperature-dependent damping factor, Γ. This was in contrast with the Raman results.[24] The collective mode analysis provides a more consistent interpretation of the T_1 results.

2. ^{133}Cs *Spin-Lattice Relaxation*

A further example of such critical relaxation behavior in KDP-type crystals has recently been observed for the ^{133}Cs nucleus in CsD_2AsO_4.[63]

[63] R. Blinc, M. Mali, J. Slak, J. Stepisnik, and S. Zumer, *J. Chem. Phys.* **56**, 3566 (1972).

Modulation of the ^{133}Cs EFG tensor by what may be described (for information available at NMR frequencies) equally well as an overdamped "soft" ferroelectric lattice mode which condenses at the Curie temperature, or relaxation behavior with correlation time approaching infinity, provides the dominant relaxation mechanism for a relatively large temperature interval above T_C. Deuteron interbond motion which dominates the deuteron T_1 masking effects of the "soft mode" at high temperature is ineffective in relaxing the ^{133}Cs spin. The results in Fig. 12 were taken at an orientation

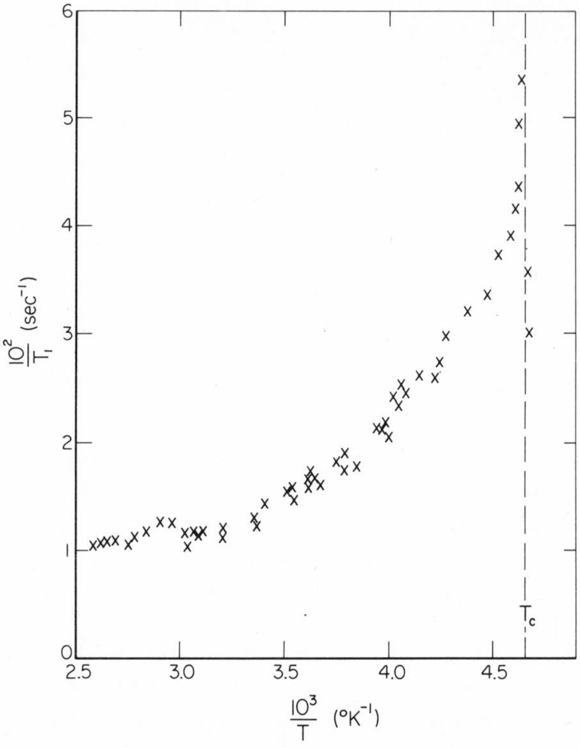

FIG. 12. Temperature dependence of the ^{133}Cs relaxation rate in CsD*A.[63] $\nu = 8.4$ MHz; $\bar{a} \perp \bar{H}_0$; $\measuredangle\ \bar{c}, \bar{H}_0 = 55$.

where all quadrupole perturbed Zeeman levels of the $I = 7/2$ nucleus are equidistant. In that case a common spin temperature is established, and the return to equilibrium may be described by a single time constant.

The EFG values at the ^{133}Cs sites are assumed to depend upon the instantaneous state of order of the entire collection of local order parameters $p_1, p_2, \ldots, p_N$ of the N elementary electric dipole moments in the crystal.

If the lattice contributions to p_i are tightly coupled to the collective O–D$\cdots$O bond deuteron motion, (T_1^{-1}) should exhibit the same temperature dependence as observed for the deuterons. Within approximations discussed previously for the deuteron and ^{31}P, the final result for ^{133}Cs above T_C is found to be [63]

$$(T_1^{-1}) = (32/147)(e^2Qq/h)^2\,\tau_c\,B(T/T_C)^2\,\{\cdots\}. \tag{84}$$

The parameter $(e^2Qq/h) = 72\,\text{kHz}$ is evaluated from the quadrupole coupling constants below and above T_C. As for the deuteron case, the time averaged EFG tensor in the paraelectric phase is approximately an average of the two ferroelectric tensors from oppositely polarized domains, demonstrating the essentially order–disorder nature of the transition.

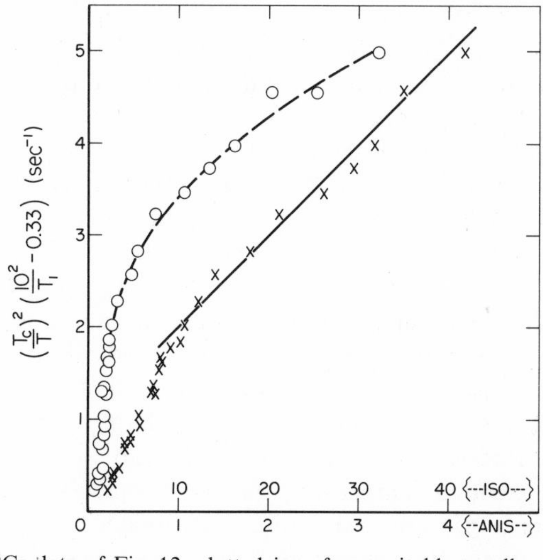

FIG. 13. The ^{133}Cs data of Fig. 12, plotted in a form suitable to allow comparison with Eq. (84), for either anisotropic (solid line) or isotropic (dashed line) dipolar interaction. An assumed background contribution of $[10^2/(T_1)_B] = 0.33\,\text{sec}^{-1}$ has been subtracted.

The data of Fig. 12 is replotted in Fig. 13 in a form suitable for the extraction of τ_c. As suggested in the original reference,[63] a "background contribution" of $[10^2/(T_1)_B] = 0.33\,\text{sec}^{-1}$ has been subtracted. While a constant background contribution of this magnitude (presumed due to paramagnetic impurities[52]) is apparent below T_C for the deuteron data of Fig. 6, it is not so clear from Fig. 12 just what background is present. Fortunately the slope of a straight line through the data of Fig. 13 just above T_C is very insensitive to the choice of $(T_1)_B$. The value of τ_c, as extracted from this slope, varies by $\lesssim 5\%$ for $0 \leqslant [10^2/(T_1)_B] \leqslant 1$.

It is apparent from Fig. 13 that, over an $\simeq 30°C$ temperature range above T_C, the assumption of anisotropic dipolar interaction gives a considerably better fit to a straight line. Assuming the same parameters for KD*P and CsD*A, the slope of the solid straight line in this figure, together with Eq. (84), gives $\tau_c = 2 \times 10^{-12}$ sec. While the τ_c values obtained from the KD*P deuteron data and ^{133}Cs data in CsD*A do bracket the related KD*P dielectric relaxation parameter, there is an order of magnitude difference.

In addition to the possibility that the parameters α, δ may be sufficiently different for KD*P and CsD*A to account for an appreciable difference in τ_c, other origins of this discrepancy are as follows: (1) Evaluation of the numerical factor in Eq. (84) is not as straightforward as for the deuteron case. (2) The assumption discussed in connection with Eq. (81), that all collective modes involve the same interaction constant, may not be as well satisfied for ^{133}Cs as for deuterons. This follows since fluctuations in the EFG tensor at the ^{133}Cs site depend upon the dipole order parameter over a macroscopic region of the crystal, while for the deuteron the EFG tensor is essentially of intrabond origin.

It would be worthwhile conducting similar experiments on the ^{87}Rb and deuterons in RbD*P to see if the same τ_c value is obtained from both. An advantage in using the phosphate is the absence of quadrupole interaction for the $I = 1/2$, ^{31}P nucleus. Cross-relaxation between the quadrupole perturbed ^{75}As spectrum and other nuclei can cause an appreciable change in T_1.[52] The cesium phosphates are not ferroelectric.

As mentioned above, with the assumption of an anisotropic dipolar interaction, Eq. (84) gives a good fit to the data, with constant τ_c, over a temperature interval of $\simeq 30°C$ above T_C. To obtain a reasonable fit over a more extended temperature range it is necessary to make some assumptions concerning the temperature dependence of τ_c. Any reasonable choice for the functional dependence, such as $\exp(W/kT)$ (or this factor divided by T), requires a considerable sacrifice in the goodness of fit over the important temperature range just above T_C in order to obtain a good fit at higher temperatures. This seems to be the case for any choice of background, $0 \leqslant [10^2/(T_1)_B] \leqslant 1\ \text{sec}^{-1}$. Since the measured relaxation rate drops below the predicted critical contribution at higher temperatures, it is possible that decoupling occurs between the deuteron and ^{133}Cs motion for $(T - T_C) \gtrsim 30°C$.

In concluding this section it is worth noting that critical relaxation has been observed for ^{39}K in KDP using a double resonance technique.[64] The very large uncertainty near T_C would make the results of a detailed numerical

[64] D. Stehlik and P. Nordal, *Proc. XVI Colloq. Ampère, Bucharest, 1970.*

comparison for τ_c rather questionable. An additional contribution to the ^{39}K relaxation rate, proportional to T^2, is quite apparent for temperatures $\gtrsim 80°C$ above the 123°K Curie temperature. This contribution arises from two-phonon Raman processes. If the quadrupole Hamiltonian is expanded to second order in the lattice displacement, and the lattice mode density approximated by a Debye spectrum, the contribution to spin-lattice relaxation takes the form

$$(T_1)^{-1} \propto (eQV_{zz}/h)^2 (T/\theta)^2,$$

where θ_1 is the Debye temperature. Such a contribution is not obvious in the ^{133}Cs data of Fig. 12.

3. *^{75}As Rotating Frame, Dipolar Relaxation, and Linewidth*

The most recent experiments which exhibit critical relaxation effects in KDP-type crystals are those on ^{75}As in the ^{39}K, ^{87}Rb, and ^{133}Cs arsenates.[65] Since the ^{75}As Zeeman spin-lattice relaxation time is so short it was necessary to use a method based upon proton–arsenic cross-relaxation in the dipolar rotating frame. Because of the somewhat greater complexity of the experiment, evaluation of the proportionality factor between the measured relaxation rate $(T_{1D})^{-1}$ and the integral $\mathscr{I}$ which contains the dynamics of the interaction is less straightforward. The relaxation rate shows the sharp increase as $T \rightarrow T_C$ observed for ^{31}P, the deuteron, and ^{133}Cs. While the data shows somewhat greater scatter than for these other nuclei, it does seem to obey the same dynamics.

The ^{75}As $(1/2) \leftrightarrow (-1/2)$ resonance linewidth also exhibits a rapid increase as $T \rightarrow T_C$.[65,66] Even though $\eta = 0$ so that in $c(Z)$-rotation the resonance frequency is independent of orientation, the linewidth is not. In addition, the functional dependence of linewidth upon temperature seems to vary markedly with θ_Z.[66] A more complete investigation would provide further information on critical dynamics at the ^{75}As site.

C. Some Comments on Dynamic Scaling

Under appropriate circumstances, singular behavior such as that displayed by $(T_1)^{-1}$ in KDP may be described in terms of critical indices. An advantage of this method of analysis is that a useful parameter may be extracted from the data without detailed evaluation of the interaction constants.

Consider a positive function $f(\varepsilon)$, continuous for sufficiently small values of the parameter $\varepsilon \equiv |T - T_c|/T_c$. The critical point exponent λ, associated

[65] R. Blinc, M. Mali, J. Pirs, and S. Zumer, *J. Chem. Phys.* **58**, 2262 (1973).
[66] R. Blinc and J. Bjorkstam, *Phys. Rev. Lett.* **23**, 788 (1969).

with $f(\varepsilon)$, is defined by

$$\lambda \equiv \lim_{\varepsilon \to 0} [\ln f(\varepsilon)/\ln \varepsilon].$$

This implies that

$$f(\varepsilon) = A\varepsilon^{\lambda}[1 + B\varepsilon^{y} + \cdots]; \quad y > 0.$$

A plot of $\ln f(\varepsilon)$ versus $\ln \varepsilon$ will, in the limit $\varepsilon \to 0$, have a slope given by the "critical-point exponent" λ. One can, from fundamental thermodynamic and statistical mechanical arguments, relate such critical-point exponents for a variety of measurable parameters.

While it is customary to make precise measurements to within millidegrees of the transition temperature in hopes of fulfilling the condition $\varepsilon \to 0$, constant power law behavior over a temperature range $0.02°\text{C} \lesssim T - T_N \lesssim 10°\text{C}$ has been observed in studies of the ^{19}F nuclear resonance linewidth in the uniaxial antiferromagnet, FeF_2.[67] Since the temperature range over which true critical effects should be observable is inversely proportional to the sixth power of the effective range of interaction, one expects a much smaller critical temperature region for ferroelectrics where the long-range Coulomb interaction is operative, than for the case of magnetic phase transitions, which depend upon an exchange interaction.[68] With the possible exception of SbSI, there seem as yet to be no conclusive results for ferroelectrics which display a transition between the classical critical region, where molecular field theory analysis is valid, and the true critical region.[68]

In the case of deuteron spin-lattice relaxation, evaluation of the fluctuating Hamiltonian was particularly simple because of the intrabond origin of EFG fluctuations. For ^{31}P–proton coupling, the ^{31}P spin was assumed stationary so that fluctuations resulted from the hydrogen motion. More generally, fluctuations in interaction between a spin at $\bar{r}_i$ and one at $\bar{r}_j$ will result from fluctuation of both spins about their equilibrium locations. In evaluating $\mathscr{H}_1(t)$ one must consider all allowed configurations of the two spins. The matrix element of $\mathscr{H}_1(t)$ responsible for $\Delta m = \pm K$ transitions may be written as $C_{ij}^{(K)} \Delta \mathscr{F}^{(K)}(\bar{r}_{ij}, t)$, where $\bar{r}_{ij} = \bar{r}_i - \bar{r}_j$. Here, $\Delta \mathscr{F}^{(K)}(\bar{r}_{ij}, t)$ represents temporal fluctuation of the interaction between the two spins, and $C_{ij}^{(K)}$ is an interaction constant. By decomposing $\Delta \mathscr{F}^{(K)}(\bar{r}_{ij}, t)$ into the product of a time independent magnitude $\Delta F^{(K)}(\bar{r}_{ij})$, and an appropriate correlation function, $\sigma(\bar{r}_{ij}, t)$, which is independent of the magnitude of the fluctuation (as well as of K), the relaxation rate may be expressed as,

$$(T_1)^{-1} = \sum_K \sum_{i,j} |C_{ij}^{(K)}|^2 \int_{-\infty}^{\infty} dt\, \overline{\Delta F^{(K)}(\bar{r}_{ij}) \sigma(\bar{r}_{ij}, t) \Delta F^{(K)*}(\bar{r}_{ij}) \sigma^*(\bar{r}_{ij}, 0)} \exp(i\omega t).$$

[67] A. Gottlieb, F. Dupre, and P. Heller, *J. Appl. Phys.* **40**, 1277 (1969); A. Gottlieb and P. Heller, *Phys. Rev. B* **3**, 3615 (1971).

[68] R. Blinc and B. Zeks, *Advan. Phys.* **21**, 693 (1972).

With the collective coordinate expansion,

$$\Delta\mathscr{F}^{(K)}(\bar{r}_{ij}, t) = \Delta F^{(K)}(\bar{r}_{ij})\sigma(\bar{r}_{ij}, t) = N^{-1/2}\sum_{\bar{q}} \Delta F_{\bar{q}}^{(K)}\sigma(\bar{q}, t)\exp(i\bar{q}\cdot\bar{r}_{ij}),$$

one may write,

$$(T_1)^{-1} = N^{-1}\sum_{K}\sum_{i,j}|C_{ij}^{(K)}|^2\sum_{\bar{q}}|\Delta F_{\bar{q}}^{(K)}|^2\int_{-\infty}^{\infty} dt\,\overline{\sigma(\bar{q}, t)\sigma(-\bar{q}, 0)}\exp(i\omega t).$$

Defining a dynamic structure factor $\mathscr{S}_{\sigma\sigma}(\bar{q}, \omega)$ as the space and time Fourier transform of the correlation function $\overline{\sigma(\bar{r}_{ij}, t)\sigma^*(\bar{r}_{ij}, o)}$,[69]

$$\mathscr{S}_{\sigma\sigma}(\bar{q}, \omega) \equiv \int d^3\bar{r}\exp(-i\bar{q}\cdot\bar{r}_{ij})\int_{-\infty}^{\infty} dt\,\overline{\sigma(\bar{r}_{ij}, t)\sigma^*(\bar{r}_{ij}, O)}\exp(i\omega t),$$

and making use of the collective coordinate expansion, one has

$$(T_1)^{-1} = \sum_{K}\sum_{i,j}|C_{ij}^{(K)}|^2\sum_{\bar{q}}|\Delta F_{\bar{q}}^{(K)}|\mathscr{S}_{\sigma\sigma}(\bar{q}, \omega).$$

If, as is the case in ferroelectrics, it is a mode condensing at the zone center which is primarily responsible for critical relaxation, one may to reasonable approximation replace all $\Delta F_{\bar{q}}^{(K)}$ by the long wavelength, $\bar{q} = 0$, value. In antiferroelectrics where condensation is at the zone boundary, $\bar{q} = \bar{q}_o$, the approximation is $\Delta F_{\bar{q}}^{(K)} = \Delta F_{\bar{q}_o}^{(K)}$ for all $\bar{q}$. As has been previously mentioned, this approximation is open to some question. In particular, it may cause overestimation of the critical relaxation rate. The approximation leads to the result,

$$(T_1)^{-1} = \left[\sum_{K}\sum_{i,j}|C_{ij}^{(K)}|^2|F_o^{(K)}|^2\right]\sum_{\bar{q}}\mathscr{S}_{\sigma\sigma}(\bar{q}, \omega).$$

All of the temperature dependence is in $\mathscr{S}_{\sigma\sigma}(\bar{q}, \omega)$. The slope of $\ln(T_1)^{-1}$ versus $\ln\varepsilon$, which gives the critical index, is thus independent of the interaction constants.

Fundamental to the static scaling hypothesis (SSH) is the existence of a correlation length κ^{-1} which diverges as $\kappa^{-1} = \kappa_o^{-1}\varepsilon^{-v}$. The critical index v will depend upon the particular static correlation function under investigation. Dependence upon temperature through ε is then replaced by dependence upon κ. With σ proportional to the order parameter of the transition, one finds for the $T > T_c$ hydrodynamic region $((q/\kappa) \ll 1)$, that the static structure factors $S_{\sigma\sigma}(\bar{q})$, defined by

$$S_{\sigma\sigma}(\bar{q}) \equiv \int_{-\infty}^{\infty} (d\omega/2\pi)\,\mathscr{S}_{\sigma\sigma}(\bar{q}, \omega),$$

[69] H. Stanley, "Introduction to Phase Transitions and Critical Phenomena." Oxford Univ. Press, London and New York, 1971. (This book provides a good introduction to the scaling hypotheses.)

may be expressed as

$$S_{\sigma\sigma}(\bar{q}) = \kappa^{v}[1 + \mathcal{O}(q/\kappa) + \cdots],$$

where

$$v = -\gamma/\nu = -2 + \eta.$$

Recognition of the dependence of both $S_{\sigma\sigma}(\bar{q})$ and $\mathcal{S}_{\sigma\sigma}(\bar{q}, \omega)$ upon temperature may be displayed through the dependence upon κ by writing, $S_{\sigma\sigma}(\bar{q}) = S_{\sigma\sigma}(\kappa, \bar{q})$ and $\mathcal{S}_{\sigma\sigma}(\bar{q}, \omega) = \mathcal{S}_{\sigma\sigma}(\kappa, \bar{q}, \omega)$. The critical exponents η, γ are defined by

$$\chi_T \propto S(\kappa, 0) = S_o \varepsilon^{-\gamma}$$

$$S(0, q) = S'_o q^{-2+\eta},$$

with S_o, S'_o constants. (Because of the difference in context, there should be no confusion between the use of γ, η as critical indices in this section, and use of these same symbols for a different purpose in the remainder of the paper.) The static susceptibility is denoted by χ_T.

The dynamic scaling hypothesis (DSH) assumes, in addition to the SSH, existence of a characteristic frequency ω_c (or inverse time) defined by

$$\int_{-\omega_c}^{\omega_c} (d\omega/2\pi)\,\mathcal{S}(\kappa, \bar{q}, \omega) = (1/2)\int_{-\infty}^{\infty} (d\omega/2\pi)\,\mathcal{S}(\kappa, \bar{q}, \omega).$$

A dimensionless shape function $f(q/\kappa, x_\omega)$ is defined by

$$\mathcal{S}(\kappa, \bar{q}, \omega) = [2\pi/\omega_c(\kappa, \bar{q})]\, S(\kappa, \bar{q})\, f(q/\kappa, x_\omega),$$

$$\int_{-\infty}^{\infty} f(q/\kappa, x_\omega)\, dx_\omega = 1; \quad x_\omega \equiv \omega/\omega_c.$$

A final assumption is that $\omega_c(\kappa, \bar{q})$ is a homogeneous function of κ and $\bar{q}$ of degree z,

$$\omega_c(\kappa, \bar{q}) = q^z G(q/\kappa).$$

Replacing the sum over $\bar{q}$ by an integral, one has finally,

$$(T_1)^{-1} \propto \int_{q=0}^{q_m} (q^2\, dq)[S(\kappa, q) f(q/\kappa, x_\omega)]/q^z G(q/\kappa).$$

With $S(\kappa, q)$ as given by the SSH, and a change of variable, $x \equiv q/\kappa$, this becomes

$$(T_1)^{-1} \propto \kappa^{d-z-(\gamma/\nu)} \int_{x=0}^{(q_m/\kappa)} x^2\, dx\, [1 + \mathcal{O}(x) + \ldots]\, f(x, x_\omega)/G(x),$$

where d is the dimensionality of the system. Since the integral converges,

$$(T_1)^{-1} \propto \kappa^{d-z-(\gamma/\nu)}.$$

The static scaling assumption, $\kappa \propto \varepsilon^{\nu}$, leads to

$$(T_1)^{-1} \propto \varepsilon^{\nu(d-z)-\gamma}.$$

Thus the critical index for $(T_1)^{-1}$ data is

$$\lambda = \nu(d-z) - \gamma.$$

In the limit $\varepsilon \to 0$, all anomalous $(T_1)^{-1}$ data published so far for ferroelectric phase transitions are, within experimental error, consistent with the behavior $(T_1)^{-1} \propto \ln \varepsilon^{-1}$. This is evidence for the fact that the true critical region has not been reached, and is consistent with dielectric experiments.[68] A possible exception to this situation is recent proton $(T_1)^{-1}$ data in KDA.[70] As is evident from Fig. 14, a substantially better fit is obtained

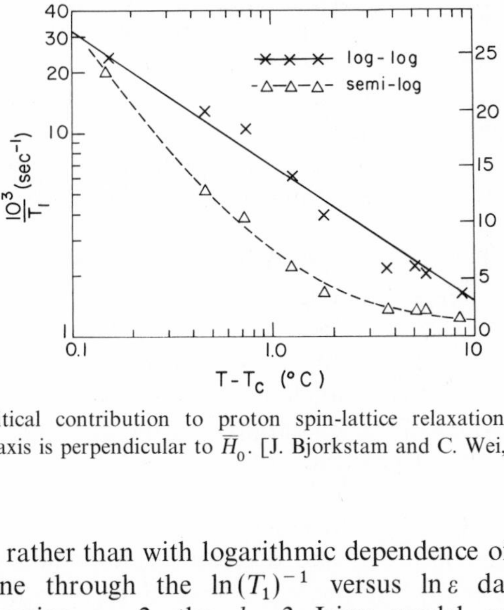

FIG. 14. The critical contribution to proton spin-lattice relaxation in KH_2AsO_4 at $\nu = 16$ MHz. The c-axis is perpendicular to $\bar{H}_0$. [J. Bjorkstam and C. Wei, *Ferroelectrics*] (to be published).

with power law, rather than with logarithmic dependence of $(T_1)^{-1}$ upon ε. The straight line through the $\ln(T_1)^{-1}$ versus $\ln \varepsilon$ data points give $\lambda = -0.67$. Assuming $z = 2$, the $d = 3$ Ising model values ($\nu \simeq 0.64$, $\gamma = 1.25$)[69] give $\lambda \simeq -0.61$. Further, more careful, experiments to check this dependence are in progress.

Attention must be called to the fact that careful dielectric experiments on KDA do not show departure from classical critical behavior.[71] If additional NMR experiments confirm the above behavior, it will be important to establish the reason for differences between the NMR and

[70] J. Bjorkstam and C. Wei, *Ferroelectrics* (to be published).

[71] R. Blinc, M. Burger, and A. Levstik, *Solid State Commun.* **12**, 573 (1973).

dielectric results. It is, of course, conceivable that NMR experiments, which do not couple directly to the order parameter in ferroelectrics, and hence, cause negligible disturbance of the system, will allow detection of true critical behavior more readily than is the case with dielectric experiments. Confirmation of power law behavior over a range of $\simeq 10°C$, as seems to be the case in Fig. 14, would point to the importance of short range, rather than Coulomb, interactions as the primary driving mechanism for this transition.

V. The Heavy Atom Spectrum

A. The ^{75}As Spectrum

Following an initial report on the $I = 3/2$, ^{75}As spectrum,[72] an anomalous dependence of the quadrupole frequency $\nu_Q \equiv |3eQV_{zz}/2hI(2I-1)|$ upon temperature above T_C, together with an estimate of 34 MHz for ν_Q below T_C, was obtained from the $1/2 \leftrightarrow -1/2$ resonance.[73] Subsequently extensive investigations have been conducted on the pure quadrupole spectrum below T_C,[74] as well as both the Zeeman perturbed quadrupole spectrum and quadrupole perturbed Zeeman spectrum, both above and below T_C.[65,75–77] Essential features, drawn from all these sources, are summarized in this section.

At room temperature the EFG tensor has cylindrical symmetry ($\eta = 0$), with the largest component along the c-axis. As the temperature is lowered toward T_C an anomalously large increase in ν_Q, illustrated in Fig. 15, occurs in the ferroelectrics. There is a rather striking difference between the antiferroelectric (NH_4) and ferroelectric crystals, just above T_C. The $1/2 \leftrightarrow -1/2$ linewidth of the quadrupole perturbed Zeeman spectrum has appreciable angular dependence, and the $\pm 3/2 \leftrightarrow \pm 1/2$ "satellites" are observable over only a small angular range where the external field $\bar{H}_0$ is nearly along the O–O bisector (Section II,A,1). Away from this orientation the satellites become broad and decrease in amplitude. The anomalous increase in $1/2 \leftrightarrow -1/2$ resonance linewidth as $T \to T_C$ is particularly striking at orientations where $\bar{H}_0$ makes an angle of 45° with respect to the tetragonal a- and b-axes. In that case the linewidth increases by more than a factor of 5 between $T_C + 30°C$ and T_C.

[72] J. Bjorkstam, *Bull. Amer. Phys. Soc.* **7**, 464 (1963).
[73] P. Kelly, M.S. Thesis, Univ. of Washington, Seattle, 1966.
[74] A. Zhukov, I. Rez. V. Pakhomov, and A. Semin, *Phys. Status Solidi* **27**, K129 (1968).
[75] W. Beezhold and E. Uehling, *Phys. Rev.* **175**, 624 (1968).
[76] G. Cinader, D. Zamir, and I. Pelah, *Solid State Commun.* **6**, 435 (1968).
[77] L. Gupta, *J. Phys. Soc. Jap.* **27**, 1229 (1969); R. Blinc, D. O'Reilly, and E. Peterson, *Phys. Rev. B* **1**, 1953 (1970).

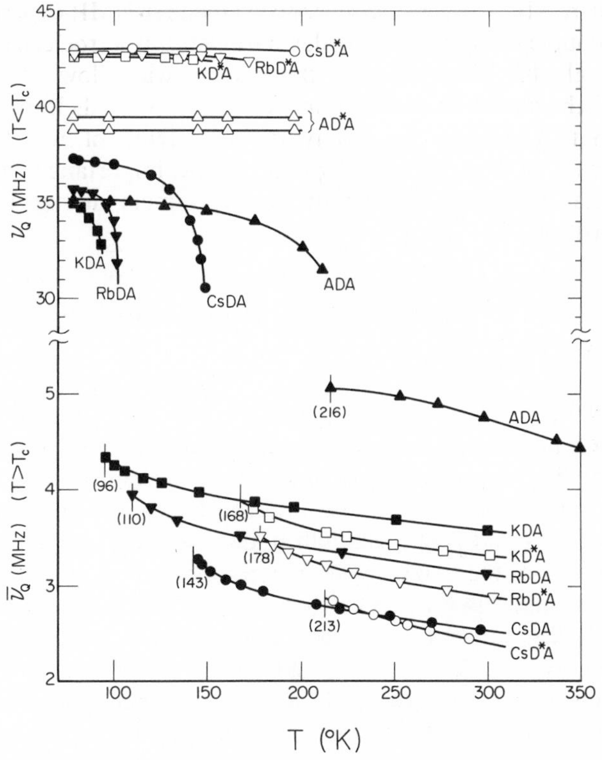

FIG. 15. Temperature dependence of the ^{75}As quadrupole frequency below T_C,[74] and above T_C.[65,66] Values of T_C, given in parentheses, depend upon the degree of deuteration. Apparent T_C values, obtained from extrapolating the $T < T_C$ data, are not fully consistent with other measurements.

Another feature of the spectrum just above T_C is the appearance of additional, relatively narrow (as compared with the $\eta = 0$ resonance) $1/2 \leftrightarrow -1/2$ resonance lines for which $\eta \neq 0$, but with V_{zz} still along the c-axis. These are superimposed upon the $\eta = 0$ line. The $\eta \neq 0$ resonances may be selectively enhanced with an applied electric field. They arise from regions of short range order in which the polarization is nonzero, but are very different from the $T < T_C$ spectrum.

Temperature dependences of the pure quadrupole frequencies below T_C are also given in Fig. 15.[74] Clearly ν_Q is an order of magnitude greater than that above T_C and, well below T_C, becomes virtually temperature independent. The largest component of the diagonalized EFG tensor V_{zz} is nearly along the "upper" O–O direction (Section II,A,1) of the distorted AsO_4 tetrahedron, with the smallest component V_{xx} along the polarization axis (c-axis).

1. *The Temperature Dependence of* ν_Q

a. *The Slater Model.* A rather successful description of the temperature dependence has been given in terms of the Slater model.[66] A slight modification of that discussion will be presented here. While the Slater model is not concerned with the question of dynamics, it continues to be a useful framework for relating equilibrium properties of KDP crystals.

In the paraelectric phase where $p = 0$, the statistical weight of nonpolar Slater groups, as given by Eq. (1), may be written

$$\delta_0 = 2[2 + \exp(\varepsilon_0/kT)]^{-1}. \quad (85)$$

The oppositely directed *polar* groups have equal weight as expressed by Eq. (2). The EFG tensor below T_C is just that of a polar group. According to the description given in the previous section, oppositely directed polar groups such as A and F of Fig. 1 will both have their smallest components V_{xx} collinear with the c-axis, while the components $|V_{zz}| > |V_{yy}|$ are along the "upper" and "lower" O–O bonds. Making use of the definition $\eta \equiv (V_{xx} - V_{yy})/V_{zz}$, together with the Laplace equation $\sum_{i=1}^{3} V_{ii} = 0$, one has $(V_{xx}/V_{zz}) = (\eta - 1)/2$. The measured pure quadrupole frequency for an $I = 3/2$ nucleus is $\nu_Q = (eQV_{zz}/2h)[1 + (\eta^2/3)]^{1/2}$.[62] With $\eta = 0.55$ and 35.5 MHz as the value which ν_Q approaches for KDA well below T_C, one finds (in units of MHz) $(eQV_{zz}/2h) = 33.7$, $(eQV_{yy}/2h) = -26.1$, and $(eQV_{xx}/2h) = -7.6$. For oppositely directed polar groups, V_{zz} and V_{yy} are interchanged but V_{xx} remains the same. Equally weighted averaging of these two possibilities gives an EFG tensor with cylindrical symmetry about the c-axis, and quadrupole frequency $\nu^{(\pm)} \equiv 7.6$ MHz.

In performing such averaging it is essential to keep in mind that the quadrupole interaction has 180° symmetry, and the absolute sign of an EFG tensor element is arbitrary. With the rapid hydrogen motion which occurs above T_C, the ^{75}As nuclear spin does not have time to quantize along the instantaneous direction of the EFG tensor, but rather only "feels" the average value. Equal weighting of the four nonpolar tensors leads to a quadrupole frequency $\nu^{(0)}$. One then expects to observe above T_C an average quadrupole frequency,

$$\bar{\nu}_Q = \delta_+ \nu^{(\pm)} + \delta_0 \nu^{(0)}. \quad (86)$$

Introducing the values of $\delta_+ = \delta_-, \delta_0$ from Eqs. (2) and (85) and rearranging gives

$$\bar{\nu}_Q = \nu^{(\pm)}/[1 + 2\exp(-\varepsilon_0/kT)] + 2\nu^{(0)}/[2 + \exp(\varepsilon_0/kT)]. \quad (87)$$

Even if the H_2AsO_4 group is the same above and below T_C so that $\nu^{(\pm)} = 7.6\,\text{MHz}$ may be used in Eq. (87), it is still necessary to make some assumption concerning $\nu^{(0)}$. One might, for example, assume the nonpolar group tensors to be identical with those of polar groups, except for orientation. The fact that $\nu_Q(T \ll T_C)$ for the antiferroelectric ADA. in which nonpolar ordering occurs below T_C,[34] is essentially the same as for the ferroelectrics[74] lends credence to such an assumption. Even with the slight distortion of AsO_4 groups from tetragonal symmetry, an equally weighted average over the four groups gives $\nu^{(0)} \ll \nu^{(\pm)}$.

The assumption $\nu^{(0)} \ll \nu^{(\pm)}$ can be validated in the following way. Equation (87) may be rewritten as

$$\ln\{2(\bar{\nu}_Q - \nu^{(0)})/(\nu^{(\pm)} - \bar{\nu}_Q)\} = \varepsilon_0/kT.$$

This result predicts that a plot of the left-hand side versus T^{-1} should give a straight line with slope ε_0/k, which will pass through the coordinate origin. Choosing $\nu^{(\pm)} = 7.6\,\text{MHz}$, and adjusting $\nu^{(0)}$ so the plot does in fact pass through the origin, leads to $\nu^{(0)} = 1.3\,\text{MHz}$ and $(\varepsilon_0/k) = 57°\text{K}$. Such a plot is indeed an excellent straight line and agreement with the Slater value of $T_C \ln 2 = 67°\text{K}$ is not unreasonable. The intercept as $(T)^{-1} \to 0$ is quite sensitive to the choice of $\nu^{(0)}$, but the important parameter (ε_0/k) is much less so. Using these parameters, Eq. (87) gives $\bar{\nu}_Q(T_C)_{\text{cal}} = 4.3\,\text{MHz}$ at $T = T_C = 96°\text{K}$. The ratio $K \equiv (\bar{\nu}_{Q_{\text{cal}}})(\bar{\nu}_{Q_{\text{exp}}})_{T=T_C}$ may be considered as a measure of the proton-lattice coupling above T_C.[66] Since $(\bar{\nu}_{Q_{\text{exp}}})_{T=T_C} = 4.33\,\text{MHz}$, within experimental error $K = 1$. This indicates that just above T_C the ^{75}As follows the hydrogen motion.

Having thus established the validity of the approximation $\nu^{(0)} \ll \nu^{(\pm)}$, the evaluation of (ε_0/k) for the other ferroelectric crystals may be accomplished in a somewhat less tedious way. Neglecting $\nu^{(0)}$ in Eq. (87) and expanding $\exp(-\varepsilon_0/kT)$ leads to

$$\bar{\nu}_Q \simeq \nu^{(\pm)}T/3[T - (2\varepsilon_0/3k)]. \tag{88}$$

A plot of $(\bar{\nu}_Q)^{-1}$ versus $(T - T_C)/T$ should then be a straight line which intercepts the abscissa at $T_0 = (2\varepsilon_0/3k)$. Such a plot is given in Fig. 16 for all the ferroelectric crystals.[65] The values of (ε_0/k) obtained from extrapolating the straight lines in Fig. 16 to their intersection with the abscissa are plotted in Fig. 17 as a function of the elementary Slater theory prediction for this parameter, i.e. $T_C \ln 2$. While a line passing through the experimental points is not in perfect agreement with the Slater theory, it does show the expected monotonic increase in (ε_0/k) with T_C and gives nearly the same slope. The fact that the approximation $(\varepsilon_0/k) \ll T$ is not so well satisfied for crystals with higher T_C is partially responsible for the difference in slope. Perhaps an even more important reason for the less than

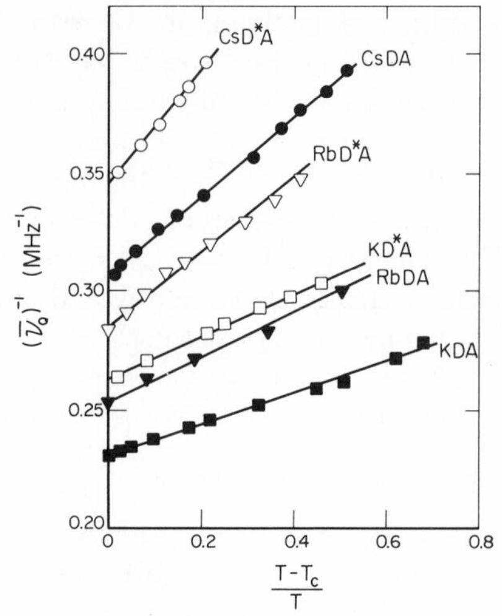

FIG. 16. Reciprocal of the ^{75}As quadrupole frequency above T_C.

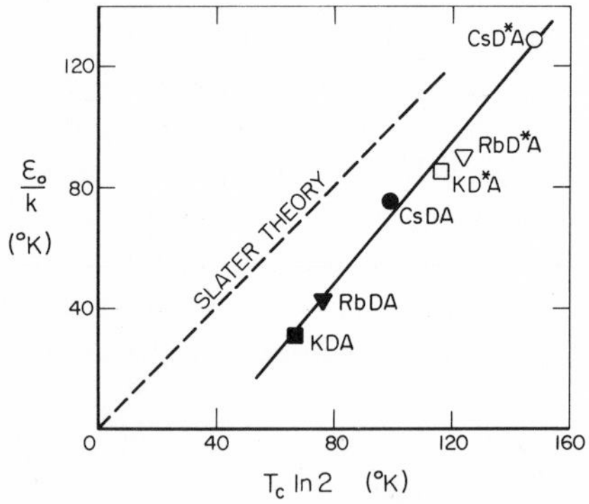

FIG. 17. Comparison of the parameter (ε_0/k) as predicted by the Slater theory, with the value obtained from the temperature dependence of the ^{75}As quadrupole frequency.

perfect agreement is incompleteness of the Slater theory itself. It should also be mentioned that for $T < T_C$ the parameter η has not been reported for any crystal except KDA. This constitutes another good reason for evaluating the parameter (ε_0/k) as indicated above.

Before leaving the discussion of temperature dependence it is important to establish the fact that "normal" variations of $\bar{\nu}_Q$ with T are indeed negligible in KDA-type crystals. It is well known that torsional oscillations can reduce $\bar{\nu}_Q$ as temperature increases.[78,79] Such oscillations tend to cause

[78] T. Das and E. Hahn, "Nuclear Quadrupole Resonance Spectroscopy." Academic Press, New York, 1958.

[79] T. Kushida, G. Benedek, and N. Bloembergen, *Phys. Rev.* **104**, 1364 (1956).

a more nearly isotropic distribution of charge about the ^{75}As nucleus, with a resulting decrease in $\bar{\nu}_Q$. If $kT \gg \omega_t$, where ω_t is the torsional oscillation frequency, the result is

$$\bar{\nu}_Q \simeq \nu_{Q_0}\{1-(3kA_t/2\omega_t^2)\,T\}.$$

The parameter A_t^{-1} has dimensions and order of magnitude of the molecular moment of inertia. Typically $(-\nu_{Q_0})^{-1}(\partial\nu_Q/\partial T) \lesssim 3\times10^{-5}\,°\mathrm{C}^{-1}$, which is an order of magnitude less than the value $\simeq 3\times10^{-4}\,°\mathrm{C}^{-1}$ observed in the KDA crystals at high temperature. It is thus a very good approximation to assume $\nu^{(\pm)}$ is constant over the temperature range of the data. While $\bar{\nu}_Q$ may also vary with lattice expansion, this contribution to the temperature dependence is normally no larger than that from torsional oscillations.

It is quite clear from Fig. 15 that for the antiferroelectric ADA, $\bar{\nu}_Q$ has a very different dependence just above T_C. Since in this crystal nonpolar ordering occurs, one might expect that simply changing the sign of the energy parameter ε_0, so that nonpolar groups are preferred, would give agreement between the above model and experiment. Such is not the case. No reasonable fit of Eq. (87) to the ADA data can be accomplished by simply changing the sign of ε_0. The qualitative features of $\bar{\nu}_Q$ versus T are consistent with a model in which, well above T_C, the crystal behaves as one of the ferroelectrics. As $T \to T_C$ the energy parameter is temperature dependent and changes sign. Since at present no theory exists for such a temperature-dependent ε_0 it does not seem worthwhile to invent an *ad hoc* dependence to fit the data.

b. *The Soft Mode Model.* As discussed in Appendix B, it is difficult at NMR frequencies to distinguish between relaxation and overdamped resonant behavior for lattice modes. In fact the critical relaxation behavior near T_C could be interpreted in terms of either model. It is thus worthwhile to consider whether the dependence of $\bar{\nu}_Q$ upon T may be described in terms of a lattice mode whose frequency approaches zero as T decreases.

For modes such as illustrated in Fig. 3 the ^{75}As moves away from its center position in the O_4 tetrahedron. Designating by ξ the ^{75}As displacement from its equilibrium position in the O_4 framework, the quadrupole frequency may be written as

$$\bar{\nu}_Q \simeq \bar{\nu}_{Q_0}(1+\alpha_0\,\xi^2).$$

The contribution to $\bar{\nu}_Q$ from the linear term in ξ is zero. For $\xi = 0$, there is a contribution $\bar{\nu}_{Q_0}$ since the tetrahedron is not quite regular. If, as discussed in Section II,A,2, the mode in question is one which condenses at T_0 as $\omega^2 \propto (T-T_0)$, an elementary application of the equipartition theorem gives

$$\xi^2 \propto [T/(T-T_0)],$$

so that

$$\bar{\nu}_Q = \bar{\nu}_{Q0}[1+\alpha T/(T-T_0)]. \tag{89}$$

The usual situation is that the condensation temperature is no more than a few ($\lesssim 5°C$) degrees below the free crystal transition temperature.[24] If that is the case no choice of parameters $\bar{\nu}_{Q0}$, α will give a reasonable fit to the data. In fact, even if one allows T_0 to take *any* value, the agreement between Eq. (89) and experiment does not begin to compare with that of the Slater model.

2. *Electric Field Effects and Polar Clusters*

As discussed in Section V, A, 1, a, an $\eta = 0$ spectrum above T_C is to be expected from rapid, *equally weighted*, averaging of the $T < T_C$ polar-group spectra. Such an $\eta = 0$ spectrum is observable at all temperatures $T > T_C$. However, as $T \to T_C$ additional resonance lines appear, superimposed upon the $\eta = 0$ spectrum.[66,80] These additional "sharp" resonance lines do *not* correspond to an EFG tensor with $\eta = 0$.

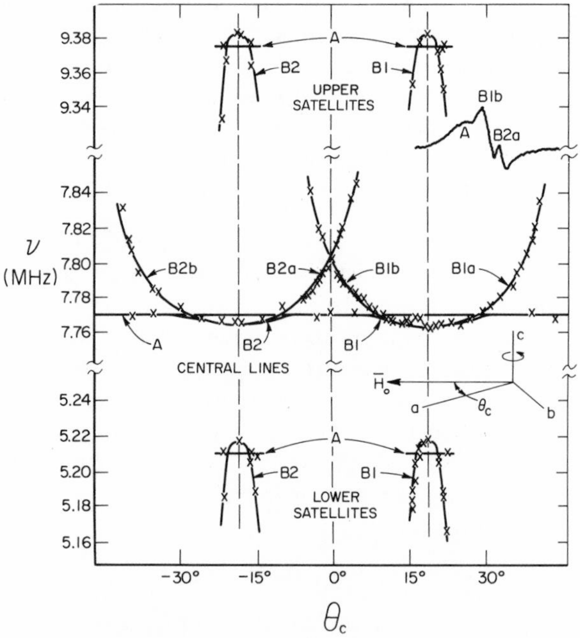

FIG. 18. Angular dependence of the ^{75}As frequency spectrum in KH_2AsO_4 at $H_0 = 10$ kG, $T = (T_C + 1)°C$. The pattern repeats every 90° in θ_c. The inset shows a trace of the first derivative of the $\frac{1}{2} \leftrightarrow -\frac{1}{2}$ dispersion mode resonance line.[80]

[80] G. Adriaenssens, J. Bjorkstam, and J. Aikins, *J. Magn. Resonance* 7, 99 (1972).

Figure 18 shows angular dependence of the observed ^{75}As NMR spectrum at $T = T_C + 1°C$ for a $c(Z)$-axis rotation with $H_0 = 10$ kG. The $\eta \neq 0$ resonance lines observed near T_C only are labeled "B." The $\eta = 0$ resonances, observed up to high temperatures, are designated "A." The B lines can be clearly identified, over the angular range indicated, up to $T \simeq T_C + 15°C$. While the B line intensity increases markedly as $T \to T_C$, the angular range over which they are observable remains essentially constant.

The B1 line pattern is symmetric with respect to the rotation of H_0 away from the orientation $\theta_c = +18.5°$, at which H_0 bisects the angle formed by the upper and lower O–O directions (i.e., the O–O bisector) for groups oriented such as A,F in Fig. 1. The B2 line pattern is symmetric about the orientation $\theta_c = -18.5°$, at which H_0 is parallel to the O–O bisector of groups such as A′,F′ in Fig. 1. The pattern of B resonances repeats every 90° in θ_c.

For a discussion of the electric field effects it will be convenient to divide the B spectrum further into B1a, B1b, B2a, B2b as shown in Fig. 19.[80] By vapor depositing silver electrodes, an electric field may be applied along the crystal c-axis, i.e. the polarization axis. An electric field $(+E)$ along the $(+c)$-axis enhances the B1b and B2b resonances while

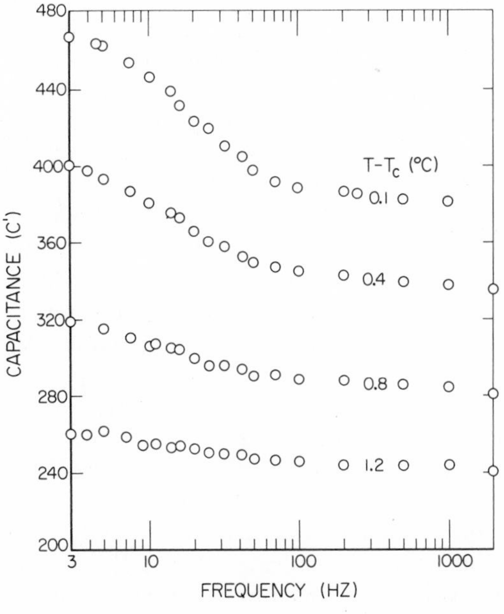

FIG. 19. Frequency dependence of the real part of the KDA dielectric constant C' (in arbitrary units), for several temperatures just above T_C.[80]

reducing B1a, B2a. The opposite effect is observed with $(-E)$. An electric field in either direction reduces the A resonance amplitude. While the position of the A resonance is, within experimental error, independent of θ_c, as should be the case for $\eta = 0$, the linewidth is dependent upon angle. At $\theta_c = 18.5°$, where A, B1a, and B1b essentially coincide, the A resonance is narrow and cannot really be distinguished from the B resonance. At this orientation an applied electric field in either direction has no effect upon the amplitude. In none of the electric field experiments was an actual shift in position of any resonance line observed. Finally, hysteresis was evident in that, when a field was applied in such a direction as to enhance a B resonance, the enhancement was virtually undiminished for at least times of the order of an hour after removal of the field. All experiments were conducted with fields $\lesssim 1000$ V/cm.

It is worth noting that the A, B1a, and B1b resonances become essentially superimposed at that orientation of H_0 where, below T_C, the A and F polar group resonances coincide. The same relationship exists between resonances A, B2a, B2b, and the A′, F′ polar groups. While the $\eta = 0$, $1/2 \leftrightarrow -1/2$ resonance is observable over the entire $c(Z)$-axis rotation, all other resonances, including the $\eta = 0$, $\pm 3/2 \leftrightarrow \pm 1/2$ satellites, broaden and disappear as the crystal is rotated away from these special orientations. This effect may be understood from the following oversimplified model. Consider an ensemble of nuclei each of which jump between two sites #1, #2 at which the frequencies are $\nu_1 > \nu_2$, respectively. If the correlation time τ for jumping is such that $\tau \gg (\nu_1 - \nu_2)^{-1}$, a separate spectrum is observed from each site. As $\tau \rightarrow (\nu_1 - \nu_2)^{-1}$, the separate resonances broaden and are "smeared" over the entire spectral range $(\nu_1 - \nu_2)$. However, with a further decrease of τ so that $\tau \ll (\nu_1 - \nu_2)^{-1}$, a single sharp resonance may be observed at the average frequency. If $\nu_1 = \nu_2$, the resonance is observable for any τ.

The above experimental results are entirely consistent with the phenomenon of long-lived, polar clusters above T_C, as previously discussed in connection with the Matsubara model of Section IV,A,2,b. The B1a resonance arises from rapid averaging of polar groups A, F in a cluster with an excess bias toward A groups, while the B1b resonance arises from clusters in which F groups dominate. Similarly the B2a resonance arises from rapid averaging of the A′, F′ groups with excess A′ groups, while B2b results from such regions where F′ groups dominate. An electric field $(+E)$ increases the fractional volume of the crystal in which F, F′ groups are dominant and hence will enhance B1b, B2b. An oppositely directed field increases the volume occupied by polar clusters in which A, A′ groups are in excess so that B1a, B2a are enhanced. An electric field in either direction will favor appropriately oriented clusters, at the expense of both oppositely directed clusters, as well as small short-lived clusters responsible for the A resonance. While an applied field does change the average cluster size it has no observable effect upon the relative bias between polar groups within a cluster. Otherwise, the field would influence the B resonance frequencies.

Since rapid averaging of oppositely directed polar clusters would lead to an A resonance, the mean lifetime $\tau_{\pm}$ of polar clusters must be large compared with the reciprocal of the minimum frequency difference between A and B resonances, for which the resonances can be resolved. This implies $\tau_{\pm} \gg 2 \times 10^{-4}$ sec.

a. *Polar Clusters and Dielectric Relaxation.* Subsequent to the NMR evidence for such long-lived polar fluctuations, dielectric relaxation experiments were conducted in the frequency range $\nu < (\tau_{\pm})^{-1} \simeq 5\,\text{kHz}$. The data of Fig. 19, obtained using c-axis plates with vapor deposited indium electrodes, clearly shows the low frequency dielectric relaxation to be expected from such slow fluctuations in polarization.[80] Expressing the dielectric constant as[81]

$$\varepsilon(\omega) = \varepsilon_{\infty} + \{(\varepsilon_0 - \varepsilon_{\infty})/[1 + (i\omega\tau_0)^{\beta}]\}$$

one finds at $T - T_C = 0.1°\text{C}$ that $\beta = 0.98 \pm 0.01$ ($\beta = 1$ corresponds to monodispersive behavior), with $\tau_0 = 8.9 \pm 0.7$ msec.

B. The Spectra of ^{39}K, ^{87}Rb, and ^{133}Cs

The quadrupole coupling tensor at the "cation" site (X = K, Rb, Cs) has been investigated in KDP,[82] RbDP,[83] RbDA,[84] and CsDA,[63] over the entire temperature range from room temperature to below T_C. Here again the temperature dependence above T_C is anomalous in that ν_Q for ^{133}Cs is temperature independent, and for the other two nuclei *decreases* monotonically with decreasing temperature. For all three nuclei $\eta = 0$ above T_C, with the largest principal axis of the diagonalized EFG tensor along the $c(Z)$-axis. Below T_C there are two physically inequivalent sites (A, B), with $\eta \neq 0$ in each domain. Both above and below T_C, the largest component of the diagonalized EFG tensor is along the $c(Z)$-axis. Below T_C the A and B site tensors have their smallest components nearly along the tetragonal $a(X)$-, $b(Y)$-axes. Polarization reversal results in the transformation A $\rightleftarrows$ B.

[81] K. Cole and R. Cole, *J. Chem. Phys.* **9**, 341 (1941).
[82] M. Kunitomo, T. Terao, and T. Hashi, *Phys. Lett. A* **31**, 14 (1970).
[83] R. Blinc, D. O'Reilly, and E. Peterson, *Phys. Rev. B* **1**, 1953 (1970).
[84] R. Blinc and M. Mali, *Phys. Rev.* **179**, 552 (1969).

A brief summary of the resulting EFG tensors has been given.[85] In all cases the quadrupole frequency is lower below T_C than just above, with the ratio between 0.93 and 0.96. The fact that the EFG tensor just above T_C is nearly an average of that at the A and B sites below T_C is further evidence for the essentially order–disorder nature of the transition.

In contrast to the ^{75}As site, the EFG tensor at the "cation" position is primarily ionic in origin, and thus depends upon the charge distribution in the entire crystal. The unusual temperature dependence therefore cannot be so easily described in terms of the Slater model, as was the case for ^{75}As. Nevertheless, it is completely consistent with that model. Since for the polar charge distribution below T_C, ν_Q is less than above T_C, the implication is that nonpolar arrangement of H_2PO_4 dipoles contribute a greater magnitude to the "cation" EFG tensor than do polar arrangements. As the temperature increases the fractional population of nonpolar groups will increase and ν_Q should increase, as observed. The temperature independence of the ^{133}Cs ν_Q above T_C probably results from an accidental balance between the dipolar and normal contributions, since these will have opposite signs.

The lack of any additional structure in the "cation" line shape as $T \to T_C$ is also to be expected from both the primarily ionic origin of the EFG tensor as well as the small change which occurs upon passing through T_C. The average "cation" EFG tensor is determined not only from charges within its own polar cluster, but from other clusters as well. Averaging over a macroscopic crystal volume, which contains many clusters, will give the same average environment for all nuclei. With the very small change in EFG tensor which occurs upon cooling through T_C, even the slightly unequal averaging of A and B site tensors within a cluster of macroscopic size should cause at most a slight line broadening.

It is thus evident that all the features of the heavy atom spectra may be reasonably well described in terms of the equilibrium Slater model. Limitations of the model in describing dynamic properties of the crystal are equally evident from the spin-lattice relaxation results.

VI. Concluding Comments

Similar collective mode descriptions have been used to discuss anomalous T_1, resonance linewidth, or rotating frame ($T_{1\rho}$) behavior near phase transitions in other ferroelectrics, as well as in antiferromagnets, liquid crystals, superconductors, and solids which undergo structural phase transitions. The method of expanding fluctuations of an appropriate order

[85] R. Blinc and M. Mali, *Solid State Commun.* **7**, 1413 (1969).

parameter in collective coordinates and utilizing the FDT or DSH has general applicability. Future NMR experiments should continue to provide important information on the dynamics of phase transitions.

Rather extensive measurements of T_1, linewidth, and T_{1_ρ} have been reported in liquid crystals which undergo isotropic-nematic phase transitions. As in KDP-type ferroelectrics, a proper interpretation of the results depends upon separating the intermolecular and intramolecular contributions,[86] as well as the critical relaxation from that due to other mechanisms.[87,88] For example, a recent comparison of the proton T_1 results just below T_c in the ordered phase of *p*-azoxyanisole (PAA) and *p*-methyoxybenzylidene *p-n* butylaniline (MBBA) suggests that, while order fluctuations are the dominant relaxation mechanism in PAA, in MBBA modulation of the intermolecular interactions either by molecular diffusion or "wagging" of the flexible alkyl end chains provides the more important mechanism.[88] An extensive guide to the literature on this subject is available in the papers referenced.

In contrast with results on KDP-type crystals, it has recently been suggested that a maximum in the *critical* contribution to the proton relaxation rate at 14 MHz in MBBA occurs several degrees ($\simeq 10°C$) above the nematic-isotropic transition temperature.[87] The resulting critical exponent for the order parameter correlation time is then found to be in agreement with mean-field theory. Similar (T_1^{-1}) maxima above T_c have been observed for ^{27}Al near the superconducting transition of Nb_2Al[89] as well as for ^{23}Na in ferroelectric Rochelle salt.[90] If the results are described in terms of a relaxation model, the approximation $\omega\tau \ll 1$ is of course inappropriate, since the maximum relaxation rate signals fulfillment of the condition $\omega\tau \simeq 1$. When the transition is weakly first order, as seems to be the case in at least most of the KDP-type crystals, the approach of (τ^{-1}) to the NMR frequency is intercepted by the first-order transition. In such a case the critical relaxation rate may increase all the way down to the transition temperature.

Similar studies on other ferroelectrics include some work on perovskites,[91] ^{23}Na in $NaNO_2$,[57,92] and protons in dicalcium strontium

[86] E. Samulski, C. Dybowski, and C. Wade, *Phys. Rev. Lett.* **29**, 340 (1972).

[87] S. Ghosh, E. Tettamanti, and P. Indovina, *Phys. Rev. Lett.* **29**, 634 (1972).

[88] M. Vilfan, R. Blinc, and J. Doane, *Solid State Commun.* **11**, 1073 (1972).

[89] M. Weger, T. Maniv, Amiram Ron, and K. Bennemann, *Phys. Rev. Lett.* **29**, 584 (1972).

[90] G. Bonera, F. Borsa, and A. Rigamonti, *Phys. Rev. Lett. A* **29**, 88 (1969); J. Willmorth, Ph.D. Thesis, Univ. of Washington, Seattle, 1972.

[91] F. Bonera, *Proc. Colloq. AMPERE XVIIth Turku, Finland, 1972* to be published (1973).

[92] A. Avogadro, E. Cavelius, D. Muller, and J. Petersson, *Phys. Status Solidi* (*B*) **44**, 639 (1971).

propionate (DSP).[58] While no ferroelectric transition has been demonstrated in the lead isomorph of DSP, $Ca_2Pb(C_2H_5COO)_6$ (DLP), this crystal does have a prominent dielectric anomally at 60°C. The proton relaxation rate exhibits a logarithmic type singularity at this temperature.[93] All ferroelectrics for which critical relaxation data have been reported show a relaxation rate near T_c which is a reasonable fit to the logarithmic type singularity observed in KDP-type crystals.

A collective mode description of NMR relaxation phenomena seems to have been first used in connection with antiferromagnets. Most of the work prior to 1967 has been summarized, or referenced, in a review article by Heller.[94] An even more recent collection of references appears in connection with the report on ^{19}F line broadening which occurs near the Neel temperature of the anisotropic antiferromagnetic crystal, FeF_2.[67] In this case the order parameter is the staggered electron magnetic moment. Well above T_N the electron-nuclear hyperfine interaction is modulated at a rapid rate by electron spin fluctuations. The correlation time for fluctuations in nuclear Larmor frequency is so short compared with the nuclear spin precession frequency that the ^{19}F linewidth is relatively small. As $T \to T_N$ in the paramagnetic phase the electron spin fluctuations increase in amplitude and the correlation times increase. The resulting modulation of the nuclear Larmor frequency causes a more than twentyfold increase in NMR linewidth over an $\simeq 10°C$ temperature interval. Such results have been analyzed in terms of both the FDT[95] and the DSH.[67] Electron spin fluctuations in several antiferromagnets have also been studied extensively, using neutron scattering techniques. In particular, the ^{19}F NMR line-broadening in MnF_2[94] and FeF_2[67] may be compared with recent neutron scattering experiments on these same crystals.[99]

Recent magnetic resonance experiments which demonstrated clearly the transition between classical and true critical behavior were conducted on the structural phase transitions in $SrTiO_3$ and $LaAlO_3$.[100,101] While these were electron spin resonance studies, NMR experiments on similar phase transitions should be equally productive.

[93] I. Tatsuzaki, K. Sakata, I. Todo, and M. Tokunaga, *J. Phys. Soc. Jap.* **33**, No. 2, 438 (1972).

[94] P. Heller, *Rep. Progr. Phys.* **30**, 731 (1967).

[95] T. Moriya, *Progr. Theor. Phys.* **28**, 371 (1962).

[96] H. Thomas, *IEEE Trans. Magn.* **5**, 874 (1969).

[97] L. Landau and E. Lifshitz, "Statistical Physics," Sect. 123. Pergamon, Oxford, 1958.

[98] R. Kubo, *Rep. Progr. Phys.* **29**, 255 (1966).

[99] P. Heller, M. Schulhof, R. Nathans, and A. Ling, *J. Appl. Phys.* **42**, 1259 (1971); M. Schulhof, M. Hutchings, and H. Guggenheim, *J. Appl. Phys.* **42**, 1376 (1971).

[100] K. Müller and W. Berlinger, *Phys. Rev. Lett.* **26**, 13 (1971).

[101] T. von Waldkirch, K. Müller, and W. Berlinger, *Phys. Rev. Lett.* **28**, 503 (1972).

Appendix A. Biased Ising Spin Model for Deuteron Motion

As discussed in Section III,A,2, orientation of the EFG tensor at the deuteron site of an O–D---O bond does depend upon the location of atoms bonded to the oxygens. The dependence is not strong, however,[51] so that small displacements which occur in these locations do not appreciably influence the elements V_K, defined in Eq. (22). This is quite evident from the nearly identical deuteron EFG tensors for all the KD*P-type crystals.[52] For a bond of sufficient length to allow deuteron motion between two off-center sites, the values of V_K at these sites may be designated $V_K(1)$, $V_K(2)$. The probability of finding a deuteron at either site will, however, depend strongly upon the surrounding atomic positions; in particular, the surrounding configuration of neighboring deuterons.

Denoting by δ_1, δ_2 the weighting factors for the two sites, the average of V_K, i.e. $\overline{V_K} = \delta_1 V_K(1) + \delta_2 V_K(2)$, is assumed equal for all bonds. That is to say, the fluctuations responsible for relaxation are sufficiently rapid so that over a mean spin-state lifetime the average environment of each deuteron is the same. Of course $\delta_1 + \delta_2 = 1$.

The fluctuation in V_K when the deuteron is at site 1 is then $\Delta V_K = V_K(1) - \overline{V_K}$, etc. A fluctuation in location of the deuteron in bond m may be described in terms of the Ising spin variable $\sigma_m(t)$ which takes the value $(+1)$ with the deuteron at site 1, and (-1) at site 2. The time-dependent fluctuation in ΔV_K for the mth deuteron is then

$$\Delta V_K(t) = (1/2)[(1 + \sigma_m(t))\Delta V_K(1) - (1 - \sigma_m(t))\Delta V_K(2)].$$

Thus,

$$\Delta V_K(t) = (1/2)[(1 + \sigma_m(t))(V_K(1) - \overline{V_K}) + (1 - \sigma_m(t))(V_K(2) - \overline{V_K})],$$

which is Eq. (41).

Appendix B. Collective Coordinates and the Fluctuation Dissipation Theorem

A brief introduction to the Fluctuation Dissipation Theorem (FDT) will be presented with the notation used by Thomas.[96] If to a static input F (magnetic field, electric field, mechanical force, etc.) one adds a small time-dependent contribution $f\exp(-i\omega t)$, the single particle static response X (magnetic moment, electric moment, spatial displacements, etc.) has an additional small contribution $x\exp(-i\omega t)$. The dynamic response and force amplitudes are related by a generalized susceptibility $\chi_s(\omega)$ as

$$x = \chi_s(\omega) f,$$

where $\chi_s(\omega)$ is the Fourier transform of the system impulse response.

It is true that interparticle interactions within a correlation range add coherently, and may lead to large fluctuations near a phase transition. In that case the single particle response may be driven into the nonlinear range, leading to breakdown of the mean field approximation. The relaxation results presented in this paper probably do not correspond to this extreme critical range, so it will be assumed that the linear response theory is valid. The dynamic response amplitude of particle m may then be expressed as

$$x_m = \chi_s(\omega)\left[f_m^{\text{ext}} + \sum_n \alpha_{mn} x_n\right] \equiv \chi_s(\omega)[f_m], \tag{B.1}$$

with f_m^{ext} the external dynamic input amplitude, and α_{mn} the instantaneous interaction coefficient between particles m and n.

For a system of N particles with translational symmetry, the input to, and single particle response of particle m may be expressed in terms of Fourier components (normal modes) as

$$\begin{aligned} x_m &= N^{-1/2} \sum_{\bar{q}} x_{\bar{q}} \exp(i\bar{q}\cdot\bar{r}_m), \\ f_m^{\text{ext}} &= N^{-1/2} \sum_{\bar{q}} f_{\bar{q}} \exp(i\bar{q}\cdot\bar{r}_m), \end{aligned} \tag{B.2}$$

with $\bar{q}$ the wave vector and $\bar{r}_m$ the particle location. The spatial Fourier transform of Eq. (B.1) gives

$$x_{\bar{q}} = \chi_s(\omega)[f_{\bar{q}} + \alpha_{\bar{q}} x_{\bar{q}}], \tag{B.3}$$

where

$$\alpha_{\bar{q}} \equiv \sum_n \alpha_{mn} \exp[-i\bar{q}\cdot(\bar{r}_m - \bar{r}_n)].$$

Defining a generalized collective susceptibility $\chi_{\bar{q}}$ by $x_{\bar{q}} = \chi_{\bar{q}}(\omega) f_{\bar{q}}$, Eq. (B.3) takes the form characteristic of a feedback control system,

$$\chi_{\bar{q}}(\omega) = \chi_s(\omega)/[1 - \alpha_{\bar{q}}\chi_s(\omega)]. \tag{B.4}$$

The result given in Eq. (B.4) for a particular system depends upon the single particle dynamics characterized by $\chi_s(\omega)$. Commonly observed behaviors include:

1. Relaxation

$$\chi_{\bar{q}}(\omega) = \chi_{\bar{q}}(0)/(1 - i\omega\tau_{\bar{q}}) \tag{B.5}$$

with static susceptibility $\chi_{\bar{q}}(0)$ and relaxation time $\tau_{\bar{q}}$.

2. Harmonic oscillator

$$\chi_{\bar{q}}(\omega) = \chi_{\bar{q}}(0)/[1 - (\omega^2 + i\omega\Gamma_{\bar{q}})/\Omega_{\bar{q}}^2] \tag{B.6}$$

with damping parameter $\Gamma_{\bar{q}}$ and resonance frequency $\Omega_{\bar{q}}$. Since NMR experiments cover such a small range on a lattice vibration frequency scale, $(\omega/\Omega_{\bar{q}})^2 \ll 1$ (except perhaps within millidegrees of the phase transition), and both forms are equivalent if one makes the connection $\tau_{\bar{q}} \leftrightarrow (\Gamma_{\bar{q}}/\Omega_{\bar{q}}^2)$. As yet there seem to be no reported results where NMR experiments have distinguished between relaxation and harmonic oscillator behavior.

That natural fluctuations occurring in an equilibrium system are related to dissipation effects, which result from interaction with external forces, has been known for a long time in the form of the Nyquist theorem.[97] Recent clarification and generalization has placed the FDT on a more substantial theoretical base.[98] Briefly stated, the noise spectrum of the frequency component of the time-dependent collective coordinate $x_{\bar{q}}(t)$ is defined by

$$S_{\bar{q}}(\omega) \equiv \int_0^\infty \overline{x_{\bar{q}}^*(t)\, x_{\bar{q}}(0)} \exp(i\omega t)\, dt + \text{c.c.}$$

The bar denotes an ensemble average. With the choice of sign

$$\chi_{\bar{q}}(\omega) \equiv \chi'_{\bar{q}}(\omega) + i\chi''_{\bar{q}}(\omega),$$

the FDT is

$$\chi''_{q}(\omega) = (S_{\bar{q}}(\omega)/\hbar)\tanh(\hbar\omega/2kT).$$

For experiments discussed in the text, $\hbar\omega \ll 2kT$, and the classical form

$$\int_0^\infty \overline{x_{\bar{q}}^*(t)\, x_{\bar{q}}(0)} \exp(i\omega t)\, dt + \text{c.c.} \simeq (2kT/\omega)\, \chi''_{\bar{q}}(\omega) \tag{B.7}$$

will be adequate. Equation (B.7) plays a central role with respect to interpretation of these NMR relaxation results.

There is a temptation to include, for reasons of completeness, a summary of the quantum mechanical form of the FDT.[98] Since the classical form is completely adequate for use in the NMR frequency range, that temptation will be resisted.

Appendix C. Ising Model Susceptibility

To complete the discussion of Section IV,A,3, an outline of the Suzuki and Kubo Calculation of $\chi_{\bar{q}}(\omega)$ for the dynamic Ising model is presented.[56,68] Beginning with the Glauber master equation introduced in Section IV,A,2,b, the transition probability is chosen to bring the stochastic system to the same equilibrium as given by the static Ising model. Under equilibrium conditions ($\dot{P}(\{\sigma\}; t) = 0$), the principle of detailed balance is used to equate terms on the right side to zero in pairs:

$$W_l(\sigma_l;\{\sigma\})\,P_0(\{\sigma\}) = W_l(-\sigma_l;\{\sigma\}')\,P_0(\{\sigma\}'). \tag{C.1}$$

The zero subscript on P designates equilibrium.

The complete Ising Hamiltonian for N dipoles, of moment p, in applied field E, may be expressed as

$$\mathscr{H} = -\sum_{l=1}^{N} \sigma_l \mathscr{E}_l \tag{C.2}$$

with the local field defined by

$$\mathscr{E}_l \equiv \sum_{m=1}^{N} J_{lm}\sigma_m + pE_l. \tag{C.3}$$

In the case of homogeneous applied fields, the field E_l acting on spin σ_l becomes the same for all spins; i.e., $E_l = E$.

With $P_0(\{\sigma\}) \propto \exp(-\beta\mathscr{H})$, Eqs. (C.1) and (C.2) give

$$\begin{aligned} W(\sigma_l;\{\sigma\})/W(-\sigma_l;\{\sigma\}') &= \exp(-\beta\mathscr{E}_l\sigma_l)/\exp(\beta\mathscr{E}_l\sigma_l) \\ &= (1-\sigma_l\tanh\beta\mathscr{E}_l)/(1+\sigma_l\tanh\beta\mathscr{E}_l. \end{aligned}$$

A form for W_l consistent with this equation is

$$W_l(\sigma_l;\{\sigma\}) = (1-\sigma_l\tanh\beta\mathscr{E}_l)/2\tau_c. \tag{C.4}$$

The noninteracting dipolar relaxation time τ_c is that discussed in Section IV,A,3. While the form of Eq. (C.4) is assumed constant, W_l may show time dependence through E_l of Eq. (C.3).

Though the calculation of $P(\{\sigma\};t)$ is in general prohibitively difficult it is possible to determine expectation values such as

$$\langle\sigma_l\rangle = \sum_{\{\sigma\}} \sigma_l P(\{\sigma\};t), \tag{C.5}$$

with the summation taken over all 2^N configurations. The equation

$$\tau_c(d\langle\sigma_l\rangle/dt) = -\langle\sigma_l\rangle + \langle\tanh\beta\mathscr{E}_l\rangle \tag{C.6}$$

then follows from the original master equation and Eq. (C.4).

The molecular field approximation involves the assumption that, in Eq. (C.3),

$$\sigma_m = \langle\sigma_m\rangle = \langle\sigma\rangle, \qquad m \neq l,$$

which means that the average behavior of each spin is the same. With this assumption, Eqs. (C.6) and (C.3) become

$$\tau_c(d\langle\sigma_l\rangle/dt) = -\langle\sigma_l\rangle + \tanh\beta\langle\mathscr{E}_l\rangle \tag{C.7}$$

$$\langle\mathscr{E}_l\rangle = pE + \sum_m J_{lm}\langle\sigma_m\rangle. \tag{C.8}$$

For homogeneous applied fields $\langle\sigma_l\rangle$ may be replaced by $\langle\sigma\rangle$ in Eq. (C.7). To terms of order $\langle\sigma\rangle^3$ this becomes the Bernoulli differential equation with solution

$$\langle\sigma\rangle = [(Q_0 + \sigma_0^{-2})\exp(t/\tau) - Q_0]^{-1/2}, \tag{C.9}$$

where

$$\begin{aligned} Q_0 &\equiv \varepsilon T\left(\beta\sum_{m=1}^{N} J_{lm}\right)^3 3T_C, \\ \varepsilon &\equiv (T - T_C)/T_C, \\ T_c &\equiv \left(\sum_{m=1}^{N} J_{lm}\right)\Big/k, \\ \sigma_0 &\equiv \langle\sigma\rangle_{t=0}, \\ \tau &\equiv [T/2(T - T_C)]\,\tau_c. \end{aligned} \tag{C.10}$$

Divergence of τ as $T \to T_C$ is responsible for the large spectral density of fluctuations at low frequencies, which leads to the enhanced NMR relaxation rate. Under equilibrium conditions $[(d\langle\sigma\rangle/dt) = 0]$ Eq. (C.9) has nonzero solutions only for $T < T_C$.

Evaluation of the $T > T_C$ small signal susceptibility $\chi_{\bar{q}}(\omega)$ for the $\bar{q}$th collective mode begins with the approximation $\tanh\beta\langle\mathscr{E}_l\rangle \simeq \beta\mathscr{E}_l$. Equation (C.7) is then

$$\tau_c[d(p\langle\sigma_l\rangle)/dt] = -\left\{p\langle\sigma_l\rangle - \beta p^2 E_l(t) - \beta\sum_m J_{lm}\,p\langle\sigma_m\rangle\right\}, \tag{C.11}$$

where each term has been multiplied by the dipole moment per spin, p. An inhomogeneous applied field, of frequency ω, may be expanded in Fourier components to give

$$E_l(t) = N^{-1/2}\left[\sum_{\bar{q}} E_{\bar{q}}\exp(i\bar{q}\cdot\bar{r}_l)\right]\mathscr{R}\exp(-i\omega t).$$

Substituting this field, together with the Fourier expansions

$$\langle\sigma_n\rangle = N^{-1/2}\sum_{\bar{q}}\langle\sigma_{\bar{q}}\rangle\exp(i\bar{q}\cdot\bar{r}_n), \tag{48}$$

into Eq. (C.11) gives the equation for $\langle\sigma_{\bar{q}}(t)\rangle$;

$$\tau_c[d(p\langle\sigma_{\bar{q}}\rangle)/dt] + (1-\beta J_{\bar{q}})p\langle\sigma_{\bar{q}}\rangle = \beta p^2 E_{\bar{q}} \mathscr{R} \exp(-i\omega t),$$

where

$$J_{\bar{q}} \equiv \sum_l J_{lm} \exp[i\bar{q}\cdot(\bar{r}_l - \bar{r}_m)].$$

Solving for $p\langle\sigma_{\bar{q}}\rangle$, and using the definition for $\chi_{\bar{q}}(\omega)$ gives finally

$$p\langle\sigma_{\bar{q}}\rangle \equiv \mathscr{R}[\chi_{\bar{q}}(\omega) E_{\bar{q}} \exp(-i\omega t)],$$

$$\chi_{\bar{q}}(\omega) = \beta p^2/(1-\beta J_{\bar{q}} - i\omega\tau_c).$$

For unit dipole moments ($p = 1$) this is Eq. (51) of Section IV,A,3.

Acknowledgments

It is a pleasure to express thanks to Prof. R. Blinc for making certain data available prior to publication, as well as for other helpful comments. The organization of Section II was greatly facilitated by the availability of the Ph.D. thesis of Dr. G. Adriaenssens. Partial support of the NASA under grant NGL 48-002-004 is gratefully acknowledged.

Spin Relaxation Theory in Terms of Mori's Formalism[*]

DANIEL KIVELSON AND KENNETH OGAN[†]

DEPARTMENT OF CHEMISTRY, UNIVERSITY OF CALIFORNIA, LOS ANGELES, CALIFORNIA

* Supported in part by a grant from the National Science Foundation.

† Present address: Department of Physiology, Boston University School of Medicine, Boston, and Department of Physics, M.I.T., Cambridge, Massachusetts.

I. Introduction

The tendency to reformulate an existing body of information every time a new theoretical approach is proposed is one that is not always fruitful, although usually, with each reformulation some further insight is developed. The relevant considerations in a new formulation should be its value in interconnecting diverse phenomena and disciplines, the physical insights which it yields, the simplicity of the presentation, the reduction in the number of special approximations and subtle mathematical procedures, and, of course, above all, its usefulness in obtaining new results. In this article we reformulate the study of spin relaxation in liquids in terms of Mori's statistical mechanical theory[1] of transport phenomena, and we believe that to some extent this presentation satisfies the criteria above. As with most new formulations, in applying the Mori theory one must first learn a new mathematical language, which in this case involves projection operators. It takes some effort to attain fluency in this new language, but thereafter the theory can provide a simple, physical approach to the study of many transport processes, in particular, to spin relaxation. A number of authors[1-5] have already looked at spin relaxation in terms of projection operator theories, and some have actually used the Mori formulation[1,3,4]; nevertheless, we believe that this presentation is useful because it is more complete and more directly aimed at people with interest in magnetic resonance, and also because it contains a number of applications and a few results which we believe to be new.

In recent years a number of different approaches have been developed for the study of time-dependent processes in liquids[1,6-10]; these theories have

[1] H. Mori, *Progr. Theor. Phys.* **33**, 423 (1965).

[2] P. N. Argyres and P. L. Kelley, *Phys. Rev. A.* **134**, 98 (1964); see also R. I. Cukier, Ph.D. Thesis, Princeton Univ., Princeton, New Jersey, (1969).

[3] J. M. Deutch, in "Electron Spin Relaxation in Liquids" (L. T. Muus and P. W. Atkins, eds.), p. 127. Plenum, New York, 1972.

[4] J. T. Hynes and J. M. Deutch, *in* "Mathematical Methods" (H. Eyring, W. Jost, and D. Henderson, eds.). Academic Press, Physical Chemistry, Vol. 11, New York, 1974; see also references cited therein.

[5] B. N. Provotorov, *Zh. Eksp. Teor. Fiz.* **41**, 1582 (1961) [*Sov. Phys.*—JETP **14**, 1126 (1962)]; R. H. Terwiel and P. Mazur, *Physica* **32**, 1813 (1966); T. Shimizu, *J. Phys. Soc. Jap.* **29**, 74 (1970); K. Kometani and H. Shimizu, *J. Phys. Soc. Jap.* **30**, 1036 (1971); O. Platz, *in* "Electron Spin Relaxation in Liquids" (L. T. Muus and P. W. Atkins, eds.), Ch. 4, p. 89. Plenum, New York, 1972.

[6] R. Kubo, *J. Phys. Soc. Jap.* **12**, 570 (1957); see also R. Kubo, *Lect. Theor. Phys.* **1**, 120 (1959); R. Kubo, *in* "Fluctuation Relaxation and Resonance in Magnetic Systems" (D. Ter Haar, ed.), p. 23. Oliver & Boyd, Edinburgh, 1962.

[7] L. P. Kadanoff and P. C. Martin, *Ann. Phys.* **24**, 419 (1963).

[8] R. Kubo, *J. Phys. Soc. Jap.* **17**, 1100 (1967).

[9] R. Kubo, *Advan. Chem. Phys.* **16**, 101 (1969); R. Kubo, *J. Phys. Soc. Jap.* **26**, Suppl., 1 (1969).

[10] B. U. Felderhof and I. Oppenheim, *Physica* **31**, 1441 (1965); P. A. Selwyn and I. Oppenheim, *Physica* **54**, 161 (1971).

systematized the study of transport and relaxation processes and have served to underscore the similarities between various relaxation phenomena. In particular, each of these formalisms has been used to study transport in liquids as well as magnetic spin relaxation.[3,11–13] The Mori theory is one such theory which can be cast into a form that yields considerable physical insight. It can be used to describe a wide range of transport processes in a systematic way. The application to magnetic resonance problems provides a concise and direct exposition of the Mori theory,[1] and this approach has the potential for obtaining new results in spin relaxation analyses.

The original theories used to develop the equations for spin relaxation in the "fast molecular motion" limit (i.e., those of Bloembergen *et al.*,[14] Redfield,[15] and Bloch[16]), are based upon time-dependent perturbation theory, or equivalently, upon expansions in the time.[11,17] For long times, where these theories are most applicable, the convergence of such expansions, and especially the validity of the necessary expedient of truncating the series after two terms, are not immediately obvious and must be justified by subtle arguments on the evaluation of various limits.[13,17] In the appropriate limits the truncated perturbation series corresponds to the first terms in the expansion of an exponential, but these theories do not provide for "slow motion" extensions since it is difficult to study long time functional behavior from a short time expansion. Other theories, such as the cumulant expansion and statistical Langevin approaches, which cast the equations into more obviously convergent form have been presented,[8,9] but in many ways the Mori approach seems to be the most direct of these approaches; it also circumvents many of the arguments involving coarse graining by a straightforward division of the problem into different time scales. One can think of the "convergent" theories, including the Mori theory, as expansions in terms of exponentials rather than in powers of time; an exponential expansion insures proper long time behavior for each term.

In this article we present detailed applications of the Mori theory to spin relaxation in liquids. We try to give simple physical connections

[11] R. Kubo, *J. Phys. Soc. Jap.* **12**, 570 (1957).

[12a] J. Freed, *J. Chem. Phys.* **49**, 376 (1968); J. H. Freed, *in* "Electron Spin Relaxation in Liquids" (L. T. Muus and P. W. Atkins, eds.), Ch. 8, p. 165. Plenum, New York, 1972.

[12b] J. H. Freed, *in* "Electron Spin Relaxation in Liquids" (L. T. Muus and P. W. Atkins, eds.), Ch. 14, p. 387. Plenum, New York, 1972.

[13] J. M. Deutch and Irwin Oppenheim, *Advan. Magn. Resonance* **3**, 43 (1968).

[14] N. Bloembergen, E. M. Purcell, and R. V. Pound, *Phys. Rev.* **73**, 679 (1948).

[15] A. G. Redfield, *Advan. Magn. Resonance* **1**, 1 (1965); A. G. Redfield, *IBM J. Res. Develop.* **1**, 19 (1963); A. G. Redfield, *Advan. Magn. Resonance* **1**, 1 (1965).

[16] F. Bloch, *Phys. Rev.* **70**, 460 (1946); F. Bloch, *Phys. Rev.* **102**, 104 (1956).

[17] A. Abragam, "The Principles of Nuclear Magnetism." Oxford Univ. Press, London and New York, 1961.

between the theory and the systems under study, and we discuss the mathematical procedures, as well as the physical interpretation, of extensions of the theory to situations where the molecular motions are both slow and nondiffusional. Because of the authors' specific interests, the presentation is given in a form most suitable for electron spin relaxation studies, but it is equally applicable to nuclear spin relaxation phenomena. Electron spin resonance with nuclear hyperfine interactions and Heisenberg spin exchange are investigated in some detail. Besides the conventional results, a procedure for handling nonsecular isotropic hyperfine interactions in the presence of exchange is outlined, and an analysis of linewidths for systems in which the paramagnetic molecules remain in contact and "exchange" spins for relatively long times is developed. Considerable discussion is given to the various diffusion constants determined from magnetic resonance experiments; it is one of the merits of the Mori theory that these diffusion constants appear explicitly in the same formalism used to study spin relaxation. In particular, it is shown that diffusion constants obtained from spin exchange broadening or narrowing may often have but little relationship to the usual particle or self-diffusion constants; a modification of the spin relaxation equation for diffusing molecules in an inhomogeneous magnetic field is presented for the "very rapid diffusion" limit.

The Mori theory makes use of a projection operator which is useful in obtaining equations of motion for averaged dynamical quantities, such as the magnetization, and we therefore expect results analogous to those in the Bloch theory.[16] Other projection operators[18a,b] could be used which would yield results more analogous to those of Redfield[15] where the equations of motion for the density matrix are obtained. This has been thoroughly discussed by Deutch.[3,4]

Outline of Article

In Section II we discuss some well-known phenomenological magnetic relaxation relations and formulate them in a manner most suitable for comparison with Mori's theory; this analysis is quite detailed because the more general and formal presentations in later sections are referred back to the examples in Section II. In Section III we give a description of relevant Hamiltonians and well-known molecular expressions for relaxation times. In Section IV we present the general Mori theory with considerable discussion of the features, both mathematical and physical, of particular interest here; we derive the general Langevin equation obtained by Mori in

[18a] R. Zwanzig, *Lect. Theor. Phys.* **3**, 106 (1961); B. J. Berne, J. P. Boon, and S. A. Rice, *J. Chem. Phys.* **45**, 1086 (1966).

[18b] G. D. Harp and B. J. Berne, *Phys. Rev. A* **2**, 975 (1970).

TABLE Ia

MORI FORMULATION

$\mathbf{A}$ = Set of slow variables $\{A_m\}$

$A_m^+ \equiv$ Hermitian conjugate of A_m

$\langle \mathbf{B} \rangle \equiv$ Equilibrium ensemble average of $\mathbf{B}$

$\overline{\mathbf{B}(t)} \equiv$ Instantaneous, nonequilbrium ensemble of $\mathbf{B}$

$\langle \mathbf{B}(t)\mathbf{B}^+ \rangle \equiv$ Correlation function matrix of $\mathbf{B}$

$\mathsf{G}(t) \equiv$ Normalized correlation function matrix of $\mathbf{A}$

$$\mathsf{G}(t) \equiv \langle \mathbf{A}(t)\mathbf{A}^+ \rangle \langle \mathbf{A}\mathbf{A}^+ \rangle^{-1}$$

$$\hat{\mathbf{A}}(\omega) = \int_0^\infty \overline{\mathbf{A}(t)}\, e^{i\omega t}\, dt$$

$\P \equiv$ projection operator

$$\P\mathbf{B} = \langle \mathbf{B}\mathbf{A}^+ \rangle \langle \mathbf{A}\mathbf{A}^+ \rangle^{-1} \mathbf{A}$$

$$\dot{\mathbf{A}} = i[\mathscr{H}, \mathbf{A}]$$

$$\dot{\boldsymbol{\alpha}} = (1 - \P)\dot{\mathbf{A}}$$

$i\Omega = \langle \dot{\mathbf{A}}\mathbf{A}^+ \rangle \langle \mathbf{A}\mathbf{A}^+ \rangle^{-1}$(precessional frequency uncorrected for relaxation shifts)

$\mathsf{K}(t) \equiv$ Memory function matrix

$$\mathsf{K}(t) = \langle \dot{\boldsymbol{\alpha}}(t_p)\dot{\boldsymbol{\alpha}}^+ \rangle \langle AA^+ \rangle^{-1}$$

$$\dot{\boldsymbol{\alpha}}(t_p) \equiv \{\exp[i(1-\P)\mathscr{H}^x t]\}\dot{\boldsymbol{\alpha}}$$

$\mathscr{K}(t) \equiv$ Effective memory function matrix

$$\mathscr{K}(t) = \mathsf{K}(t)e^{-i\Omega t}$$

$$\hat{\mathscr{K}} \equiv \hat{\mathscr{K}}(0) \equiv \lim_{\omega \to 0} \int_0^\infty dt\, \mathscr{K}(t) e^{-i\Omega t} e^{-i\omega t} \quad \text{(relaxation matrix)}$$

$$\hat{\mathscr{K}} = \hat{\mathscr{K}}_{RE} - i\hat{\mathscr{K}}_{IM}$$

$\mathscr{T}^{-1} \equiv$ Transport matrix

$$\mathscr{T}^{-1} = -i\Omega + \hat{\mathscr{K}}$$

$\Lambda \equiv$ Eigenvalue matrix of $\mathscr{T}^{-1}$ (Diagonal matrix)

$$\Lambda = \Gamma - i\bar{\omega}$$

Γ is a diagonal matrix of Lorentzian linewidths

$\bar{\omega}$ is a diagonal matrix of Larmor frequencies

$\tau_I \equiv$ Correlation time associated with $\dot{\boldsymbol{\alpha}}$

$$\tau_I \equiv \mathrm{Re} \int_0^\infty dt\, \langle \dot{\boldsymbol{\alpha}}(t_p)\dot{\boldsymbol{\alpha}}^+ \rangle \langle \dot{\boldsymbol{\alpha}}\dot{\boldsymbol{\alpha}}^+ \rangle^{-1} e^{-i\Omega t}$$

Generalized Langevin equation

$$d\mathbf{A}(t)/dt = i\Omega\mathbf{A}(t) - \int_0^t dt\, \mathsf{K}(t-\tau)\mathbf{A}(t) + \dot{\boldsymbol{\alpha}}(t_p)$$

Generalized transport equation [also for $\mathsf{G}(t)$]

$$d\overline{\mathbf{A}(t)}/dt = i\Omega\overline{\mathbf{A}(t)} - \int_0^t d\tau\, \mathsf{K}(t-\tau)\overline{\mathbf{A}(t)}$$

Linear transport equation [also for $\mathsf{G}(t)$]

$$d\overline{\mathbf{A}(t)}/dt = -\mathsf{T}^{-1}\overline{\mathbf{A}(t)} = -[-i\Omega + \hat{\mathscr{K}}]\overline{\mathbf{A}(t)}$$

TABLE Ib

EXTENDED MORI FORMULATION[a]

$\mathbf{A}^{(\mathrm{o})} \equiv$ Original set of slow variables

$\mathbf{A} \equiv$ Extended set of slow variables

$\mathbf{A} \equiv \{\mathbf{A}^{\mathrm{o}}, \dot{\mathsf{A}}\}$

$\dot{\mathsf{A}} \equiv (1 - \P_{\mathbf{A}^{\mathrm{o}}})\,\dot{\mathbf{A}}^{\mathrm{o}}$ Same theory as for unextended set

$\P\mathbf{B} \equiv \P_{\mathbf{A}^{\mathrm{o}}\dot{\mathsf{A}}}\mathbf{B}$

$\mathbf{A} \equiv$ Further extended set of slow variables

$\mathbf{A} \equiv \{\mathbf{A}^{\mathrm{o}}, \dot{\mathsf{A}}, \ddot{\mathfrak{A}}\}$

$\dot{\mathsf{A}} \equiv (1 - \P_{\mathbf{A}^{\mathrm{o}}})\,\dot{A}^{\mathrm{o}}$

$\ddot{\mathfrak{A}} \equiv (1 - \P_{\mathbf{A}^{\mathrm{o}}\dot{\mathsf{A}}})\,\ddot{\mathbf{A}}^{\mathrm{o}}$

$\P\mathbf{B} \equiv \P_{\mathbf{A}^{\mathrm{o}}\dot{\mathsf{A}}\ddot{\mathfrak{A}}}\mathbf{B}$

[a] Same theory as in Table Ia with new definitions of A and ¶.

TABLE Ic

SPIN RELAXATION (SINGLE VARIABLE)[a]

$\mathbf{A} = S_+$

$S_+ = \sum_j S_{+j}$ where S_{+j} is spin on jth molecule

$\mathscr{H} = \mathscr{H}_{\mathrm{S}} + \mathscr{H}_{\mathrm{L}} + \mathscr{H}_{\mathrm{SL}}$

$\mathscr{H}_{\mathrm{S}} \equiv$ Pure spin Hamiltonian

$\mathscr{H}_{\mathrm{L}} \equiv$ Pure lattice Hamiltonian

$\mathscr{H}_{\mathrm{SL}} \equiv$ Spin lattice Hamiltonian (perturbation)

$\langle \mathscr{H}_{\mathrm{SL}} \rangle = 0$

$\mathscr{H}_{\mathrm{SL}}^{(\mathrm{sec})} \equiv$ Secular part of $\mathscr{H}_{\mathrm{SL}}$, i.e. $[\mathscr{H}_{\mathrm{SL}}^{(\mathrm{sec})}, S_z] = 0$

$\omega_0 \equiv$ Larmor frequency (corrected for spin-lattice shifts)

$\Omega =$ Larmor frequency (uncorrected for spin-lattice shifts) (also use Ω_0)

$\omega_0 - \Omega = \hat{\mathscr{K}}_{\mathrm{IM}} \equiv$ Frequency shift due to spin-lattice interaction

$T_2^{-1} = \hat{\mathscr{K}}_{\mathrm{RE}} = \omega_\lambda^2 \tau_{\mathrm{I}}$

$\hat{\mathscr{K}} \equiv \hat{\mathscr{K}}_{\mathrm{RE}} - i\hat{\mathscr{K}}_{\mathrm{IM}} \equiv$ Relaxation matrix

$$\hat{\mathscr{K}} = \left\{ \int_0^\infty dt e^{-i\omega_0 t} \langle e^{i(1-\P)\mathscr{H}^{\times} t} [\mathscr{H}_{\mathrm{SL}}, S_+], [S_-, \mathscr{H}_{\mathrm{SL}}] \rangle \right\} \Big/ \langle S_+ S_- \rangle$$

$\omega_\lambda =$ Spin-lattice interaction frequency

$\omega_\lambda^2 \equiv \langle [\mathscr{H}_{\mathrm{SL}}, S_+][S_-, \mathscr{H}_{\mathrm{SL}}] \rangle / \langle S_+ S_- \rangle$

$\tau_{\mathrm{I}} \equiv$ Correlation time associated with $[\mathscr{H}_{\mathrm{SL}}, S_+]$, i.e. with molecular motion

$\tau_{\mathrm{I}} \equiv \hat{\mathscr{K}}_{\mathrm{RE}} / \omega_\lambda^2$

[a] See also Table Ia—General Mori Theory.

TABLE Id

SPIN RELAXATION (EXTENDED THEORY FOR ONE FREQUENCY)[a]

$\tilde{A} \equiv (S_+, \dot{\mathscr{S}}_+)$

$\dot{\mathscr{S}}_+ = i[\mathscr{H}_{SL}, S_+]$

$\P \mathbf{B} = \P_{S_+ \dot{\mathscr{S}}_+} \mathbf{B}$

$\tilde{\ddot{\alpha}} = (0, \ddot{\mathfrak{S}}_+)$

$\ddot{\mathfrak{S}}_+ = \ddot{\mathscr{S}}_+ - \langle \ddot{\mathscr{S}}_+ S_- \rangle [\langle \ddot{\mathscr{S}}_+ S_- \rangle]^{-1} S_+ - \langle \ddot{\mathscr{S}}_+ \dot{\mathscr{S}}_- \rangle [\langle \dot{\mathscr{S}}_+ \dot{\mathscr{S}}_- \rangle]^{-1} \dot{\mathscr{S}}_+$

$\hat{\mathscr{K}} = \hat{\mathscr{K}}'' \begin{bmatrix} 0 & 0 \\ 0 & 1 \end{bmatrix} \equiv$ Relaxation matrix

$$\hat{\mathscr{K}}'' = \left\{ \int_0^t dt \langle e^{i(1-\P)\mathscr{H}^x t} \ddot{\mathfrak{S}}_+, \ddot{\mathfrak{S}}_- \rangle e^{-i\Omega t} \right\} \Big/ \langle [\mathscr{H}_{SL}, S_+][S_-, \mathscr{H}_{SL}] \rangle$$

$\omega_L^2 = \langle \ddot{\mathfrak{S}}_+ \ddot{\mathfrak{S}}_- \rangle / \omega_\lambda^2 \langle S_+ S_- \rangle$

$\tau_{II} \equiv$ Correlation time associated with $\mathscr{K}(t)$ when $\mathscr{K}(t)$ arises from two-variable extended theory, i.e. associated with $\ddot{S}_+$

$\tau_{II} = \mathscr{K}''_{RE} / \omega_L^2$

$\omega_l =$ Lattice frequency or molecular rotational frequency

$\omega_l = [\mathscr{H}_L, [\mathscr{H}_{SL}^{(sec)}, S_+]] / \omega_\lambda$

$\omega_l^2 \equiv \langle |[\mathscr{H}_L, [\mathscr{H}_{SL}^{(sec)}, S_+]|^2 \rangle / \omega_\lambda^2$

$\omega_l \approx \omega_L$

For rotations: $\omega_l^2 = l(l+1) k_B T / I$

$T_\pm \equiv$ Lorentzian width

$\sigma_\pm \equiv$ Lorentzian frequency shifts

$\beta \equiv$ dimensionless coupling parameter

$\beta = \langle [\mathscr{H}_{SL}, [\mathscr{H}_{SL}, S_+]][S_-, \mathscr{H}_{SL}^+] \rangle / \omega_\lambda \langle [\mathscr{H}_{SL}, S_+][S_-, \mathscr{H}_{SL}^+] \rangle$

---

[a] Same theory as in Table Ic with some new definitions.

TABLE Ie

SPIN RELAXATION; HYPERFINE INTERACTION[a]

$\tilde{\mathbf{A}} = (S_+^{(1)}, S_+^{(2)}, S_+^{(3)}, \ldots)$

$S_+^{(m)} = \sum_j S_{+j} P_{mj}$ where the nuclear spin projection operator is defined as

$P_{mj} = |mj\rangle \langle mj|$

and $|mj\rangle$ represents the jth nuclear spin with specific quantum number, $I_{zj} = m$

Ω is diagonal and

$\Omega_n = \omega_0 + a m_n$

$$\hat{\mathscr{K}}_{mm} \approx [2(2I+1)/2N] \int_0^\infty dt \langle [e^{it(\mathscr{H}_S^x + \mathscr{H}_L^x)t} \mathscr{H}_{SL}, S_+^{(m)}][S_-^{(m)}, \mathscr{H}_{SL}] \rangle$$

Relaxation $T_2^{-1}(m) = (\hat{\mathscr{K}}_{RE})_{mm}$

Frequency shift: $\sigma_m = (\hat{\mathscr{K}}_{IM})_{mm}$

---

[a] Same theory as in Table Ic with new definitions.

TABLE If

SPIN EXCHANGE

$$\mathscr{H}_{SL} = \sum_{jk} J_{jk}(\mathbf{r}_{jk})\mathbf{S}_j \cdot \mathbf{S}_k$$

$$\tilde{\mathbf{A}} \equiv (S_+^{(1)}, S_+^{(2)}, \ldots)$$

$$K_{mm'} = \int_0^\infty dt \sum_k \langle J_{lk}(t) J_{lk} \rangle [I\delta_{mm'} - \tfrac{1}{2}(1-\delta_{mm'})]$$

$$\tau_J = \left(\int_0^\infty \langle J_{lk}(t) J_{lk} \rangle \, dt\right) \Big/ \langle J_{lk}^2 \rangle$$

$$\tau_J = \tau_I$$

$$\mathscr{J}^2 = \sum_k \langle J_{lk}^2 \rangle; J_{lk} = J_{lk}(r_{lk})$$

$$\omega_{ex} \equiv \tfrac{1}{2}\mathscr{J}^2 \tau_J$$

$$\mathscr{K}_{mm'} = \omega_{ex}[I\delta_{mm'} - \tfrac{1}{2}(1-\delta_{mm'})]$$

$$\mathscr{J}^2 \equiv J^2 X;$$

J^2 has only slight concentration dependence

Model: $J_{12} = 0$

$J_{12} = J_0 \quad r < r_0$

$$\mathscr{J}^2 = J_0^2(\tau_{co}/\tau_{en}); \tau_{co} \approx \tau_J$$

τ_{co} is the time of "contact"

τ_{en} is the time between molecular "encounters"

Appendix A, but refer the reader to Mori's excellent article on the subject for a more comprehensive idea of the general theory.[1] In Section V we obtain simple Bloch equations[16] by the Mori method, and we then extend the treatment to a time domain not adequately described by the Bloch equations, i.e. we obtain results where the relaxation is nonexponential and the lineshapes nonLorentzian. In Section VI, problems involving hyperfine splittings are analyzed, and in Section VII we study spin exchange. A study of the various diffusion constants that one determines in magnetic resonance experiments is presented in Sections V and VII; Section VIII is devoted to diffusion constants measured in spin-echo experiments. In some ways these results are disappointing since the diffusion constants, particularly those measured by spin exchange measurements, may differ considerably from the usual diffusion constants derived from momentum autocorrelation functions.

Tables Ia–Ih contain many of the symbols used throughout this article.

The references to the work of others is by no means comprehensive but include many of the important original articles, some books and review papers, and a number of articles with illustrate specific points.

TABLE Ig

SPIN EXCHANGE: EXTENDED THEORY

$$\tilde{\mathbf{A}} \equiv (S_+^{(1)}, \dot{\mathscr{S}}_+^{(1)}, S_+^{(2)})$$

$$\P\mathbf{B} = \P_{S_+^{(1)} \dot{\mathscr{S}}_+^{(1)} S_+^{(2)}}\mathbf{B}$$

$$\tilde{\dot{\boldsymbol{\alpha}}} = (0, \ddot{\mathfrak{S}}_+, 0); \ \ddot{\mathfrak{S}}_+ \text{ defined in Eq. (274)}$$

$$\widehat{\mathscr{K}}''\left(\int 8\langle e^{i(1-\P)\mathscr{H}^x t}\ddot{\mathfrak{S}}_+, \ddot{\mathfrak{S}}_-\rangle e^{-i\Omega t}\, dt\right) N\mathscr{J}^2$$

$$\omega_{\mathrm{L}}^2 = 8\langle \ddot{\mathfrak{S}}_+ \ddot{\mathfrak{S}}_- \rangle / N\mathscr{J}^2$$

$$\omega_{\mathrm{L}}^2 \approx \omega_J^2$$

$$\omega_J^2 \equiv \sum \langle \dot{J}_{lm}^2 \rangle / \sum \langle J_{lj}^2 \rangle$$

τ_{II} = Correlation time associated with $\ddot{S}_+$,

$$\tau_{\mathrm{II}} = \mathscr{K}''_{\mathrm{RE}}/\omega_{\mathrm{L}}^2$$

$$\tau_{\mathrm{II}} \approx \sum_{ml} \int_0^\infty dt\, \langle \dot{J}_{jl}(t)\, \dot{J}_{lm} \rangle / \omega^2 \mathscr{J}^2$$

$$\omega_{\mathrm{ex}} = \mathscr{J}^2 / 2\omega_J^2 \tau_{\mathrm{II}}$$

$$\tau_J = (\omega_J^2 \tau_{\mathrm{II}})^{-1}$$

TABLE Ih

DIFFUSION IN INHOMOGENEOUS FIELD

$$\mathbf{A}(\mathbf{R}) = \sum_j S_{+j}\delta(\mathbf{R} - \mathbf{r}_j) \quad \text{(Spin density)}$$

$$\mathbf{A}(\mathbf{k}) = \int d\mathbf{R}\, e^{i\mathbf{k}\cdot\mathbf{R}} \mathbf{A}(\mathbf{R}) = \sum S_{+j} e^{i\mathbf{k}\cdot\mathbf{r}_j}$$

Let $A = A(\mathbf{k})$

$G = (1/B_0)\, dB/dZ$ $\quad$ (dB/dZ = field gradient)

Do not confuse with correlation function $\mathrm{G}(t)$

D = Translational diffusion constant

$$D = \int_0^\infty dt\, \langle x_j(t)\, x_j \rangle$$

II. Bloch Equations

In this section, we develop some well-known results in the theory of magnetic relaxation in order to motivate the discussion of Mori's statistical mechanical theory of time-dependent processes discussed in later sections. We formulate a few simple problems in a manner that makes the connection with the Mori theory rather obvious. Furthermore, we show how the simple Bloch equations can be generalized to describe more complex situations, and we also point out the need for a more general theory such as that of Mori. In Section II, A we develop the simplest case, the one in which the system consists of spins with a single Larmor frequency and a single constant relaxation time; in Section II, B we extend the treatment to include a frequency-dependent relaxation time; in Section II, C a treatment is presented for systems in which the spins have a choice of two distinct Larmor frequencies and the spin relaxation times are constants; and in Section II, D the two-frequency problem is extended to include frequency-dependent spin relaxation times.

A. Bloch Equations. One Variable

The fundamental equations that describe the time evolution of ensemble-averaged magnetic quantities are the Bloch equations.[16,17] In their simplest form they are

$$\overline{dS_z(t)/dt} = -(1/T_1)[\overline{S_z(t)} - \langle S_z \rangle], \tag{1}$$

$$\overline{dS_\pm(t)/dt} = [\pm i\omega_0 - (1/T_2)]\,\overline{S_\pm(t)}, \tag{2}$$

where $\mathbf{S}_j$ is the spin angular momentum of the jth molecule in a system with N such molecules,

$$\mathbf{S} = \sum_j^N \mathbf{S}_j, \tag{3}$$

$$S_\pm = S_x \pm iS_y, \tag{4}$$

the bar over a quantity indicates an instantaneous, nonequilibrium ensemble average, the pointed brackets indicate an equilibrium ensemble average,[18c] ω_0 is the Larmor precessional angular frequency for these spins in an applied magnetic field along the z-axis, and T_1 and T_2 are constants known as the

[18c] Note that in a dc magnetic field along the z-axis, $\langle S_z \rangle \neq 0$ but $\langle S_\pm \rangle = 0$.

longitudinal and transverse spin relaxation times, respectively. These equations are simple transport equations. The solutions to Eqs. (1) and (2) are exponentials; in particular

$$\overline{S_+(t)} = \exp\,[(i\omega_0 - T_2^{-1})t]\,\overline{S_+(0)}. \tag{5}$$

The Fourier transform of the Bloch equations is also of interest; from Eq. (2) we obtain

$$i\omega\hat{S}_+(\omega) - \overline{S_+(0)} = [i\omega_0 - T_2^{-1}]\,\hat{S}_+(\omega), \tag{6}$$

where

$$\hat{S}_+(\omega) = \int_0^\infty \overline{S_+(t)}e^{-i\omega t}\,dt. \tag{7}$$

It can readily be shown[1,19] that the same equations of motion [Eqs. (2), (5), (6)] hold for the *autocorrelation function*, $\langle S_+(t)S_-\rangle$, as for $\overline{S_+(t)}$. Since the line-shape, $I(\omega)$, for a magnetic resonance absorption is connected to this autocorrelation function by the relation

$$I(\omega) = \mathrm{Re}\int_0^\infty dt\,e^{-i\omega t}\langle S_+(t)S_-\rangle, \tag{8}$$

it follows that

$$I(\omega) = \{T_2/[1+(\omega-\omega_0)^2T_2^2]\}\,\langle S_+S_+\rangle. \tag{9}$$

This is a Lorentzian line centered at $\omega = \omega_0$ with half-width at half-height equal to T^{2-1}.

B. Extended Bloch Equations. One Variable

We now wish to extend the Bloch equations and to discuss these extensions in a manner calculated to give insight into the development of the Mori theory presented in Section IV. A formal extension of the results above can be made by assuming that T_2^{-1} in Eq. (6) is a function of frequency, i.e. $T_2^{-1} \to \hat{K}(\omega)$, and for reasons to be explained below, we now denote the Larmor frequency by Ω_0 rather than by ω_0:

$$i\omega\hat{S}_+(\omega) - \overline{S_+(0)} = [i\Omega_0 - \hat{K}(\omega)]\,\hat{S}_+(\omega). \tag{10}$$

[19] L. Onsager, *Phys. Rev.* **37**, 405 (1931); **38**, 2265 (1931).

This equation can be Fourier transformed back to the time domain:

$$\overline{dS_+(t)/dt} = i\Omega_0 \overline{S_+(t)} - \int_0^t K(t-\tau)\overline{S_+(\tau)}\,d\tau, \tag{11}$$

where $K(t)$ is defined by

$$K(t) = (1/2\pi)\int_0^\infty d\omega\, e^{i\omega}t\,\hat{K}(\omega). \tag{12}$$

Equation (11) is a generalized variant of the Bloch equation in Eq. (2), and $K(t)$ is called a *memory function*[18b] since it correlates events that occurred at time τ with events occurring at time t, where $t \geqslant \tau$. In order to consider under what conditions the extended Bloch equations reduce to the ordinary Bloch equations, we rewrite Eq. (11) in the equivalent form,

$$\overline{dS_+(t)/dt} = i\Omega_0 \overline{S_+(t)} - \int_0^t d\tau\, K(\tau)\overline{S_+(t-\tau)}. \tag{13}$$

If $K(\tau)$ has a short memory, i.e. if it is a rapidly decaying function with characteristic decay time, τ_l, then if $S_+(t)$ is a slow function of time we would be interested in times t such that

$$t \gg \tau_l. \tag{14}$$

Under these conditions we could replace $S_+(t-\tau)$ by $S_+(t)$ in the integrand in Eq. (13) and we could extend the limits of integration to $t \to \infty$. However, this procedure is not justified because for high Larmor frequencies, i.e. for $\Omega_0 T_2 \gg 1$, $\overline{S_+(t)}$ may be a rapidly precessing function as exhibited in Eq. (5). On the other hand, under these conditions the quantity $\exp(-i\Omega_0 t)\,S_+(t)$ should be a slowly varying function; in fact, from Eq. (5) we see that $\exp(-i\Omega_0 t)\,S_+(t)$ should be a decaying function with relaxation time T_2. S_+ expressed this way is said to be in the rotating coordinate system[17] or in the interaction representation.[11] Thus if we rewrite Eq. (13) as

$$dS_+(t)/dt = i\Omega_0 \overline{S_+(t)} - e^{i\Omega_0 t}\int_0^t d\tau\,[e^{-i\Omega_0\tau}K(\tau)]\,[e^{-i\Omega_0(t-\tau)}\overline{S_+(t-\tau)}], \tag{15}$$

and if $(t, T_2, T_1 \gg \tau_l)$, we can replace $[e^{-i\Omega_0(t-\tau)}\overline{S_+(t-\tau)}]$ by $[e^{-i\Omega_0 t}\overline{S_+(t)}]$, and we can let the upper limit of the integral be $t \to \infty$. See Fig. 1. It is convenient to define a complex parameter, $\hat{\mathscr{K}}$, which we can call the *relaxation parameter*:

$$\hat{\mathscr{K}} \equiv \int_0^\infty dt\, e^{-i\Omega_0 t}K(t). \tag{16}$$

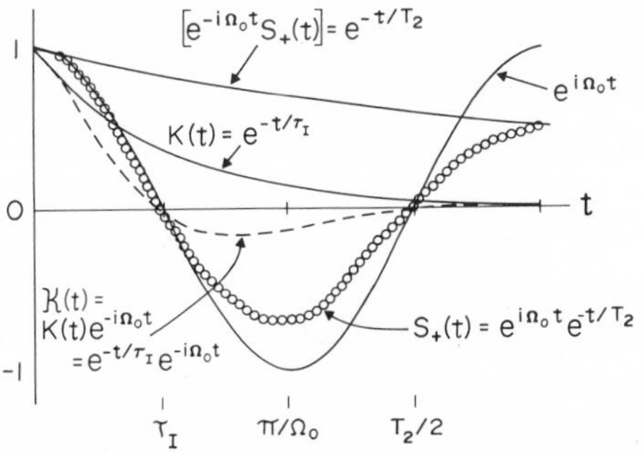

FIG. 1. Explanation of time scales: $\tau_{\mathrm{I}} \ll T_2$ and $\tau_{\mathrm{I}}\Omega_0 \approx 1$. (Real part of functions are plotted.)

In terms of the notation introduced above, $\hat{\mathscr{K}} = \hat{K}(\Omega_0)$. The *generalized transport equation* in Eq. (15) can now be rewritten in the approximate form

$$\overline{dS_+(t)/dt} = [i\Omega_0 - \hat{\mathscr{K}}]\,\overline{S_+(t)}; \tag{17}$$

we call this a *linear transport equation.* If we now define the relaxation time, T_2, by the relation

$$T_2^{-1} = \mathrm{Re}\,\hat{\mathscr{K}} \tag{18}$$

and the "shifted" or corrected Larmor frequency, ω_0, by the relation

$$\omega_0 = -\,\mathrm{Im}\,\hat{\mathscr{K}} + \Omega_0, \tag{19}$$

then Eq. (15) reduces to the Bloch equation, Eq. (2). $(\omega_0 - \Omega_0)$ is the "non-secular" or "dynamic frequency shift."[27] Because of this shift, even in the coordinate system rotating with frequency Ω_0, the quantity $\exp(-i\Omega_0 t)\,\overline{S_+(t)}$ precesses slowly with frequency $(\omega_0 - \Omega_0)$; the arguments presented above concerning the slow variation of $\exp(-i\Omega_0 t)\,\overline{S_+(t)}$ hold, therefore, only if the nonsecular shift $(\omega_0 - \Omega_0)$ is small, i.e. if $\omega_0 - \Omega_0|\tau_{\mathrm{I}} \ll 1$.

We have somewhat overmanipulated the equations above and have defined a somewhat excessive number of related parameters with the purpose of studying a simple case from numerous viewpoints so that our handling of the Mori theory in Section IV will be more meaningful. In summary, we note that a generalized transport equation of the form given in Eq. (13) can be reexpressed as a linearized transport equation, provided appropriate conditions on times and frequencies are satisfied. In particular, the generalized transport equations for the case of spins reduce to the Bloch equations; the appropriate conditions on times and frequencies are

$$t \gg \tau_{\mathrm{I}}, \tag{20}$$

and specifically for Eq. (2) to be valid

$$|\omega \mp \omega_0|\,\tau_{\mathrm{I}} \ll 1, \tag{21a}$$

$$\tau_{\rm I}/T_2 \ll 1, \tag{21b}$$

$$|\Omega_0 - \omega_0|\,\tau_{\rm I} \ll 1, \tag{21c}$$

while for Eq. (1) to be valid,

$$\omega\tau_{\rm I} \ll 1, \tag{22a}$$

$$\tau_{\rm I}/T_1 \ll 1. \tag{22b}$$

Thus the Bloch equations are valid only at frequencies low compared to $\tau_{\rm I}^{-1}$ *and times long compared to* $\tau_{\rm I}$. The simple Lorentzian lineshape, given in Eq. (9), is not valid if these conditions break down, since $\overline{S_+(t)}$ and $\langle S_+(t)\,S_-\rangle$ are then determined not by Eq. (17), but by Eq. (15). We shall say more about this later.

From the discussion above we see that the Bloch equations are valid provided $\overline{S_z(t)}$ and $(e^{\mp i\omega_0 t}\overline{S_\pm(t)}]$ vary slowly compared to all other relevant time-dependent quantities. For example, in a liquid the time dependences of $\{\overline{S_z(t)}, \overline{S_\pm(t)}\}$ may arise from the modulation of the magnetic dipolar interactions among spins by the rotational motions of the molecules, and for the Bloch equation to be applicable it is necessary that the rotational relaxation times, $\tau_{\rm I}$, be short compared to the spin relaxation times, T_2 and T_1, i.e. the conditions in Eqs. (21) and (22) must hold. This is usually the case in normal liquids where the molecular rotational correlation times are of order $\tau_{\rm I} \approx 10^{-11}$ sec and the spin relaxation times are of the order of 10^{-7} sec for electron and 10^{-1} sec for nuclear spins. Various statistical theories have been employed to derive the Bloch equations and to obtain molecular expressions for the spin relaxation times, T_1 and T_2. Most of these theories are based upon time-dependent perturbation theory and the extension to situations in which Eqs. (21) and (22) do not apply is not always straightforward.[11,14,15–17,20] One of our goals is to seek a formulation which allows for extension to time domains where Eqs. (21) and (22) are not fulfilled.

C. Bloch Equations. Two Variables

We next look at a slightly more complex situation involving two coupled Bloch equations. This problem will also be useful as a model for our discussion of the Mori theory. Let the system contain two sets of spins, $\mathbf{S}^{(1)}$ and $\mathbf{S}^{(2)}$, which, in the absence of exchange, are characterized by distinct Larmor frequencies, ω_1 and ω_2, respectively; the Larmor frequencies may be determined by different electronic environment or by nuclear hyperfine interactions with nuclei in different spin states. If some mechanism of exchange exists which can convert a member of one set, e.g. $\mathbf{S}_j^{(1)}$, to a

[20] C. Schlicter, "Principles of Magnetic Resonance." Harper, New York, 1963.

member of the other set, $\mathbf{S}_j^{(2)}$, then one has the coupled Bloch equations[21,22a]

$$\frac{d}{dt}\begin{pmatrix}\overline{S_+^{(1)}(t)}\\ \overline{S_+^{(2)}(t)}\end{pmatrix} = -\mathsf{T}^1\begin{pmatrix}\overline{S_+^{(1)}(t)}\\ \overline{S_+^{(2)}(t)}\end{pmatrix}, \tag{23}$$

where the *transport matrix*, $\mathscr{T}^{-1}$, is

$$\mathscr{T}^{-1} \equiv (-i\boldsymbol{\omega}+ \hat{\mathscr{K}}_{\mathrm{RE}}), \tag{24}$$

the *precessional frequency matrix*, $\boldsymbol{\omega}$, is

$$\boldsymbol{\omega} \equiv \begin{bmatrix}\omega_1 & 0\\ 0 & \omega_2\end{bmatrix} \tag{25}$$

and the real *relaxation matrix*, $\hat{\mathscr{K}}_{\mathrm{RE}}$, is

$$\hat{\mathscr{K}}_{\mathrm{RE}} = \begin{bmatrix} T_2^{(1)^{-1}} + \frac{1}{2}\omega_{\mathrm{ex}} & -\frac{1}{2}\omega_{\mathrm{ex}}\\ -\frac{1}{2}\omega_{\mathrm{ex}} & T_2^{(2)-1} + \frac{1}{2}\omega_{\mathrm{ex}}\end{bmatrix}. \tag{26}$$

The relaxation times, $T_2^{(n)}$, are real constants as is the *exchange rate*, ω_{ex}, which might be due to either chemical or spin exchange. We will discuss this more thoroughly in Section VII. Note that if $\omega_{\mathrm{ex}} = 0$, Eq. (23) reduces to two uncoupled Bloch equations of the form given in Eq. (17), and the relaxation time, $T_2^{(n)}$, for the spins $\overline{S_+^{(n)}(t)}$, is given by the relation, $T_2^{(n)} = (\hat{\mathscr{K}}_{\mathrm{RE}})_{nn}^{-1}$.

The coupled Bloch equations form a set of coupled linear transport or relaxation equations, in this case two coupled equations. The solutions to these equations, in the time domain, can be represented as a linear combination of two exponentials, i.e.

$$\begin{aligned}\overline{S_+^{(n)}(t)} = [1/(\Lambda_n - \Lambda_{n'})]\,\{[(-\Lambda_{n'} + \mathscr{T}_{nn}^{-1})\,\overline{S_+^{(n)}(0)} + \mathscr{T}_{nn'}^{-1}\,\overline{S_+^{(n')}(0)}]\exp(-\Lambda_n t)\\ + [(\Lambda_n - \mathscr{T}_{nn}^{-1})\,\overline{S_+^{(n)}(0)} - \mathscr{T}_{nn'}^{-1}\,\overline{S_+^{(n')}(0)}]\exp(-\Lambda_{n'} t)\},\end{aligned} \tag{27}$$

where $n = 1, 2$; $\overline{S_+^{(n)}(0)}$ is the initial value of $\overline{S_+^{(n)}(t)}$; $n \neq n'$; and Λ_1 and Λ_2 are the complex eigenfunctions of the transport matrix, $\mathscr{T}^{-1}$.

The spectrum, $I(\omega)$, corresponding to the solutions above, is given by a

[21] H. M. McConnell, *J. Chem. Phys.* **28**, 430 (1958).

[22a] H. S. Gutowsky and A. Saika, *J. Chem. Phys.* **21**, 1688 (1953); H. S. Gutowsky, D. M. McCall, and C. P. Slichter, *J. Chem. Phys.* **21**, 279 (1953).

relation analogous to that in Eq. (8), i.e. [22b]

$$I(\omega) = \text{Re} \int_0^\infty \langle [S_+^{(1)}(t) + S_+^{(2)}(t)]\,[S_-^{(1)} + S_-^{(2)}] \rangle e^{-i\omega t} dt. \tag{28}$$

In most cases of interest, $\langle S_-^{(n)} S_+^{(n')} \rangle = \langle S_-^{(1)} S_+^{(1)} \rangle \delta_{nn'}$, and it then follows that

$$I(\omega) = \text{Re} \left\{ \left[\frac{1 - (\mathscr{T}_{12}^{-1} + \mathscr{T}_{21}^{-1})/(\Lambda_2 - \Lambda_1)}{i\omega + \Lambda_1} + \frac{1 + (\mathscr{T}_{21}^{-1} + \mathscr{T}_{12}^{-1})/(\Lambda_2 - \Lambda_1)}{i\omega + \Lambda_2} \right] \right\}$$
$$\times \langle S_+^{(1)} S_-^{(1)} \rangle. \tag{29}$$

In the next two paragraphs we study the behavior of $I(\omega)$ and in the third paragraph below we do the same for $\overline{S_+^{(n)}(t)}$.

In analyzing the spectral formula in Eq. (28), it is convenient to separate the eigenvalues Λ_n into real and imaginary parts:

$$\Lambda_n = \Gamma_n - i\bar{\omega}_n. \tag{30}$$

The formula for $I(\omega)$, given in Eq. (29), then becomes

$$I(\omega) = \sum_{n=1}^{2} \left[\frac{\sigma_n \Gamma_n}{\Gamma_n^2 + (\omega - \bar{\omega}_n)^2} + \frac{\eta_n(\omega - \bar{\omega}_n)}{\Gamma_n^2 + (\omega - \bar{\omega}_n)^2} \right] \langle S_-^{(1)} S_+^{(1)} \rangle, \tag{31}$$

where the amplitudes σ_n and η_n, for the specific problem in Eq. (26), are determined by the relation

$$\sigma_n + i\eta_n = 1 - \{\omega_{\text{ex}}[(\Gamma_2 - \Gamma_1) + i(\bar{\omega}_2 - \bar{\omega}_1)]/[(\Gamma_2 - \Gamma_1)^2 + (\bar{\omega}_2 - \bar{\omega}_1)^2]\}(-1)^n, \tag{32}$$

and the corresponding spectrum, presented in Eq. (31), is the sum of two "generalized" Lorentzians. The first term in the square brackets in Eq. (31) is a *normal Lorentzian* at frequency $\bar{\omega}_n$ and with half-width Γ_n, and the second term might be called a *dispersive Lorentzian term.* The total spectrum, therefore, is a *generalized Lorentzian*, consisting of a normal and a dispersive contribution, and it is not symmetric about its peak, i.e. its first moment is nonvanishing. This generalized Lorentzian is generated by the correlation function in Eq. (28) which is the sum of two complex exponentials with complex coefficients.

If we now turn, for simplicity, to the case where $T_2^{(1)} = T_2^{(2)} = T_2$, we

[22b] Actually we should have

$$I(\omega) = \text{Re} \int_0^\infty \langle [\gamma_1 S_+^{(1)}(t) + \gamma_2 S_+^{(2)}(t)]\,[\gamma_1 S_+^{(1)} + \gamma_2 S_+^{(2)}] \rangle e^{i\omega t} dt,$$

where γ_1 and γ_2 are gyromagnetic ratios. If $\gamma_1 \approx \gamma_2$, then Eq. (28) is adequate, but if the two classes of spins are very different, e.g. one an electron and one a nuclear spin of a proton and the other that of a fluorine nucleus, then the equation in this footnote must be used.

see that

$$\Gamma_n - i\bar{\omega}_n = T_2^{-1} + \tfrac{1}{2}\omega_{ex} - \tfrac{1}{2}i(\omega_1 + \omega_2) - (-1)^n \tfrac{1}{2}[\omega_{ex}^2 - (\omega_2 - \omega_1)^2]^{1/2} \quad (33)$$

and

$$\sigma_n + i\eta_n = 1 + \{\omega_{ex}(-1)^n / [\omega_{ex}^2 - (\omega_2 - \omega_1)^2]^{1/2}\}. \quad (34)$$

ω_n are the Larmor frequencies in the absence of coupling terms, i.e. for $\omega_{ex} = 0$, whereas $\bar{\omega}_n$ are the precessing frequencies in the presence of exchange.

With these formulas we can readily study the interesting behavior of $I(\omega)$ in various limiting cases. For slow exchange, i.e.

$$(\omega_2 - \omega_1)^2 \gg \omega_{ex}^2, \quad (35)$$

the imaginary and real parts of the eigenvalues of Λ_n are

$$\bar{\omega}_n \cong \omega_n + [(-1)^n \omega_{ex}^2 / (\omega_2 - \omega_1)] \quad (36)$$

and

$$\Gamma_n \cong T_2^{-1} + \omega_{ex}/2, \quad (37)$$

respectively; the intensity coefficients in Eq. (34) are then

$$\sigma_n = 1 \quad (38)$$

$$\eta_n \cong -(-1)^n \omega_{ex} / (\omega_2 - \omega_1). \quad (39)$$

Thus as $\omega_{ex} \to 0$, i.e. the limit of no exchange, the spectrum consists of two symmetrical "normal" Lorentzians of equal intensity centered at ω_1 and ω_2. As ω_{ex} increases the lines shift towards each other and become asymmetric because of the presence of the dispersive components with intensity factor, η_n. See Fig. 2. For rapid exchange, i.e.

$$(\omega_2 - \omega_1)^2 \ll \omega_{ex}^2, \quad (40)$$

the two Lorentzians are both located at frequency $(\omega_1 + \omega_2)/2$, and the dispersive contributions vanish because $\eta_n = 0$. Furthermore, from Eqs. (33) and (34) we find

$$\Gamma_n = T_2^{-1} + \tfrac{1}{2}\omega_{ex} - (-1)^n \tfrac{1}{2}\omega_{ex}\{1 - [(\omega_2 - \omega_1)/\omega_{ex}]^2\}^{1/2} \quad (41)$$

and

$$\sigma_n = 1 + ((-1)^n / \{1 - [(\omega_2 - \omega_1)/\omega_{ex}]^2\}^{1/2}). \quad (42)$$

Thus the sharper of the two Lorentzians is the more intense and has positive intensity whereas the other has negative intensity; as ω_{ex} increases the sharper line becomes narrower and more intense whereas the broader negative line becomes wider and less intense. See Fig. 3.

We can gain additional insight by expressing the solution for $\overline{S_+^{(1)}(t)}$ more explicitly than in Eq. (21). If we assume, as in the discussion above,

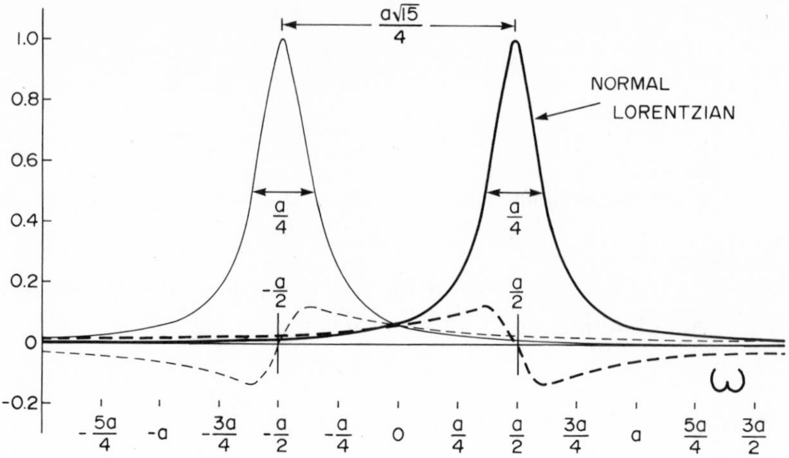

FIG. 2. Slow exchange. Let $\omega_n = (-1)^n a/2$; $\omega_{ex} = a/4$, $T_2^{-1} = 0$; then, $\Gamma_n = a/8$, $\bar{\omega}_n = (-1)^n a(15)^{1/2}/8$, $\sigma = 1$; $\eta_n = -(-1)^n/(15)^{1/2}$. See Eqs. (33), (34), and (31). We have plotted the separate Lorentzian but have not summed to obtain net lineshapes.

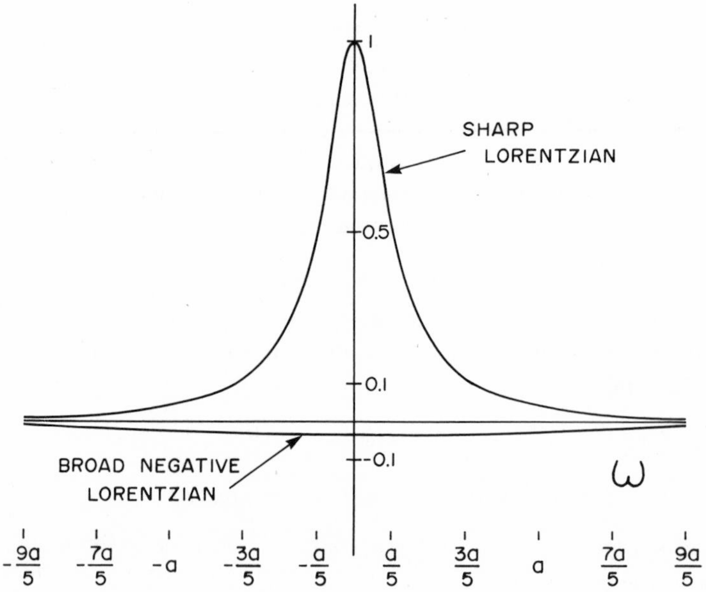

FIG. 3. Fast exchange. Let $\omega_n = (-1)^n a/2$; $\omega_{ex} = a\sqrt{2}$; $T_2^{-1} = 0$; then, $\Gamma_2 = \frac{1}{2}a[\sqrt{2}-1]$; $\Gamma_1 = \frac{1}{2}a[\sqrt{2}+1]$; $\bar{\omega}_1 = \bar{\omega}_2 = 0$; $\sigma_2 = 1+\sqrt{2}$; $\sigma_1 = -(-1+\sqrt{2})$; $\eta_1 = \eta_2 = 0$. See Eqs. (33), (34), and (31).

that $T_2^{(1)} = T_2^{(2)} = T_2$, and, furthermore, if we set $\overline{S_+^{(1)}(0)} = \overline{S_+^{(2)}(0)}$, then

$$\overline{S_+^{(1)}(t)} = \overline{S_+^{(1)}(0)}\{[1+\tfrac{1}{2}\delta+\tfrac{1}{2}i\omega_{\text{ex}}(1+\delta)/(\omega_2-\omega_1)]\,e^{-\Gamma_1 t}e^{i\bar{\omega}_1 t} + [-\tfrac{1}{2}\delta-\tfrac{1}{2}i\omega_{\text{ex}}(1+\delta)/(\omega_2-\omega_1)]\,e^{-\Gamma_2 t}e^{i\bar{\omega}_2 t}\}, \tag{43}$$

where

$$\delta = [1-\omega_{\text{ex}}^2/(\omega_2-\omega_1)^2]^{-1/2}-1. \tag{44}$$

In the slow exchange limit, i.e. $(\omega_1-\omega_2)^2 > \omega_{\text{ex}}^2$, the parameter δ is real and positive, and $\Gamma_1 = \Gamma_2$. Furthermore, if $(\omega_1-\omega_2)^2 \gg \omega_{\text{ex}}^2$, then $0 < \delta \ll 1$, and

$$\overline{S_+^{(1)}(t)} = \overline{S_+^{(1)}(0)}\,e^{-\Gamma_1 t}\{[1+\tfrac{1}{2}\delta+\tfrac{1}{2}i\omega_{\text{ex}}/(\omega_2-\omega_1)]\,e^{i\bar{\omega}_1 t} + [-\tfrac{1}{2}\delta-\tfrac{1}{2}i\omega_{\text{ex}}/(\omega_2-\omega_1)]\,e^{i\bar{\omega}_2 t}\}. \tag{45}$$

We note the following interesting properties of this solution. (i) $\overline{S_+^{(1)}(t)}$ is a sum of two oscillating terms so that one cannot find a rotating frame which cancels out both precessional terms. (ii) In the limit that $\omega_{\text{ex}} \to 0$, and the solution reduces to the simple solution in Eq. (5) with a single precessional frequency. (iii) For finite ω_{ex} both precessional terms have complex amplitudes, and the imaginary part of the amplitude, $\pm i\omega_{\text{ex}}(\omega_2-\omega_1)^{-1}/2$, is an out-of-phase contribution which gives rise to a dispersive Lorentzian in frequency space. (iv) In the fast exchange limit, $\omega_{\text{ex}}^2 > (\omega_2-\omega_1)^2$, and we must revert back to Eq. (43); in this case, $\Gamma_1 \neq \Gamma_2$, δ is imaginary, and $\bar{\omega}_1 = \bar{\omega}_2 = \tfrac{1}{2}(\omega_1+\omega_2)$, so that a rotating coordinate system which cancels out the precessional motion can be found. The most important conclusion that we draw from this discussion is that *in the slow exchange limit, a coordinate frame rotating with frequency* $\bar{\omega}_1$ *cancels the precessional motion of the dominant contribution to* $S_+^{(1)}(t)$, *but the small contribution with frequency* $\bar{\omega}_2$ *precesses with frequency* $(\bar{\omega}_2-\bar{\omega}_1)$ *in this frame.* This precessional frequency is "slow" if

$$(\bar{\omega}_2-\bar{\omega}_1)^2\tau_{\text{I}}^2 \ll 1; \tag{46}$$

where τ_{I} is the correlation time characteristic of the molecular motions, or as we shall see, characteristic of $\dot{S}_+$. We can also analyze the precessional motion in a coordinate system rotating with frequency ω_1; the precessional frequencies of the two components are then $(\omega_1-\bar{\omega}_1)$ and $(\omega_1-\bar{\omega}_2)$. These precessional frequencies are "slow" if

$$(\omega_1-\bar{\omega}_2)^2\tau_{\text{I}}^2 \ll 1, \tag{47a}$$

$$(\omega_2-\bar{\omega}_1)^2\tau_{\text{I}}^2 \ll 1. \tag{47b}$$

In the fast exchange limit the precessional frequency in the coordinate frame rotating with frequency ω_1 is $(\omega_2-\omega_1)/2$ and this is "slow" if

$$(\omega_2-\omega_1)^2\tau_{\mathrm{I}}^2 \ll 4. \tag{48}$$

Of course, in the fast exchange limit the precessional motion can be completely canceled if the rotating frame rotates with frequency $(\omega_1+\omega_2)/2$.

D. Extended Bloch Equations. Two Variables

We can also seek an extension of the coupled Bloch equations by the same procedure that we followed in Section II, B for the simple uncoupled Bloch equations; however, additional difficulties arise in the present case. The generalized form of Eqs. (23)–(26) is

$$\frac{d}{dt}\begin{pmatrix}\overline{S_+^{(1)}(t)}\\ \overline{S_+^{(2)}(t)}\end{pmatrix} = i\Omega\begin{pmatrix}\overline{S_+^{(1)}(t)}\\ \overline{S_+^{(2)}(t)}\end{pmatrix} - \int_0^t d\tau \begin{pmatrix}K_{11}(\tau) & K_{12}(\tau)\\ K_{21}(\tau) & K_{22}(\tau)\end{pmatrix}\begin{pmatrix}\overline{S_+^{(1)}(t-\tau)}\\ \overline{S_+^{(2)}(t-\tau)}\end{pmatrix}, \tag{49}$$

where

$$\Omega = \begin{pmatrix}\Omega_1 & 0\\ 0 & \Omega_2\end{pmatrix}; \tag{50}$$

these equations bear the same relation to Eqs. (23)–(26) that Eq. (11) bears to Eq. (2). Note that we have replaced ω_1 and ω_2 by Ω_1 and Ω_2, respectively. Again we ask under what conditions the generalized Bloch equations reduce to the simple one, i.e. when does Eq. (49) reduce to Eqs. (23)–(26). For this reduction to take place the *memory function matrix*, $K_{nm}(\tau)$, must decay rapidly compared to the characteristic times of $\overline{S_+^{(1)}(t)}$ and $\overline{S_+^{(2)}(t)}$, but as we saw above, these mean spins can have rapidly precessing components with frequency, ω_n. (See Eq. (43).) Once again we can transform to a rotating coordinate system, but now both $\overline{S_+^{(1)}(t)}$ and $\overline{S_+^{(2)}(t)}$ have components which precess with characteristic frequencies approximately equal to $\bar{\omega}_1$ and $\bar{\omega}_2$.[22c] In order to handle this difficulty, we rewrite Eq. (49) as

$$\frac{d}{dt}\begin{pmatrix}\overline{S_+^{(1)}(t)}\\ \overline{S_+^{(2)}(t)}\end{pmatrix} = i\Omega\begin{pmatrix}\overline{S_+^{(1)}(t)}\\ \overline{S_+^{(2)}(t)}\end{pmatrix} - \int_0^t d\tau\,[\mathsf{K}(\tau)e^{-i\Omega\tau}]\,e^{i\Omega t}\left[e^{-i\Omega(t-\tau)}\begin{pmatrix}\overline{S_+^{(1)}(t-\tau)}\\ \overline{S_+^{(2)}(t-\tau)}\end{pmatrix}\right], \tag{51}$$

and the elements of the matrix $\mathsf{K}(\tau)$ are $K_{nm}(\tau)$. Equation (51) is analogous to Eq. (15). The functions $[e^{-i\Omega_1 t}\overline{S_+^{(1)}(t)}]$ and $[e^{-i\Omega_2 t}\overline{S_+^{(2)}(t)}]$ are more slowly

[22c] $\overline{S_+^{(1)}(t)}$ has components precessing with frequencies ω_1 and ω_2 in the simple case described by Eq. (23); the behavior of $\overline{S_+^{(1)}(t)}$ in Eq. (49) may be somewhat different.

varying than are $\overline{S_+^{(1)}(t)}$ and $\overline{S_+^{(2)}(t)}$, respectively, since the exponential factors cancel part of the precessional motion in $\overline{S_+^{(1)}(t)}$ and $\overline{S_+^{(2)}(t)}$; however, since the precessional frequencies are approximately $\bar{\omega}_1$ and $\bar{\omega}_2$ rather than Ω_1 and Ω_2, and since $\overline{S_+^{(1)}(t)}$ and $\overline{S_+^{(2)}(t)}$ are each composed of the sum of two terms, one with precessional frequency $\bar{\omega}_1$ and the other with precessional frequency $\bar{\omega}_2$, it follows that the precessional effects are not completely canceled by the exponential factors. (See the discussion in Section II, C.) If τ_{I} again characterizes the decay time of the elements of $\mathsf{K}(\tau)$, and if the characteristic times involved in $[e^{-i\Omega_n t}\overline{S_+^{(n)}(t)}]$ are long compared to τ_{I}, then, following the procedure used in obtaining Eq. (17) from Eq. (15), we can replace $[e^{-i\Omega_n(t-\tau)}\overline{S_+^{(n)}(t-\tau)}]$ in Eq. (51) by $[e^{-i\Omega_n t}\overline{S_+^{(n)}(t)}]$ and we can let the upper limit on the integral be $t \to \infty$. These changes are valid provided the conditions in Eqs. (47a) and (47b) apply, or in the most extreme case, if the condition in Eq. (48) applies (of course, we replace ω_n by Ω_n). If these approximations are made, Eq. (51) reduces to

$$\frac{d}{dt}\begin{pmatrix} \overline{S_+^{(1)}(t)} \\ \overline{S_+^{(2)}(t)} \end{pmatrix} = -\mathscr{T}^{-1}\begin{pmatrix} \overline{S_+^{(1)}(t)} \\ \overline{S_+^{(2)}(t)} \end{pmatrix}, \tag{52}$$

where

$$\mathscr{T}^{-1} = -i\Omega + \hat{\mathscr{K}}, \tag{53}$$

and

$$\hat{\mathscr{K}} \equiv \int_0^\infty d\tau\ \mathsf{K}(t)\exp[-i\Omega t]. \tag{54}$$

Equation (52) is the simple two-variable Bloch equation presented in Eqs. (23)–(26) provided the following associations are made:

$$\hat{\mathscr{K}} = \hat{\mathscr{K}}_{\mathrm{RE}} - i\hat{\mathscr{K}}_{\mathrm{IM}} \tag{55}$$

and

$$\Omega + \hat{\mathscr{K}}_{\mathrm{IM}} = \omega \tag{56}$$

and $\hat{\mathscr{K}}_{\mathrm{RE}}$ is given by Eq. (26).

We have written down a generalized two-variable Bloch equation, Eq. (51), and have seen that under appropriate conditions the memory function matrix, $\mathsf{K}(t)$, can be replaced by $\hat{\mathsf{K}}(\Omega) = \hat{\mathscr{K}}$, and the generalized transport equations reduce to the simple two-variable Bloch equations in Eqs. (23)–(26). This simplification can be made if $[\exp(-i\Omega_n t)\overline{S_+^{(n)}(t)}]$ varies slowly compared to $[\mathsf{K}(t)\exp(-i\Omega t)]$; the appropriate conditions are

$$\tau_{\mathrm{I}}/T_2^{(n)} \ll 1, \tag{57a}$$

$$\tau_{\mathrm{I}}/t \ll 1, \tag{57b}$$

and

$$(\Omega_2-\Omega_1)^2\tau_{\rm I}^2/4 \ll 1, \tag{58a}$$

$$(\omega-\bar{\omega}_n)^2\tau_{\rm I}^2 \ll 1. \tag{58b}$$

These discussions should serve as aids in understanding Section IV. We shall return to the examples discussed above later in this article and shall treat them more thoroughly.

III. Molecular Considerations

The spin relaxation times, T_1 and T_2, have been calculated in terms of molecular parameters by a variety of different approaches.[10,11,14–17] Detailed information concerning the time evolution of a system of spins requires detailed solutions of the molecular equations of motion of the individual molecules. This is, of course, impossible to achieve but even approximate calculations based on molecular descriptions require knowledge of the appropriate Hamiltonians. In Section III, A we shall discuss, in general terms, Hamiltonians appropriate for the study of magnetic relaxation. In Section III, B we comment briefly on some molecular expressions for T_2^{-1}; this section is a review of existing formulations which can be compared with those obtained in ensuing sections.

In this section we consider Hamiltonians, $\mathscr{H}$, which give rise to magnetic resonance spectra and account for the relaxation of the spins. The spatial electronic coordinates must be properly "averaged" in the ground electronic states, and the spin-dependent parts of the Hamiltonian constitute what is call a spin Hamiltonian.[23–25] The Hamiltonian, $\mathscr{H}$, is also dependent on the spatial coordinates of the nuclei; we refer to the system described by nuclear spatial coordinates as the *lattice*.

A. Hamiltonian

We will consider a Hamiltonian, $\mathscr{H}$, in units of $\hbar$:

$$\mathscr{H} = \mathscr{H}_{\rm S} + \mathscr{H}_{\rm L} + \mathscr{H}_{\rm SL}, \tag{59}$$

where $\mathscr{H}_{\rm S}$ depends only upon spin variables, $\mathscr{H}_{\rm L}$ represents the molecular motions and interactions which are independent of the spins, and $\mathscr{H}_{\rm SL}$

[23] G. E. Pake, "Paramagnetic Resonance." Benjamin, New York, 1962.

[24] A. Abragam and B. Bleaney, "Electron Paramagnetic Resonance of Transition Ions." Oxford Univ. Press (Clarendon), London and New York, 1970.

[25] A. Carrington and A. D. McLachlan, "Introduction to Magnetic Resonance." Harper, New York, 1967.

involves those interactions which involve both spins and nuclear spatial coordinates. $\mathscr{H}_S$ is known as the *pure spin term*, $\mathscr{H}_L$ the *lattice interaction* (where by lattice we mean all molecular behavior other than spin) and $\mathscr{H}_{SL}$ is the *spin-lattice interaction*. It is clear that

$$[\mathscr{H}_S, \mathscr{H}_L] = 0. \tag{60}$$

In all cases of interest, the equilibrium average of $\mathscr{H}_{SL}$ vanishes, i.e.

$$\langle \mathscr{H}_{SL} \rangle = 0. \tag{61}$$

Most schemes for studying spin relaxation in liquids are based upon the separation of the Hamiltonian into a zeroth-order term, $\mathscr{H}_S + \mathscr{H}_L$, and a perturbation which is $\mathscr{H}_{SL}$. It is usually a simple matter to obtain solutions to $\mathscr{H}_S$, and in liquids one usually assumes that $\mathscr{H}_L$ can be replaced by the classical equivalent. In the limits usually studied, $\mathscr{H}_{SL}$ is a time-dependent perturbation *whose lattice variables evolve classically in time and whose spin variables must be treated quantum mechanically* and precess with a time dependence determined by $\mathscr{H}_S$.

The simplest pure spin Hamiltonian is the Zeeman interaction

$$\mathscr{H}_S = \omega_0 S_z, \tag{62}$$

where S_z is the sum of z-components of spins described in Eq. (3), ω_0 is the Larmor frequency,

$$\omega_0 = (e/4\pi mc)\, gB, \tag{63}$$

e is the electronic charge in esu, m is the electronic (or protonic) mass, c is the speed of light, B is the dc magnetic field applied along the laboratory z-axis, and g is the isotropic g-factor or, in nuclear resonance experiments, the isotropic chemical shift. The terms in $\mathscr{H}_{SL}$ for which

$$[S_z, \mathscr{H}_{SL}^{(\mathrm{sec})}] = 0 \tag{64}$$

are called the *secular terms*, whereas those for which

$$[S_z, \mathscr{H}_{SL}^{(\mathrm{nsec})}] \neq 0 \tag{65}$$

are called the *nonsecular terms*.[26a–27] One example of $\mathscr{H}_{SL}$ is the anisotropic g-tensor interaction or, equivalently in nuclear resonance experiments, the

[26a] D. Kivelson, *J. Chem. Phys.* **33**, 1094 (1960).

[26b] D. Kivelson, *in* "Electron Spin Relaxation in Liquids" (L. T. Muus and P. W. Atkins, eds.), Ch. 10, p. 213. Plenum, New York, 1972.

[26c] R. Wilson and D. Kivelson, *J. Chem. Phys.* **44**, 154 (1966).

[27] J. Freed and G. Fraenkel, *J. Chem. Phys.* **39**, 326 (1963).

anisotropic chemical shift interaction. This interaction has the form

$$\mathscr{H}_{\text{SL}} = \sum_j B[\Gamma_{zz}^{(j)} S_{z_j} + \Gamma_{z+}^{(j)} S_{+_j} + \Gamma_{z-}^{(j)} S_{-j}], \tag{66}$$

where $\Gamma^{(j)}$ is a traceless second rank tensor representing the anisotropic g-tensor or chemical shift of the jth molecule; it is a function of the Eulerian angles of the molecule which in turn are functions of time since the molecules rotate. The first term in Eq. (66) is the secular term. Another spin-lattice interaction is the spin-spin dipolar interaction,

$$\mathscr{H}_{\text{SL}} = \sum_{\alpha,\beta} \sum_{j,k} S_{\alpha j} \mathscr{D}_{\alpha\beta}^{(j\ k)} S_{\beta k}, \tag{67}$$

where ($\alpha, \beta = z, +, -$) and $\mathscr{D}^{(jk)}$ is a second rank traceless tensor representing the dipolar interaction between the jth and the kth spin; $\mathscr{D}_{\alpha\beta}^{(jk)}$ depends upon the orientations of both the jth and the kth molecules and upon the intermolecular separation which also varies with time as the molecules move. The interaction in Eq. (66) is intermolecular; an intramolecular variant has $j = k$. This intramolecular interaction arises only if the spin $S \geqslant 1$; for electron spins this is the spin-spin interaction and for nuclear spins it is the quadrupolar coupling interaction.

A somewhat more complicated pure spin Hamiltonian is

$$\mathscr{H}_{\text{S}} = \omega_1 S_z^{(1)} + \omega_2 S_z^{(2)}, \tag{68}$$

where $S_z^{(1)}$ and $S_z^{(2)}$ are sums of spins with Larmor frequencies ω_1 and ω_2, respectively. The different Larmor frequencies correspond to different g-values in ESR and different chemical shifts in NMR experiments. If the two spins are on different molecules, one form of $\mathscr{H}_{\text{SL}}$ that is interesting is

$$\mathscr{H}_{\text{SL}} = \sum_{\alpha,\beta} \sum_{jk} S_{\beta j}^{(1)} \mathscr{D}_{\beta\alpha}^{(jk)} S_{\alpha k}^{(2)}, \tag{69}$$

where $\mathscr{D}^{(jk)}$ has already been described as the traceless dipolar interaction tensor. If the two spins are on the same molecule, we must set $j = k$ in Eq. (69). A closely related intermolecular interaction is that of spin exchange,[28]

$$\mathscr{H}_{\text{SL}} = \tfrac{1}{2} \sum_{j,k} J_{jk} S_j^{(1)} \cdot S_k^{(2)}; \tag{70}$$

this is a scalar interaction and J_{jk} can be treated as the missing trace of $\mathscr{D}^{(jk)}$ in Eq. (69). The parameter J_{jk} is twice the exchange integral between

[28] J. H. Van Vleck, "The Theory of Electric and Magnetic Susceptibilities," Sect. 76. Oxford Univ. Press, London, 1932.

electrons j and k; it is a function of intermolecular separation and relative orientation, and it is, therefore, time-dependent in a liquid. We will discuss it in detail in Section VII.

Finally, we wish to consider the pure spin Hamiltonian

$$\mathscr{H}_S = \omega_0 S_z + \omega_N I_z + a \sum_j \mathbf{I}_j \cdot \mathbf{S}_j \tag{71}$$

where S_z has already been defined as the sum of spins S_{z_j} with Larmor frequency ω_0,

$$I = \sum I_j, \tag{72}$$

where I_{z_j} is the z-component of angular momentum (in units of $\hbar$) for a spin with Larmor frequency ω_N, also on the jth molecule, and a is a scalar coupling constant between the two spins. We have used a notation most appropriate for ESR where S_j is an electron spin, I_j a nuclear spin, and a the Fermi contact or isotropic hyperfine coupling constant. In the NMR counterpart both I_j and S_j are nuclear spins, and a is the spin-spin coupling constant.

The spin-lattice interactions of most importance to the ESR hyperfine problem are the anisotropic g-factor interaction [Eq. (66)], the exchange interaction [Eq. (70)], and the anisotropic hyperfine interaction

$$\mathscr{H}_{SL} = \sum_{\alpha,\beta} \sum_{j,k} I_{\alpha_j} a_{\alpha\beta}^{(jj)} S_{\beta_j}, \tag{73}$$

where $\boldsymbol{a}^{(jk)}$ is the second rank traceless tensor which represents the intramolecular dipolar interaction between nuclear and electronic spins on the jth molecule. In the NMR counterpart, $\mathscr{H}_{SL}$ is the anisotropic spin-spin dipolar interaction.

There are many other interactions and complications that occur, but the examples given above cover many of the situations encountered in both ESR and NMR. One additional spin-lattice interaction of considerable interest is the spin-rotational interaction.[28-32] Note that intermolecular spin-lattice interactions depend upon the translational motion of the molecules and

[29] P. S. Hubbard, *Phys. Rev.* **131**, 1155 (1963).

[30] P. W. Atkins and D. Kivelson, *J. Chem. Phys.* **44**, 169 (1966).

[31] P. Atkins, *Mol. Phys.* **12**, 201 (1967); P. W. Atkins, *in* "Electron Spin Relaxation in Liquids" (L. T. Muus and P. W. Atkins, eds.), Ch. 11, p. 279. Plenum, New York, 1972; P. W. Atkins, *Advan. Mol. Relaxation Processes* **2**, 121 (1972).

[32] J. M. Deutch and J. S. Waugh, *J. Chem. Phys.* **43**, 1914 (1965); C. S. Johnson and J. S. Waugh, *J. Chem. Phys.* **35**, 2020 (1961).

hence give information concerning molecular diffusion. One of our objects will be to study the significance of the various diffusion constants that can be determined.

B. Molecular Expression for T_2^{-1}

Starting with the classic work of Bloembergen *et al.*,[14] numerous authors have developed techniques for calculating spin relaxation times. Simple exponential relaxation of spin is observed only if the Bloch equations apply, or equivalently, as discussed in Section II, if $\tau_I \ll T_2, T_1$ where T_1 and T_2 are spin relaxation times and τ_I is a correlation time characteristic of molecular motions.[17,20,23] This condition holds if $\mathscr{H}_{SL}$ is a small perturbation on $\mathscr{H}_S + \mathscr{H}_L$, and under these conditions, expressions for T_2^{-1} and T_1^{-1} can be obtained by second-order time-dependent perturbation calculations. Such a calculation can be considered an expansion in powers of $\mathscr{H}_{SL}$, and only terms through second order are kept. It can also be considered an expansion in the time t, and only terms through first order in t are kept; the terms in this expansion are then assumed to be the first two terms in the expansion of an exponential in time, and the coefficient of t is associated with the appropriate spin relaxation frequency, T_1^{-1} or T_2^{-1}. For a single uncoupled Bloch equation in the absence of hyperfine interactions, it can readily be shown that [11,26a–c]

$$T_2^{-1} = \int_0^\infty dt \, \langle [\mathscr{H}_{SL}(t), [\mathscr{H}_{SL}(0), S_+]] \, S_- \rangle / \langle S_+ S_- \rangle, \tag{74}$$

where $[\mathbf{A}, \mathbf{B}]$ is the commutator, $\mathbf{AB} - \mathbf{BA}$, and

$$\mathscr{H}_{SL}(t) = e^{-i(\mathscr{H}_S + \mathscr{H}_L)t} \, \mathscr{H}_{SL} \, e^{i(\mathscr{H}_S + \mathscr{H}_L)t}. \tag{74a}$$

In Section V we derive these expressions by means of Mori's formulism. For completeness, we also include a molecular expression for T_1^{-1}:

$$T_1^{-1} = \int_0^\infty dt \, \langle [\mathscr{H}_{SL}(t), [\mathscr{H}_{SL}(0), S_z]] \, S_z \rangle / \langle S_z S_z \rangle. \tag{75}$$

IV. Mori's Theory

Because of the enormous number of molecules involved, an exact description of the motion of a molecule in a liquid is beyond hope. However, for a Brownian particle, i.e. a particle that is large compared to the solvent molecules, we can write an equation of motion which is valid for times much longer than molecular times. This equation of motion is the Langevin equation,

$$m\,\mathbf{dv}/\mathbf{dt} = \mathscr{F} - \zeta\mathbf{v} + \mathbf{F}(t), \tag{76}$$

where $\mathbf{v}$ is the velocity of the Brownian particle, $\mathscr{F}$ is the force on it due to an externally applied field, $\zeta\mathbf{v}$ represents the slow, frictional force where ζ is the friction constant, and $\mathbf{F}(t)$ is a force which is a rapidly varying random function which averages to zero, i.e. for which

$$\overline{\mathbf{F}(t)} = 0. \tag{77}$$

The key point in the development of the Langevin equation, and of the Bloch equations described in Section II, is the separation of time scales into a slow time scale associated with the motion of the Brownian particle and a fast time scale associated with colecular motion. We might ask then if it is possible, with the right choice of variables, to divide the time evolution of molecular variables into a slowly varying part and a rapidly varying part.

Mori[1] has developed a "generalized Langevin equation" which provides this division of time scales. This theory has been used to study hydrodynamics and a variety of relaxation and transport relaxation and transport phenomena, and in this article we will apply it to the study of spin relaxation. We will not discuss the derivation of Mori's theory here; in Appendix A we give a brief summary of the derivation. An interpretive description of the theory is given in Section IV, A. For details and a more thorough interpretation, the reader is referred to the original article by Mori. Solutions to the transport equations are discussed in Section IV, B with emphasis on results that are useful to the analysis of spin relaxation in succeeding sections. In Section IV, C we discuss extensions of the simple theory.

A. Formalism of Mori

Mori chooses a set of dynamical variables, $A_i(t)$, which hopefully form a complete set in the sense that they describe the relevant slow variations in the system. We shall specify this aspect of the theory, which constitutes the physical input, in more detail below, but we might expect $[S_z - \langle S_z \rangle]$ to contribute the complete set in the spin problem giving rise to the simple Bloch equation in Eq. (1), whereas $S_+^{(1)}$ and $S_+^{(2)}$ might constitute the complete set in the spin problem giving rise to the coupled Bloch equations in Eq. (23). The variables, $A_i(t)$, can be represented as a column vector,

$$\mathbf{A}(t) = \begin{pmatrix} \vdots \\ A_i(t) \\ \vdots \end{pmatrix}. \tag{78}$$

The variables, $\mathbf{A}(t)$, are always taken such that

$$\langle \mathbf{A}(t) \rangle = 0, \tag{79}$$

i.e. $\mathbf{A}(t)$ describes a displacement from equilibrium. A projection operator, ¶, is introduced which projects an arbitrary vector **B**, with the same dimensionality as **A**, onto **A**, i.e.

$$\P\mathbf{B} = \langle \mathbf{B}\mathbf{A}^{+} \rangle \cdot \langle \mathbf{A}\mathbf{A}^{+} \rangle^{-1} \mathbf{A}, \tag{80}$$

where $\mathbf{A}^{+}$ is the adjoint of **A**. (¶ is idempotent, i.e. $\P^2 = \P$.) In the appropriate Hilbert space where the set of variables, **A**, constitutes *a complete set of slow variables,* any function which is orthogonal to all the A_i's should vary rapidly; thus for an arbitrary vector **B**, its orthogonal component, $(1-\P)\mathbf{B}$, should consist only of rapidly varying terms. It is in this manner that the Mori theory provides a division of time scales between fast and slow motions, and it is for this reason that it can be successful only if a complete set of slow variables, **A**, is selected.

The fundamental equation of motion in this theory is the generalized Langevin equation

$$d/dt\,\mathbf{A}(t) = i\boldsymbol{\Omega}\cdot\mathbf{A}(t) - \int_0^t \mathsf{K}(\tau)\,\mathbf{A}(t-\tau)\,d_\tau + \dot{\boldsymbol{\alpha}}(t_\mathrm{p}), \tag{81}$$

where the time derivative and time dependence of $\mathbf{A}(t)$ can be expressed in terms of the superoperator, $\mathscr{H}^x$, of the Hamiltonian as[33]

$$\dot{\mathbf{A}} = i\mathscr{H}^x\mathbf{A}, \tag{82}$$

$$\mathbf{A}(t) = e^{i\mathscr{H}^x t}\mathbf{A}, \tag{83}$$

and the superoperator notation indicates that if $\mathscr{H}^x$ acts on an arbitrary operator, **0**,

$$\mathscr{H}^x\mathbf{0} = [\mathscr{H}, \mathbf{0}]; \tag{84}$$

Ω is a set of oscillatory frequencies specified by

$$i\Omega \equiv \langle \dot{\mathbf{A}}\mathbf{A}^{+} \rangle \cdot \langle \mathbf{A}\mathbf{A}^{+} \rangle^{-1}; \tag{85}$$

$\dot{\boldsymbol{\alpha}}$ represents components of the time derivative of **A** that are orthogonal to **A**, and are therefore, rapidly varying, i.e.

$$\dot{\boldsymbol{\alpha}} \equiv (1-\P)\,\dot{\mathbf{A}}; \tag{86}$$

the time evolution of $\dot{\boldsymbol{\alpha}}(t_p)$ is determined by a propogator composed only of those components of the Hamiltonian, $\mathscr{H}$, which lie outside the subspace determined by the slow variables, i.e.

$$\dot{\boldsymbol{\alpha}}(t_\mathrm{p}) \equiv \exp[t(\mathbf{1}-\P)\,i\mathscr{H}^x]\,\dot{\boldsymbol{\alpha}}, \tag{87}$$

[33] L. T. Muus, *in* "Electron Spin Relaxation in Liquids" (L. T. Muus and P. W. Atkins, eds.), Ch. 1, p. 1. Plenum, New York, 1972.

and the subscript on t_p indicates this projected time dependence; $\mathsf{K}(t)$ is a memory function matrix, not unlike that introduced in Eqs. (13) and (49), which is defined as

$$\mathsf{K}(t) \equiv \langle \dot{\boldsymbol{\alpha}}(t_p)\dot{\boldsymbol{\alpha}}^+ \rangle \cdot \langle \mathbf{A}\mathbf{A}^+ \rangle^{-1}. \tag{88}$$

The generalized Langevin equation in Eq. (81) is an identity but it is only in the linear response limit, i.e. for small displacements from equilibrium, that one can assign to it the property [specified in Eq. (77) for the usual Langevin relation]

$$\overline{\dot{\boldsymbol{\alpha}}(t_p)} = 0, \tag{89}$$

where the bar indicates an instantaneous average in contrast to $\langle\rangle$ which indicates an equilibrium average. We shall henceforth assume this linear response result. We can, therefore, obtain the general transport equation

$$d/dt\,\overline{\mathbf{A}(t)} = i\boldsymbol{\Omega}\cdot\overline{\mathbf{A}(t)} - \int_0^t \mathsf{K}(\tau)\cdot\overline{\mathbf{A}(t-\tau)}\,d\tau \tag{90}$$

from Eq. (81). The analogy between Eq. (90) and both Eqs. (13) and (49) is striking.

If the variable set, $\mathbf{A}(t)$, is properly chosen to contain all slowly varying terms, then $\dot{\boldsymbol{\alpha}}(t_p)$ will be a rapidly varying function of time, and the memory function matrix, $\mathsf{K}(t)$, should decay rapidly to zero. Indeed, if the choice of $\mathbf{A}(t)$ is made carefully, one might expect $\mathsf{K}(t)$ to decay to zero before $\overline{\mathbf{A}}(t)$ changes very much, and we could then simplify the equation of motion in Eq. (90) by replacing $\mathbf{A}(t-\tau)$ by $\mathbf{A}(t)$ and by replacing the upper limit of integration by $t \to \infty$. However, it is sometimes convenient to choose a set of variables, $\mathbf{A}(t)$, that decay slowly but that have rapidly oscillating or precessing contributions, and we cannot then make the simplifying changes in Eq. (90). Such is the case for $S_+(t)$ as discussed in Section II, A. The frequency matrix, $\boldsymbol{\Omega}$, in Eq. (90) represents such oscillating contributions, and it is then $[\exp(-i\Omega t)\mathbf{A}(t)]$, rather than $\mathbf{A}(t)$, that is a slowly varying quantity. The quantity $[\exp(-i\Omega t)\overline{\mathbf{A}}(t)]$ corresponds, in spin relaxation theory, to choosing components of the spin in the rotating frame or in the interaction representation. *Note that since* $\mathsf{K}(t)$ *is a function of* $\dot{A}$, *the transport equation relates the relaxation of* A *to the appropriately projected fluctuations of* $\dot{A}$, *and often these fluctuations of* A *are more rapid than the relaxation of* A.

If the variable set, $\mathbf{A}(t)$, is properly chosen so that $\{\exp(-i\Omega t)\mathbf{A}(t)\}$ is slowly varying, then we can write the integral in Eq. (90) as

$$\int_0^t \mathsf{K}(\tau)\overline{\mathbf{A}(t-\tau)}\,d\tau = \int_0^t \mathsf{K}(\tau)\exp[i\quad(t-\tau)]\{\exp[-i\Omega(t-\tau)]\overline{\mathbf{A}(t-\tau)}\}. \tag{91}$$

It is convenient to define an *effective* memory function matrix, $\mathcal{K}(t)$:

$$\mathcal{K}(t) \equiv \mathsf{K}(t)e^{-i\Omega t}. \tag{92}$$

Then, if $\overline{\mathbf{A}(t)}$ is properly chosen, $\mathcal{K}(t)$ will decay to zero before $\{\exp[-i\Omega t]\overline{\mathbf{A}(t)}\}$ changes much, and we can approximate

$$\{\exp[-i\Omega(t-\tau)]\overline{\mathbf{A}(t-\tau)}\}$$

in Eq. (91) by $\{\exp(-i\Omega t)\overline{\mathbf{A}(t)}\}$, and allow the upper limit of integration to go to $t \to \infty$. Equation (90) thus becomes

$$d/dt\,\overline{\mathbf{A}(t)} = [i\Omega - \hat{\mathcal{K}}]\,\overline{\mathbf{A}(t)} \tag{93}$$

where the *relaxation matrix*, $\hat{\mathcal{K}}$, is defined by

$$\hat{\mathcal{K}} \equiv \int_0^\infty \mathcal{K}(t)\,dt. \tag{94}$$

The relaxation matrix may be complex:

$$\hat{\mathcal{K}} = \hat{\mathcal{K}}_{\mathrm{RE}} - i\hat{\mathcal{K}}_{\mathrm{IM}}; \tag{95}$$

the real part is associated with relaxation times and the imaginary part with frequency shifts. Thus Eq. (93) becomes

$$(d/dt)\,\overline{\mathbf{A}(t)} = -[-i(\Omega + \hat{\mathcal{K}}_{\mathrm{IM}}) + \hat{\mathcal{K}}_{\mathrm{RE}}]\cdot\overline{\mathbf{A}(t)} \tag{96}$$

which is quite analogous to the coupled Bloch equations in Eqs. (23) and (52)–(56). The coupled Mori transport equations are most useful when they reduce to the form in Eq. (96), i.e. when the transport matrix $[-i(\Omega + \hat{\mathcal{K}}_{\mathrm{IM}}) + \hat{\mathcal{K}}_{\mathrm{RE}}]$ is constant. Equation (96) is our fundamental relation upon which all further discussions are based.

The discussion above can also be carried out in the frequency domain. The Fourier transform of Mori's equation, Eq. (90), is

$$i\omega\hat{\mathbf{A}}(\omega) - \mathbf{A}(0) = i\Omega\cdot\hat{\mathbf{A}}(\omega) - \hat{\mathcal{K}}(\omega-\Omega)\cdot\hat{\mathbf{A}}(\omega) \tag{97}$$

where we have used Eqs. (91) and (92), and where we have also defined the half-transform,

$$\mathcal{K}(\omega) \equiv \int_0^\infty e^{-i\omega t}\mathcal{K}(t)\,dt. \tag{98}$$

As discussed in the next section, $\hat{\mathcal{K}}(\omega-\Omega)$ is independent of frequency for small $|1\omega-\Omega|$, and we can replace $\hat{\mathcal{K}}(\omega-\Omega)$ by $\hat{\mathcal{K}}(0)$:

$$\hat{\mathcal{K}}(0) = \int_0^\infty \mathcal{K}(t)\,dt. \tag{99}$$

Comparison of Eqs. (94) and (99) shows that the respective relaxation matrixes are identical under the appropriate conditions, and we can write

$$\hat{\mathscr{K}}(0) = \hat{\mathscr{K}}_{\mathrm{RE}} - i\hat{\mathscr{K}}_{\mathrm{IM}}. \tag{100}$$

In this approximation the transport equation, Eq. (97), is,

$$i\omega\hat{\mathbf{A}}(\omega) - \overline{\mathbf{A}(0)} = [i(\Omega + \hat{\mathscr{K}}_{\mathrm{IM}}) - \hat{\mathscr{K}}_{\mathrm{RE}}]\,\hat{\mathbf{A}}(\omega), \tag{101}$$

which is just the Fourier transform of Eq. (95).

Mori has also shown that an equation of motion for the correlation function, $\langle \mathbf{A}(t)\,\mathbf{A}^{+}\rangle$, can be obtained in an analogous manner under the same conditions for which Eqs. (95) and (101) hold. (See Appendix A.) If the normalized correlation function matrix, $\mathsf{G}(t)$, is defined by

$$\mathsf{G}(t) = \langle \mathbf{A}(t)\,\mathbf{A}^{+}\rangle \langle \mathbf{A}\mathbf{A}^{+}\rangle^{-1} \tag{102}$$

then

$$d\mathsf{G}(t)/dt = i\Omega \cdot \mathsf{G}(t) - \int_0^t d\tau\, \mathsf{K}(\tau) \cdot \mathsf{G}(t-\tau) \tag{103a}$$

in analogy with Eq. (90), and under the appropriate conditions

$$d\mathsf{G}(t)/dt = [i(\Omega + \hat{\mathscr{K}}_{\mathrm{IM}}) - \hat{\mathscr{K}}_{\mathrm{RE}}] \cdot \mathsf{G}(t) \tag{103b}$$

in analogy with Eq. (95). If Eq. (103b) holds, then

$$i\omega\hat{\mathsf{G}}(\omega) - \mathsf{G}(0) = [i(\Omega + i\hat{\mathscr{K}}_{\mathrm{IM}}) - \hat{\mathscr{K}}_{\mathrm{RE}}] \cdot \hat{\mathsf{G}}(\omega) \tag{103c}$$

where $\hat{\mathsf{G}}(\omega)$ is the Fourier transform of $\mathsf{G}(t)$. The real part of $\hat{\mathsf{G}}(\omega)$ is related to the frequency spectrum, $I(\omega)$, as discussed in Section II.

B. Discussion of Solutions. Extended Solutions

In the preceding subsection we saw that the fundamental transport equations are most useful if $\hat{\mathscr{K}}(\omega - \Omega)$ is approximately independent of frequency, i.e. $\hat{\mathscr{K}}(\omega - \Omega) \approx \hat{\mathscr{K}}(0)$, This is the case when the variable set, **A**, is *complete in the sense that it describes all the slow motions.* These motions are slow in that the decay times for $[(\exp - i\Omega t)\,\overline{\mathbf{A}(t)}]$, which are associated with $\hat{\mathscr{K}}_{\mathrm{RE}}$, are long compared to the molecular times associated with $\dot{\boldsymbol{\alpha}}(t_{\mathrm{p}})$, or rather with those associated with $\mathsf{K}(t)$. Thus the set, **A**, must be chosen so that $\dot{\boldsymbol{\alpha}}(t_{\mathrm{p}})$ is a rapidly fluctuating function and $\mathsf{K}(t)$ is a rapidly decaying function with characteristic decay or correlation time of the order of τ_{I}, where

$$\tau_{\mathrm{I}} |\hat{\mathscr{K}}_{\mathrm{RF}}| \ll 1. \tag{104}$$

Furthermore, this is equivalent to restricting the times, t, of interest to

$$t \gg \tau_{\mathrm{I}} \tag{105}$$

and frequency of interest to

$$|1\omega - \Omega|\,\tau_{\mathrm{I}} \ll 1. \tag{106}$$

If we wish to extend the times downward and the frequencies upward, we must expand the set of variables, **A**, to include more rapid motions in our specification of "slow"; thus we must include a larger part of Hilbert space, specified by a larger number of basic functions, **A**. (Another approach is to assume an ω dependence for $\mathbf{K}(\omega)$.[18,34]) *The ability to choose the appropriate set for a given problem is the aspect of the approach that requires physical insight; if an appropriate set is chosen, the transport equations, Eq. (90), have constant coefficients,* $\hat{\mathscr{K}}$, *and it is only in this limit that the transport equation will be of use to us.* Remember that all these results are *valid only in the linear response region.* Note the analogies between the discussion above and the more specialized discussions in Section II.

The formal solution to Eq. (95), the coupled transport equations with constant coefficients, is

$$\overline{\mathbf{A}(t)} = \exp[i(\Omega + \hat{\mathscr{K}}_{\mathrm{IM}})\,t - \hat{\mathscr{K}}_{\mathrm{RE}}\,t]\,\overline{\mathbf{A}(0)}. \tag{107}$$

This is a sum of complex exponentials. If the set of variables, **A**, consists of n members, then the solutions for each $A_i(t)$ will consist of a sum of n such exponentials with exponents $(\Lambda_n t)$, where the Λ_n are the eigenvalues *of the transport matrix,* $[-i(\Omega + \hat{\mathscr{K}}_{\mathrm{IM}}) + \hat{\mathscr{K}}_{\mathrm{RE}}]$. The same exponentials, with different amplitudes, will enter for each $\overline{A_i(t)}$. See the discussion below Eq. (23) in Section II.

In frequency space the solution to Eq. (101) or, of more interest to us, to Eq. (103), is

$$\hat{\mathsf{G}}(\omega) = [+\hat{\mathscr{K}}_{\mathrm{RE}} + i(\omega - \Omega - \hat{\mathscr{K}}_{\mathrm{IM}}]^{-1}\;, \tag{108}$$

where we have used the fact that $\mathsf{G}(0) = 1$. The real part of $\hat{\mathsf{G}}(\omega)$ can be used, together with equations such as Eq. (8) or Eq. (28), to yield the spectral density, $I(\omega)$. It is clear that *if A is n-dimensional,* $I(\omega)$ *is a sum of n generalized Lorentzian lines* as discussed in Section II, C in Eq. (27).

We see, therefore, that a single variable leads to solutions which are represented by a single exponential in time or a single normal Lorentzian in frequency space, i.e. just the results of the Bloch theory. For two variables, as in the coupled Bloch equations presented in Eq. (23), two exponentials in time or two generalized Lorentzians in frequency are required. As the number of variables is increased to include faster (high frequency) motions, the number of exponentials or Lorentzians increases, and the net shape of $\overline{\mathbf{A}(t)}$ or $I(\omega)$ becomes more complex. When only the very slow motions are considered, we have the equivalent of statistical coarse-graining, and by

[34] R. Zwanzig, personal communication (1970).

adding more and more rapid variables, we refine the coarse-graining until we approach molecular times and behavior. Of course, we may need large numbers, (10^{23}), of variables to describe molecule motions, but, nevertheless, we can think of the addition of variables as an appropriate expansion in exponentials in time or, correspondingly, of Lorentzians in frequency, in order to describe behavior at shorter times and higher frequencies.

We see that at very long times and low frequencies, one always has exponential and Lorentzian behavior, respectively, a result that is well-known. A Lorentzian has an infinite second moment because its behavior at large ω (short t) cannot be correct. Thus at high frequencies, $\hat{\mathscr{K}}(\omega - \Omega)$ is not independent of ω; one can either assume a functional form for this ω-dependence, or we can expand our set of variables to include faster motions in the expectation that $\hat{\mathscr{K}}(\omega - \Omega)$ in this larger Hilbert space will be frequency independent out to higher ω. If one adds the set of variables, $\{\dot{\mathbf{A}}_i\}$, to the complete slow set of n variables, $\{\mathbf{A}_i\}$, we obtain an expanded set of $2n$ variables which should give a better description in terms of transport equations with constant coefficients, at shorter times and larger frequencies. In fact, this set yields spectra with finite second but infinite fourth moments.[35] If we add the variables $\{\ddot{\mathbf{A}}_i\}$ to our set, we also obtain finite fourth but infinite sixth moments. This is another example of improved descriptions with sets of increasing size. Since all moments must be finite, we would clearly need an infinite number of variables in our set to describe the spectrum and motions exactly.

C. Extended Mori Theory

The Mori approach discussed above is most useful when the behavior of interest can be described in terms of a small set of variables and transport equations with constant relaxation matrix, $\hat{\mathscr{K}}$. Extension of the analysis to shorter times is meaningful only if these shorter times in turn are long compared to most of the other molecular times, so that a relatively small set of variables, together with transport equations with a constant relaxation matrix, describes the system adequately. Spin problems are particularly well suited for treatment by this method. The spin (S_+) relaxation times are almost always very slow. The time derivatives of spin ($\dot{S}_+$) often depend upon molecular orientations which vary rapidly compared to spin relaxation, but often slowly compared to other molecular motions in liquids. Thus in many cases S_+ itself forms a complete set, and if the molecular tumbling rates are low, it may be adequate to add $\dot{S}_+$ as an additional variable; the variables S_+ and $\dot{S}_+$ form the basis of the "extended" theory presented in Section V, B. If this extended theory is not adequate, i.e. if the times

[35] D. Kivelson and T. Keyes, *J. Chem. Phys.* **57**, 4599 (1972).

of interest are very short or the frequencies large, then we can introduce $\ddot{S}_+$ as a third variable in the hope that although it fluctuates more rapidly than S_+ and $\dot{S}_+$, it varies more slowly than all other relevant variables. The physical basis of this procedure can be illustrated by considering the case where $\dot{S}_+$ is dependent upon the Eulerian angles which determine molecular orientation; then $\ddot{S}_+$ depends upon angular velocities and $\dddot{S}_+$ upon intermolecular torques. In many liquids it is reasonable to assume that the spin relaxes slowly (T_2) compared to the variation in Eulerian angle (τ_{I}) which in turn varies slowly compared to changes in the angular velocity (τ_{II}) which vary slowly compared to the variation in the intermolecular torques (τ_{III}) (see the discussion in Mori[36a]). This procedure can be extended and in numerous situations a satisfactory description for spin relaxation can be obtained with a relatively small set of variables. Eventually, by adding higher derivatives, we reach times that are so short that the times characteristic of every motion of every molecule are comparable with those being discussed, and there is no way to exclude some variables and not others. (We have used τ_{I}, τ_{II}, τ_{III} to indicate the correlation times associated with the first, second, and third derivatives, respectively, of S_+ or **A**.)

Instead of adding the variables $\dot{\mathbf{A}}^{\mathrm{o}}$ to the original set $\mathbf{A}^{\mathrm{o}}$, where the superscript (o) indicates the original set, it is convenient to choose as the additional variables, $\dot{\mathsf{A}}$,

$$\dot{\mathsf{A}} = (1 - \P_{\mathbf{A}^o})\,\dot{\mathbf{A}}^{\mathrm{o}}, \tag{109}$$

where $\P_{\mathbf{A}^o}$ is the projection operator upon the original $\mathbf{A}^o$'s. The set $\dot{\mathsf{A}}$ is *orthogonal* to the $\mathbf{A}^{\mathrm{o}}$ in that

$$\langle \dot{\mathsf{A}}\mathbf{A}^{\mathrm{o}+} \rangle = 0. \tag{110}$$

Thus the expanded set of preferred variables is[36b]

$$\tilde{\mathbf{A}} = (A_1^{\mathrm{o}}, A_2^{\mathrm{o}}, \ldots A_n^{\mathrm{o}}, \dot{\mathsf{A}}_1, \dot{\mathsf{A}}_2, \ldots \dot{\mathsf{A}}_n). \tag{111}$$

If the set is to be extended further, we may choose as our set of slow variables the combined set of $\mathbf{A}^o$, $\dot{\mathsf{A}}$, and $\mathfrak{A}$, where

$$\ddot{\mathfrak{A}} = (1 - \P_{\mathbf{A}^o} - \P_{\dot{\mathsf{A}}})\,\ddot{\mathbf{A}}^o \tag{112}$$

and $\P_{\dot{\mathsf{A}}}$ is the projection operator upon $\dot{\mathsf{A}}$. This extended approach is equivalent to Mori's continued fraction expansion.[36a, 37]

[36b] H. Mori, *Progr. Theor.* **34**, 399 (1965).

[36b] To avoid confusion, we have used the following notation: $\dot{\mathsf{A}}$ represents the set of added variables in the first extended case, but $\dot{\boldsymbol{\alpha}}$ represents the projected derivative of the entire set **A**. Thus, if **A** is defined in Eq. (111), then

$$\dot{\boldsymbol{\alpha}} = (1 - \P_{\mathbf{A}})\,\dot{\mathbf{A}} = (1 - \P_{\mathbf{A}^o\dot{\mathsf{A}}}) \begin{pmatrix} \dot{\mathbf{A}}^o \\ \ddot{\mathsf{A}} \end{pmatrix}.$$

[37] J. M. Deutch, personal communication (1971). Deutch has studied spin relaxation by means of the Mori continued fraction technique.

V. Simple Spin Resonance. One Variable

Clearly, the components of the spin (S_z, S_+, S_-) are a natural first choice of the variable set, **A**, for use in applying the Mori theory to magnetic resonance and spin relaxation problems. As we discussed in the last subsection, the time scale for spin relaxation is likely to be long compared to that for the lattice (liquid) motions, and so the spin components may well constitute a complete set in the sense described above. It should be carefully noted, however, that *additional slow variables may be necessary in some cases;* for example, for $S = 3/2$ one must consider both S_+ and S_+^3 as independent slow variables if the spin-lattice interaction is due to quadrupolar interactions in NMR problems or due to zero-field (spin-spin dipole) interactions in ESR problems. This has been discussed by Yoon and Deutch.[38a]

In Section V, A we shall use the Mori formalism to obtain the simple Bloch equations in Eqs. (1) and (2), that is, we shall restrict ourselves to a system for which each molecule of interest has a single spin $\mathbf{S}_j$ with $S_j = 1/2$, and we shall assume that the lattice correlation times (τ_l) are short compared to the spin relaxation times (T_1, T_2). Section V, B is devoted to a comparison of the Mori and Redfield or perturbation results. In Section V, C we shall extend the results for the same systems to situations where the time restrictions are somewhat relaxed; we shall make use of an extended formulation in which the variables, $\dot{S}_z$, $\dot{S}_+$ and $\dot{S}_-$ are also included in the set, **A**, of slow variables, and the resulting lineshapes will be nonLorentzian. A discussion of the nonLorentzian line shape is given in Section V, D, whereas Section V, E is devoted to a discussion of the conditions under which the extended theory applies. In Section V, F the physical consequences of the results and their relation to rotational diffusion constants are considered. Finally in Section V, G, extensions of the theory to yet shorter times are considered with special attention both to nondiffusive lattice motions and to slow diffuse tumbling.

The reader who is not interested in the extended formulations but wishes only to follow the derivation of the Bloch equations by means of Mori's theory, can, after reading Section V, A, proceed directly to Section VI and the discussion of spectra with hyperfine structure. However, we emphasize that one of the principal features of the Mori formulation is that it provides a powerful means for constructing extended spin relaxation theories applicable to time and frequency domains inaccessible to the Bloch equations.

A. Derivation of Bloch Equations of Single Spin Systems

We shall consider a system composed of N identical spins with $S = 1/2$, and we shall derive the appropriate Bloch equation by means of Mori's

[38a] B. Yoon and J. M. Deutch, personal communication (1972).

theory. The Hamiltonian, $\mathscr{H}_S + \mathscr{H}_L + \mathscr{H}_{SL}$, is described in Eq. (59); the pure spin term is $\mathscr{H}_S = \omega_0 S_z$ where S_z is the z-component of **S**, and **S** is the spin angular momentum operator summed over all spins as specified in Eq. (3). As an example, we might think of the spin-lattice interaction as that arising from the anisotropy of the g-factor or chemical shift [see Eq. (66)]; in this case, $\mathscr{H}_L$ represents molecular tumbling or reorientation. The slow variables **A** can readily be selected as

$$\mathbf{A} = \begin{bmatrix} S_+ \\ \Delta S_z \\ S_- \end{bmatrix}, \tag{113}$$

where

$$\Delta S_z(t) = S_z(t) - \langle S_z \rangle. \tag{114}$$

These variables are discussed in Section II, A. Actually, we shall see that the three variables above enter into the transport equations in such a way that they are each determined by a single, uncoupled Bloch equation.

To lowest order in $(k_B T)^{-1}$, where k_B is the Boltzmann constant, the normalization matrix $\langle \mathbf{AA}^+ \rangle$ in Eqs. (80), (85), and (88) is

$$\langle \mathbf{AA}^+ \rangle = \tfrac{1}{4}N \begin{bmatrix} 2 & 0 & 0 \\ 0 & 1 & 0 \\ 0 & 0 & 2 \end{bmatrix} \tag{115}$$

where N is the number of spins.

In order to calculate $i\Omega$ in Eq. (85), we need the matrix elements $\langle \dot{S}_\alpha S_\beta^+ \rangle$ where $(\alpha, \beta = z, +, -)$, and hence we need to evaluate $\dot{S}_\alpha$;

$$\dot{S}_\alpha = i[\mathscr{H}, S_\alpha]. \tag{116}$$

Since $\mathscr{H}_L$ commutes with S_α, Eq. (116) becomes

$$\dot{S}_\alpha = i[\mathscr{H}_S, S_\alpha] + i[\mathscr{H}_{SL}, S_\alpha]. \tag{117}$$

If we consider as an example, $\alpha = \beta = +$, then it can readily be seen that

$$\langle \dot{S}_+ S_- \rangle = i\omega_0 \langle S_+ S_- \rangle + i\langle [\mathscr{H}_{SL}, S_+] S_- \rangle. \tag{118}$$

Similarly,

$$\langle \dot{S}_z S_z \rangle = i\langle S_z [\mathscr{H}_{SL}, S_z] \rangle. \tag{119}$$

In all cases of interest, terms linear in $\mathscr{H}_{SL}$ vanish when averaged over the ensemble, so that Eq. (119) and the second term in Eq. (118) vanish. It can then readily be shown that $i\Omega$ in Eq. (85) is

$$i\Omega = i\omega_0 \begin{bmatrix} 1 & 0 & 0 \\ 0 & 0 & 0 \\ 0 & 0 & -1 \end{bmatrix} \qquad (120)$$

Next we must calculate the memory function matrix, $\mathsf{K}(t)$, in Eq. (88), which requires the evaluation of the matrix, $\langle \dot{\boldsymbol{\alpha}}(t_p)\dot{\boldsymbol{\alpha}}^{+}(0)\rangle$. To do this we first examine $\dot{\boldsymbol{\alpha}}(0)$ which may be obtained from Eq. (86). In order to evaluate this quantity we note the following identity:

$$\P_A \dot{\mathbf{A}} = i\Omega \cdot \mathbf{A}, \qquad (121)$$

which leads to the result

$$\dot{\boldsymbol{\alpha}} = \dot{\mathbf{A}} - i\Omega \cdot \mathbf{A}. \qquad (122)$$

If we let **A** be the spins in Eq. (113), it follows that

$$i[\mathscr{H}_S + \mathscr{H}_L, \mathbf{A}] = i\Omega \cdot \mathbf{A}, \qquad (123)$$

and the quantity $\dot{\boldsymbol{\alpha}}$ is then

$$\dot{\boldsymbol{\alpha}} = i[\mathscr{H}_{SL}, \mathbf{A}]. \qquad (124)$$

Equation (124) holds if **A** is an eigenvector of $\mathscr{H}_S + \mathscr{H}_L$; it is the equation of motion for **A** in the interaction representation.

In order to obtain the relaxation matrix $\hat{\mathscr{K}}(\omega-\Omega)$, which plays an important role in the theory [see Eq. (100)], we Fourier transform $\mathsf{K}(t)$:

$$\hat{\mathscr{K}}_{\alpha\beta}(\omega-\Omega) = 2/\mathrm{N}\int_0^{\infty} dt\, e^{i\omega t}(1+\delta_{\beta z})\langle[\mathscr{H}_{SL}(t_p), S_\alpha(t_p)]\,[\mathscr{H}_{SL}, S_\beta]^{+}\rangle$$
$$\times \exp\{+i\omega_0 t[\delta_{\beta+} - \delta_{\beta-}]\}, \qquad (125)$$

where $(\alpha, \beta = z, +, -)$ and ω_0 is the Larmour frequency.[38b] Symmetry arguments in this case show that the off-diagonal elements of the relaxation matrix vanish[1]; therefore, the transport equation, Eq. (97), reduces to three uncoupled generalized relaxation equations:

$$\hat{\dot{S}}_z(\omega) = -\hat{\mathscr{K}}_{zz}(\omega)\Delta\hat{S}_z(\omega), \qquad (126)$$

$$\hat{\dot{S}}_\pm(\omega) = [i(\pm\omega_0 + \mathrm{Im}\,\hat{\mathscr{K}}_{\pm\pm}(\omega\mp\omega_0)) - \mathrm{Re}\,\hat{\mathscr{K}}_{\pm\pm}(\omega\mp\omega_0)]\,\hat{S}_\pm(\omega). \qquad (127)$$

Next we wish to study the time dependence of $\mathsf{K}(t)$, i.e. the time dependence of the integrand in Eq. (125). First of all, we shall assume that $\mathsf{K}(t)$ decays rapidly so that $\mathscr{K}(\omega-\Omega)$ can be replaced by $\hat{\mathscr{K}}_{RE} - i\hat{\mathscr{K}}_{IM}$.

[38b] $[\mathscr{H}_{SL}\, t_p), S_\alpha(t_p)] \equiv \{\exp i[1-\P]\,\mathscr{H}^{\times}t\}\,[\mathscr{H}_{SL}, S_\alpha]$.

(See the discussion in Section IV, A) The transport equations, Eqs. (126) and (127), transformed back to the time domain, then become

$$(d/dt)\overline{S_z(t)} = -T_1^{-1}[\overline{S_z(t)} - \langle S_z \rangle] \tag{128}$$

and

$$(d/dt)\overline{S_\pm(t)} = [i(\pm\omega_0 + \sigma_\pm) - T_2^{-1}]\overline{S_\pm(t)}, \tag{129}$$

where [38b]

$$T_1^{-1} \equiv \mathrm{Re}\,\hat{\mathscr{K}}_{zz}\; 4/N \int_0^\infty \langle [\mathscr{H}_{\mathrm{SL}}(t_{\mathrm{p}}), S_z(t_{\mathrm{p}})]\,[S_z, \mathscr{H}_{\mathrm{SL}}] \rangle\, dt, \tag{130}$$

$$T_2^{-1}(\hat{\mathscr{K}}_{\mathrm{RE}})_{\pm\pm} = \mathrm{Re}\,(2/N) \int_0^\infty \langle [\mathscr{H}_{\mathrm{SL}}(t_{\mathrm{p}}), S_\pm(t_{\mathrm{p}})]\,[S_\mp, \mathscr{H}_{\mathrm{SL}}] \rangle e^{\mp i\omega_0 t}\, dt, \tag{131}$$

$$\sigma = (\mathscr{K}_{\mathrm{IM}})_{\pm\pm} = \mathrm{Im}\,(2/N) \int_0^\infty \langle [\mathscr{H}_{\mathrm{SL}}(t_{\mathrm{p}}), S_\pm(t_{\mathrm{p}})]\,[S_\mp, \mathscr{H}_{\mathrm{SL}}] \rangle\, e^{\mp i\omega_0 t}\, dt. \tag{132}$$

Equations (128) and (129) are the simple Bloch equations presented in Eqs. (1) and (2), except that the $\sigma_\pm$ term which represents the so-called *nonsecular* or *dynamic frequency shift* is given explicitly. See Eq. (19). Note that $\mathrm{Im}\,\hat{\mathscr{K}}_{zz} = 0$.

Since Eqs. (126) and (127) represent three separate transport equations, we can consider the theory above as a one-variable theory in each of the variables, S_+, S_-, or S_z. The one-variable theory is particularly useful only if $\hat{\mathscr{K}}(\omega - \Omega)$ can be replaced by $\hat{\mathscr{K}}$, i.e. if the transport coefficients $T_1, T_2, \sigma_\pm$ are independent of time or frequency. This replacement can be made if the correlation time, τ_{I}, for $\mathsf{K}(t)$ is short, i.e. if

$$\tau_{\mathrm{I}}^2 \omega_\lambda^2 \ll 1 \tag{133}$$

where

$$\omega_\lambda^2 \equiv \langle [\mathscr{H}_{\mathrm{SL}}, S_\pm]\,[S_\mp, \mathscr{H}_{\mathrm{SL}}] \rangle / \langle S_\pm S_\mp \rangle, \tag{134}$$

$$\tau_{\mathrm{I}} \equiv (2/N)\,\mathrm{Re}\left(\int_0^\infty dt\, \langle [\mathscr{H}_{\mathrm{SL}}(t_{\mathrm{p}}), S_\pm(t_{\mathrm{p}})][S_\mp, \mathscr{H}_{\mathrm{SL}}] \rangle\, e^{\mp i\omega_0 t} / \omega_\lambda^2 \right) \tag{135}$$

Although we shall use this definition of correlation times, the more usual definition is

$$\tau_{\mathrm{c}} = \mathrm{Re}\, \frac{\int_0^\infty dt\, \langle [\mathscr{H}_{\mathrm{SL}}^{(\mathrm{sec})}(t_{\mathrm{p}}), S_+(t_{\mathrm{p}})]\,[S_{-\mp}, \mathscr{H}_{\mathrm{SL}}^{(\mathrm{sec})}] \rangle\, e^{\mp i\omega_0 t}}{\langle [\mathscr{H}_{\mathrm{SL}}^{(\mathrm{sec})}, S_+]\,[S_\mp, \mathscr{H}_{\mathrm{SL}}^{(\mathrm{sec})}] \rangle}, \tag{136}$$

where $\mathscr{H}_{\mathrm{SL}}^{(\mathrm{sec})}$ is discussed in Eq. (64).

We can rewrite Eq. (131) as

$$T_2^{-1} = \omega_\lambda^2 \tau_{\mathrm{I}}, \tag{137}$$

and Eq. (133) as

$$\tau_{\mathrm{I}} \ll T_2. \tag{138}$$

A similar relation could be obtained for the single variable S_z theory. We will discuss these results in more detail in Section V, B, but we note that Eqs. (137), (134), (135), and Eqs. (130)–(132) constitute microscopic expressions for the spin relaxation times and shifts.

B. Comparison with Results of Conventional Perturbation Theory

The expressions in Eqs. (130)–(132) differ slightly from the usual microscopic results obtained by time-dependent perturbation theory. [See Eq. (74).] If we evaluate the integrals in Eqs. (130)–(132) in the limit of small $\mathscr{H}_{SL}$, i.e. if we retain only the lowest order terms in $\mathscr{H}_{SL}$, we find that the projected time dependence of $\mathsf{K}(t)$ is equivalent to that in the rotating frame interaction representation, the representation used in perturbation schemes of the Redfields[15] and Bloch[16] theories.[17,20] To show this, consider $\dot{\boldsymbol{\alpha}}(t_p)$, defined in Eqs. (86) and (87);

$$\dot{\boldsymbol{\alpha}}(t_p) = \exp[it(1-\P)\mathscr{H}^x]\,\mathrm{i}[\mathscr{H}_{SL}, \mathbf{A}], \tag{139}$$

which in the limit of small $\mathscr{H}_{SL}$ becomes

$$\dot{\boldsymbol{\alpha}}(t_p) \cong \exp[it(1-\P)(\mathscr{H}_L^x + \mathscr{H}_S^x)\,i[\mathscr{H}_{SL}, \mathbf{A}] \tag{140}$$

where $\mathscr{H}_S + \mathscr{H}_L$ is the unperturbed Hamiltonian. Only the lowest order terms in $\mathscr{H}_{SL}$ have been kept in Eq. (140). In this limit the density matrix involved in the ensemble averaging procedure, i.e. in the evaluation of $\langle\,\rangle$, is independent of $\mathscr{H}_{SL}$.[38c] Therefore, the projection operator in Eq. (140) involves ensemble averages that are linear in $\mathscr{H}_{SL}$; this can be seen by expanding the exponential in powers of t and making use of the definition of $\P$ in Eq. (80). For all physical situations, averages involving terms linear in $\mathscr{H}_{SL}$ vanish.[17] Thus

$$\dot{\boldsymbol{\alpha}}(t_p) \cong \exp\{it(\mathscr{H}_S^x + \mathscr{H}_L^x)\}\,i[\mathscr{H}_{SL}, \mathbf{A}] \tag{141}$$

to lowest order in $\mathscr{H}_{SL}$. If $\mathbf{A} = \mathbf{S}_+$ and if Eq. (141) is substituted into Eq. (131), one obtains

$$T_2^{-1} = \mathrm{Re}(2/N)\int_0^\infty dt\,\langle[\exp\{it(\mathscr{H}_S^x + \mathscr{H}_L^x)\}\,\mathscr{H}_{SL}, S_+], [S_-, \mathscr{H}_{SL}]\rangle, \tag{142}$$

where the time dependence is just that obtained by conventional perturbation theories in the rotating-frame interaction representation.[17,20] In obtaining this result we note that

$$\exp\{it(\mathscr{H}_S^x + \mathscr{H}_L^x)\}\,S_+ = S_+ e^{i\omega_0 t}. \tag{143}$$

[38c] $\langle\mathbf{Q}\rangle = \mathrm{Tr}\{e^{-\beta\mathscr{H}}\mathbf{Q}\}/\mathrm{Tr}\{e^{-\beta\mathscr{H}}\}$ and to lowest order in $\mathscr{H}_{SL}$, $\langle\mathbf{Q}\rangle = \mathrm{Tr}\{e^{-\beta(\mathscr{H}_0+\mathscr{H}_L)}\mathrm{Q}\}/\mathrm{Tr}\{e^{-\beta(\mathscr{H}_0+\mathscr{H}_L)}\}$. Actually it is sufficient for us to work in the high temperature limit and to replace $\langle\mathrm{Q}\rangle$ by $\mathrm{Tr}\,\mathbf{Q}/\mathrm{Tr}\,\mathbf{1}$.

Generally, if Eq. (124) holds and if terms linear in $\mathscr{H}_{SL}$ vanish in the various equilibrium ensemble averages, then Eq. (141) holds. If Eq. (141) is substituted into Eqs. (130) and (132), we have the usual perturbation results for T_1^{-1} and $\sigma_{\pm}$.

If we do not treat $\mathscr{H}_{SL}$ as a small perturbation, then the full Hamiltonian must be used in the exponential which determines the time evolution propagator of $\dot{\boldsymbol{\alpha}}(t_p)$. See Eq. (139). In this case, in contrast to that in Eq. (141), the projection operator plays an important role. In fact, the projection operator reduces the effect of $\mathscr{H}_{SL}$ in the time evolution operator, enabling us to treat $\mathscr{H}_{SL}$ as a small quantity even if it is quite large. This reduction can be observed by expanding the expression for $\mathsf{K}(t)$ in powers of t. This in effect justifies the perturbation approach used above even for reasonably large values of $\mathscr{H}_{SL}$.

C. Two-Variable Extended Theory. Identical Spins

In Section IV, C we discussed the possibility of extending our set of slow variables, $\mathbf{A}$, to include in addition to the very slow variables, A_i, their rather more rapidly varying derivatives, $\dot{A}_i$. We shall extend our treatment of the simple spin problem discussed in the last section in just this way, and we shall obtain a set of transport equations, with constant coefficients, which constitute an extension to the Bloch equations. The new set of equations should be valid at short times, or for spin relaxation times comparable to molecular tumbling times.

Since in the last subsection we saw that there was no mixing of S_+, S_-, and S_z components, we will henceforth restrict our discussion to the S_+ variable, the spin component which determines lineshapes and transverse spin relaxation. As a second variable, we could add $\dot{S}_+$ which would, hopefully, enable us to set $\hat{\mathscr{K}}(\omega-\Omega) \cong \hat{\mathscr{K}}$ for higher frequencies than permitted in the Bloch equations; however, as discussed in Section IV, C, S_+ and $\dot{S}_+$ are not orthogonal in the sense that $\langle \dot{S}_+ S_+ \rangle \neq 0$. For ease of interpretation, we wish to choose an orthogonal set in which one element is S_+ and the other is $\dot{\mathscr{S}}_+$, defined as

$$\dot{\mathscr{S}}_+ = i[\mathscr{H}_{SL}, S_+]. \qquad (144)$$

[$\dot{\mathscr{S}}_+ = (1-¶_S)\dot{S}_+$, where $¶_S$, is the projection operator onto S_+. See Eq. (109).]

Thus

$$\mathbf{A}(t) = \begin{bmatrix} S_+(t) \\ \dot{\mathscr{S}}_+(t) \end{bmatrix}. \qquad (145)$$

As in the last section we choose our pure spin Hamiltonian to be $\mathscr{H}_S = \omega_0 S_z$, and terms linear in $\mathscr{H}_{SL}$ vanish when averaged. Thus the normalization matrix $\langle \mathbf{AA}^+ \rangle$ which enters into Eq. (80) is

$$\langle \mathbf{AA}^+ \rangle = \frac{N}{2} \begin{bmatrix} 1 & 0 \\ 0 & \omega_\lambda^2 \end{bmatrix} \tag{146}$$

where the frequency, ω_λ, is specified by an equilibrium average and is defined in Eq. (134). In order to calculate $i\Omega$ in Eq. (85) we need $\ddot{\mathscr{S}}_+$, where

$$\ddot{\mathscr{S}}_+ = -[\mathscr{H}, [\mathscr{H}_{SL}, S_+]]. \tag{147}$$

[See Eq. (116) and remember that $\mathscr{H} = \mathscr{H}_L + \mathscr{H}_S + \mathscr{H}_{SL}$.] It readily follows that

$$i\Omega = \begin{bmatrix} i\omega_0 & 1 \\ -\omega_\lambda^2 & ib\omega_0 + i\beta\omega_\lambda \end{bmatrix} \tag{148}$$

where b and β are defined as

$$b \equiv \langle [\mathscr{H}_S, [\mathscr{H}_{SL}, S_+]]\,[S_-, \mathscr{H}_{SL}^+] \rangle / \omega_0 \langle [\mathscr{H}_{SL}, S_+]\,[S_-, \mathscr{H}_{SL}^+] \rangle \tag{149}$$

and

$$\beta = \langle [\mathscr{H}_{SL}, [\mathscr{H}_{SL}, S_+]]\,[S_-, \mathscr{H}_{SL}^+] \rangle / \omega_\lambda \langle [\mathscr{H}_{SL}, S_+]\,[S_-, \mathscr{H}_{SL}^+] \rangle \tag{150}$$

respectively. We note that averages over terms linear in $\mathscr{H}_L$ vanish. b and β are dimensionless parameters which are of the order of unity. In some cases, such as Heisenberg spin exchange, $\beta = 0$.

Next we must calculate the memory function matrix, $\mathsf{K}(t)$, in Eq. (88), which requires that we evaluate $\dot{\boldsymbol{\alpha}}$. Equation (123) no longer holds and [38d]

$$\dot{\boldsymbol{\alpha}} = \begin{bmatrix} 0 \\ \ddot{\mathfrak{S}}_+ \end{bmatrix} \tag{151}$$

where

$$\ddot{\mathfrak{S}}_+ = -[\mathscr{H}, [\mathscr{H}_{SL}, S_+]] + b\omega_0 [\mathscr{H}_{SL}, S_+] + \omega_\lambda^2 S_+ + \beta\omega_\lambda [\mathscr{H}_{SL}, S_+]. \tag{152}$$

We use $\dot{\mathscr{S}}_+$ to indicate $(1 - \P_{S_+}) \dot{S}_+$, and $\ddot{\mathfrak{S}}_+$ to indicate $(1 - \P_{S_+\dot{\mathscr{S}}_+})\, \ddot{\mathscr{S}}_+$ where $\P_{S_+}$ is the projection operator which projects onto S_+ and $\P_{S_+\dot{\mathscr{S}}_+}$ projects onto both S_+ and $\dot{\mathscr{S}}_+$. For intuitive modeling, *one can associate* $\dot{\mathscr{S}}_+$ *with* $\dot{S}_+$ and $\ddot{\mathfrak{S}}_+$ *with* $\ddot{S}_+$ *where most of the processional contributions have*

[38d] $\ddot{\mathfrak{S}}_+ = \ddot{\mathscr{S}}_+ - \langle \ddot{\mathscr{S}}_+ S_- \rangle S_+ / \langle S_+ S_- \rangle - \langle \ddot{\mathscr{S}}_+ \dot{\mathscr{S}}_- \rangle \dot{\mathscr{S}}_+ / \langle \dot{\mathscr{S}}_+ \dot{\mathscr{S}}_- \rangle$.

been removed. It follows that

$$\mathsf{K}(t) = K''(t)\begin{bmatrix} 0 & 0 \\ 0 & 1 \end{bmatrix} \tag{153}$$

where

$$K''(t) = \langle \ddot{\mathfrak{S}}_+(t_{p'})\,\ddot{\mathfrak{S}}_-\rangle/\omega_\lambda^2\langle S_- S_+\rangle \tag{154}$$

and[38e]

$$\ddot{\mathfrak{S}}_+(t_{p'}) = \exp[t(1-\P)\,i\mathscr{H}^x]\,\ddot{\mathfrak{S}}_+. \tag{155}$$

The double prime on $K''(t)$ indicates that it involves second derivatives of S_+; the projection operator in Eq. (155) is $\P_{S_+\dot{\mathscr{S}}_+}$. If $\mathsf{K}(t)$ decays rapidly compared to $\{e^{-i\Omega t}\overline{\mathbf{A}(t)}\}$, where

$$e^{-i\Omega t}\mathbf{A}(t) = e^{-i\Omega t}\begin{pmatrix} \overline{S_+(t)} \\ \overline{\dot{\mathscr{S}}_+(t)} \end{pmatrix}, \tag{156}$$

then, as we have discussed previously, the transport coefficient $\hat{\mathscr{K}}(\omega-\Omega)$ can be replaced by $\hat{\mathscr{K}}_{\mathrm{RE}} - i\hat{\mathscr{K}}_{\mathrm{IM}}$. *The transport equations are useful only if this replacement can be made.* The conditions under which the replacement can be made is then the fundamental restriction on the usefulness of the two-variable $(S_+, \dot{\mathscr{S}}_+)$ theory. This condition can be stated in terms of the correlation time, τ_{II}, which is defined as

$$\tau_{\mathrm{II}} \equiv \mathrm{Re}\int_0^\infty dt\,(K''(t)\,e^{-i\Omega t})(\omega_{\mathrm{L}}^2)^{-1}, \tag{157}$$

where

$$\omega_{\mathrm{L}}^2 \equiv \langle \ddot{\mathfrak{S}}_+\,\ddot{\mathfrak{S}}_-\rangle/\omega_\lambda^2\langle S_+ S_-\rangle. \tag{158}$$

The required condition is then

$$\tau_{\mathrm{II}}^2\omega_{\mathrm{L}}^2 \ll 1; \tag{159}$$

or, equivalently, if we define $\hat{\mathscr{K}}''_{\mathrm{RE}}$ as

$$\mathscr{K}''_{\mathrm{RE}} = \omega_{\mathrm{L}}^2\tau_{\mathrm{II}}, \tag{160}$$

a definition consistent with notation used previously, the required condition can be restated as

$$\hat{\mathscr{K}}''_{\mathrm{RE}}\tau_{\mathrm{II}} \ll 1. \tag{161}$$

A comparison should be made between Eq. (159), the requirement for the two variable theory, and Eq. (133), the requirement for the one-variable

[38e] Note that $t_{p'}$ corresponds to time evolution under a projection operator which projects along both S_+ and $\dot{\mathscr{S}}_+$, while t_p corresponds to time evolution under a projection operator which projects along S_+.

theory; these conditions and the distinctions between ω_L and ω_λ and between τ_I and τ_{II} will be discussed in the next section, but in broad terms, *ω_λ is a frequency associated with the strength of $\mathscr{H}_{SL}$, i.e. $\dot{\mathscr{S}}_+$, and τ_I is the associated correlation time, whereas ω_L is a frequency associated with $\mathscr{H}_L$, i.e.* $(\ddot{\mathfrak{S}}/\omega_\lambda)$, and τ_{II} is the associated correlation time.

The choice of $[\exp(-i\Omega t)\mathbf{A}(t)]$ in Eq. (156) as the slowly varying variable of interest is somewhat arbitrary and not always convenient. The arbitrariness of this variable can be understood by referring back to Section II, C and noting that if $\mathbf{A}$ is composed of two components, e.g. $S_+^{(1)}$ and $S_+^{(2)}$, and the transport equations describing their motions are coupled, then each of the two componets of $\mathbf{A}(t)$ has a time dependence described by the sum of two exponentials, each with a different precessing or Larmor frequency. It is thus not possible to find a rotating coordinate system in which all precession of $\mathbf{A}(t)$ vanishes. The criterion for slowness is such that a rotating frame be selected so that the precession of both exponential time contributions be slow; this is discussed in Eqs. (46)–(48). If the matrix Ω is diagonal with matrix elements Ω_1 and Ω_2, then if $|\Omega_1 - \Omega_2|$ is sufficiently small we can satisfy the slowness criterion by choosing the variable $[\exp(-i\Omega t)\mathbf{A}(t)]$ as the slow variable. If Ω is not diagonal it is less clear how to choose the rotating frame. We will assume for simplicity that the slow variable can be selected as

$$e^{-i\Omega t}\overline{\mathbf{A}(t)} \approx e^{-i\Omega^{(D)}t}\overline{\mathbf{A}(t)}. \tag{162}$$

where $\Omega^{(D)}$ is the matrix composed of the diagonal elements of Ω. In Appendix B we discuss this problem further in terms of a specific example.

If we assume that the off-diagonal elements of Ω have a negligible effect upon $\hat{\mathscr{K}}(\omega - \Omega)$, we can write the transport equations as

$$\frac{d}{dt}\begin{pmatrix} \overline{S_+(t)} \\ \overline{\dot{\mathscr{S}}_+(t)} \end{pmatrix} = -\begin{pmatrix} -i\omega_0 & -1 \\ +\omega_\lambda^2 & \hat{\mathscr{K}}''_{RE} - i(\omega_0 b + \omega_\lambda\beta + \hat{\mathscr{K}}''_{IM}) \end{pmatrix}\begin{pmatrix} \overline{S_+(t)} \\ \overline{\dot{\mathscr{S}}_+(t)} \end{pmatrix}, \tag{163}$$

where

$$\hat{\mathscr{K}}'' = \int_0^\infty dt\, e^{-i(b\omega_0 + \beta\omega_\lambda)t} K''(t) \tag{164}$$

and

$$\hat{\mathscr{K}}'' = \hat{\mathscr{K}}''_{RE} + i\hat{\mathscr{K}}''_{IM}. \tag{165}$$

[See Eq. (160).] These coupled transport equations represent an extended form of the uncoupled Bloch equation for $S_+(t)$ given in Eq. (2). In the next section we shall study the implications to the magnetic resonance spectrum of the coupled equations in Eq. (163) and the conditions under which they are valid. In Section V, E we shall study the form and significance of the transport coefficient, $\hat{\mathscr{K}}''$.

D. Effect of Extended Theory on The Spectrum

The coupled transport equations in Eq. (163) can be solved; the solutions for both $\overline{S_+(t)}$ and $\overline{\hat{\mathscr{S}}_+(t)}$ are sums of two complex exponentials with arguments given by the eigenvalues of the transport matrix in Eq. (163). These are similar to the exponential solutions in Eq. (27) which correspond to the coupled transport equations in Eqs. (23). Since the precessions, as discussed in Section II, C, cannot be canceled out exactly in any rotating frame, the variables evaluated in the rotating frame, i.e. $(\exp - i\Omega t)\,\overline{\mathbf{A}(t)}$, are slowly varying only if

$$\tau_{\mathrm{II}}^2[\mathrm{Im}(\Lambda_+ - \Lambda_-)]^2 \ll 1, \tag{166}$$

where Λ_+ and Λ_- are the eigenvalues of the transport matrix, i.e. Λ_1 and Λ_2 in Eq. (27). The spectral density, $I(\omega)$, is again given by Eq. (8), i.e. it is the real part of the Fourier transform of the autocorrelation function, $\langle S_-(t)\,S_+\rangle$, and the correlation function, according to the Mori theory, obeys the same transport equation as does $S_+(t)$. $I(\omega)$ can be obtained by taking the real part of Eq. (27) with $S_+^{(1)}(0) = 1$ and $S_+^{(2)} = 0$:

$$I(\omega) = \mathrm{Re}\left\{(\Lambda_+ - \Lambda_-)^{-1}\left[\frac{(\Lambda_+ + i\omega_0)}{(i\omega + \Lambda_-)} - \frac{(\Lambda_- + i\omega_0)}{(i\omega + \Lambda_+)}\right]\right\}, \tag{167}$$

where $\Lambda_\pm$ are the eigenvalues of the transport matrix in Eq. (163). $I(\omega)$ in Eq. (167) is the sum of two generalized Lorentzians.

The procedures outlined above can readily be interpreted in the special case that the spin-lattice Hamiltonian $\mathscr{H}_{\mathrm{SL}}$ is secular, i.e. it satisfies the condition in Eq. (64). If $\mathscr{H}_{\mathrm{SL}}$ is secular, we see from an analysis of Eq. (149) that

$$b = 1, \tag{168}$$

and both Eqs. (163) and (164) are modified accordingly. As we shall show in the next section, under these conditions $\hat{\mathscr{K}}''$ is real. The eigenvalues $\Lambda_\pm$ of the transport matrix in Eq. (163) are then:

$$\Lambda_\pm = -i\omega_0 + \tfrac{1}{2}(\hat{\mathscr{K}}'' - i\beta\omega_\lambda)\left[1 \pm \{1 - [4\omega_\lambda^2/(\hat{\mathscr{K}}'' - i\beta\omega_\lambda)^2]\}^{1/2}\right]. \tag{169}$$

As we shall also show in the next section, the two-variable theory is valid only if

$$\omega_\lambda^2\,\hat{\mathscr{K}}''^{-2} \ll 1, \tag{170}$$

and

$$\beta^2 \omega_\lambda^2 \hat{\mathscr{K}}''^{-2} \ll 1; \tag{171}$$

under these conditions the requirement of Eq. (166) is also satisfied. In the limit of Eqs. (170) and (171), we can thus expand Eq. (169) in powers of $(\omega_\lambda/\hat{\mathscr{K}}'')$ and retain terms through order $(\omega_\lambda/\hat{\mathscr{K}}'')^3$; Eq. (167) then becomes[38f]

$$\begin{aligned} I(\omega) = & \{T_-^{-1}/[T_-^{-2} + (\omega_0 - \omega - \sigma_-)^2]\}[1 - 2(\omega_0 - \omega - \sigma_-)\beta T_-(\omega_\lambda/\hat{\mathscr{K}}'')^3] \\ & - (\omega_\lambda/\mathscr{K}'')^2 \{T_+^{-1}/[T_+^{-2} + (\omega_0 - \omega + \sigma_+)^2]\} \\ & \times [1 - 2(\omega_0 - \omega + \sigma_+)\beta T_+(\omega_\lambda/\mathscr{K}'')], \end{aligned} \tag{172}$$

where

$$T_-^{-1} = \omega_\lambda^2/\hat{\mathscr{K}}'', \tag{173}$$

$$T_+^{-1} \approx \hat{\mathscr{K}}'', \tag{174}$$

$$\sigma_- = \beta T_-^{-1}(\omega_\lambda/\hat{\mathscr{K}}''), \tag{175}$$

$$\sigma_+ \approx \beta\omega_\lambda. \tag{176}$$

Thus the spectral line consists of two generalized Lorentzians; a sharp, intense one with half-width, T_-^{-1}, located at $\omega = \omega_0 - \sigma_-$, and a broad, weak one with half-width, T_+^{-1}, located at $\omega = \omega_0 + \sigma_+$. These results are quite similar to those in Fig. 3. The sharp one can be called the "fundamental" while the broad one is an "auxilliary" line.[39–41a] The intensity of the "normal" part of the sharp one is unity whereas that of the broad one is approximately $(-\omega_\lambda^2/\hat{\mathscr{K}}''^2)$, i.e. small and negative. The negative broad line contributes most significantly in the spectral wings, and the resultant line shape is less intense in the wings than is a simple Lorentzian; lowered contribution in the wings assures a finite second moment and, as discussed in Section IV, B, this is a more realistic lineshape than is a simple Lorentzian. Of course, the net intensity must be positive at all frequencies. The shift, σ_-, in the central frequency of the fundamental line is called the "secular shift,"[40,41a] and it is small in the limit of Eqs. (170) and (171); the shift, σ_+, of the auxilliary line is large. The dispersive part of the fundamental becomes important when $(\omega - \omega_0 - \sigma_-)^2 \gtrsim T_-^{-2}$, whereas that of the auxilliary line becomes important when $(\omega_0 - \omega + \sigma_+)^2 \gtrsim T_+^{-2}$.

As the ratio $\omega_\lambda^2/\hat{\mathscr{K}}''^2$ decreases, the spectrum predicted by the two-variable theory approaches that obtained by the one-variable theory. In

[38f] The $\sigma_\pm$ terms in the numerators are actually of higher order in $(\omega_\lambda/\hat{\mathscr{K}}'')$ but have been kept in order to keep the formula in more symmetrical form.

[39] H. Sillescu and D. Kivelson, *J. Chem. Phys.* **48**, 3493 (1968).

[40] R. Hwang, Ph.D. Thesis, Univ. of California, Los Angeles, 1973.

[41a] R. Hwang and D. Kivelson, *J. Mag. Res.* (in press) (1974).

this limit the fundamental line gets sharper since $T_-^{-1} \to \omega_\lambda^2/\hat{\mathscr{K}}''$; the secular shift vanishes since $\sigma_- \to 0$ as $T_-^{-1}(\omega_\lambda/\hat{\mathscr{K}}'') \to 0$; and the dispersive part of the fundamental vanishes as $T_-^{-1}(\omega_\lambda/\hat{\mathscr{K}}'') \to 0$. At the same time, the broad line becomes vanishingly small as $(\omega_\lambda^2/\hat{\mathscr{K}}''^2) \to 0$. Thus in the limit of very small $(\omega_\lambda^2/\hat{\mathscr{K}}''^2)$, the spectrum consists of a single, normal Lorentzian of width $\omega_\lambda^2/\hat{\mathscr{K}}''$, and the pair of coupled equations in Eq. (163) reduces to the Bloch equation in Eq. (2); consequently, $T_2 = T$. We shall discuss this further in the next section.

E. Discussion of Extended Theory

In this subsection we discuss the conditions under which the two-variable theory holds, and we discuss the relationship between the one- and two-variable theories. In the extreme limit $\omega_\lambda^2 \hat{\mathscr{K}}''^{-2} \ll 1$, we have seen that the variable theory with $\mathbf{A}^+ = (S_-, \mathscr{S}_-)$ reduces to the single variable Bloch theory with $\mathbf{A} = S_+$, and $T_-^{-1} = T_2^{-1}$. Therefore, it is interesting to compare the transverse relaxation times obtained from the one-variable theory [Eq. (131)] with that obtained from the limit of the two variable theory. We shall do this below for the case of secular $\mathscr{H}_{\mathrm{SL}}$.

We wish to establish that the requirement under which the two-variable theory applies is

$$(\omega_\lambda^2/\hat{\mathscr{K}}''^2) \ll 1. \tag{177}$$

This condition, we shall show, follows from the requirement, Eq. (161), that

$$\hat{\mathscr{K}}''\tau_{\mathrm{II}} \ll 1, \tag{178}$$

which in turn guarantees that $K''(t)e^{-i(b\omega_0+\beta\omega_\lambda)t}$ decays rapidly compared to $\overline{S_+(t)}$ and $\overline{\dot{\mathscr{S}}_+(t)}$ in the rotating frame. τ_{II} is the correlation time for $\langle \ddot{\mathfrak{S}}_+(t_{p'}) \ddot{\mathfrak{S}}_- \rangle$, i.e. for $K''(t)$, defined in Eq. (157). This rapid decay is required in order for the coefficients in the two coupled transport equations to be independent of time or frequency, and it is only under these conditions, as we have discussed, that the two-variable theory is useful.

We shall show that if Eqs. (177) and (178) hold, then

$$\tau_{\mathrm{II}} \approx \int dt \, \langle \omega_l(t)\,\omega_l \rangle^{e^{-i\omega_0 t}} / \omega_l^2, \tag{179}$$

where ω_l is a frequency characteristic of the lattice motions and defined as

$$\omega_l = [\mathscr{H}_{\mathrm{L}}, [\mathscr{H}_{\mathrm{SL}}^{(\mathrm{sec})}, S_+]]/\omega_\lambda, \tag{180}$$

and when we write ω_l^2 we will always mean

$$\omega_l^2 = \langle |[\mathscr{H}_{\mathrm{L}}, [\mathscr{H}_{\mathrm{SL}}^{(\mathrm{sec})}, S_+]]|^2 \rangle / \omega_\lambda^2. \tag{181}$$

Under these conditions we also have

$$\omega_\lambda^2 \ll \omega_l^2, \tag{182}$$

$$\hat{\mathscr{K}}'' = \omega_l^2 \tau_{\mathrm{II}}, \tag{183}$$

and the condition in Eq. (178) can be rewritten as

$$\hat{\mathscr{K}}'' \tau_{\mathrm{II}} \ll 1 \tag{184a}$$

or

$$\omega_l^2 \tau_{\mathrm{II}}^2 \ll 1. \tag{184b}$$

Furthermore since

$$T_-^{-1} \to T_2^{-1}, \tag{185}$$

where $T_-^{-1} = \omega_\lambda^2/\hat{\mathscr{K}}''$ from the two-variable theory result [Eq. (173)], and $T_2^{-1} = \omega_\lambda^2 \tau_{\mathrm{I}}$ from the one-variable theory result [Eq. (137)], we see that in the limit of Eq. (177)

$$\tau_{\mathrm{I}} = (\hat{\mathscr{K}}'')^{-1} \tag{186}$$

or

$$\tau_{\mathrm{I}} = (\omega_l^2 \tau_{\mathrm{II}})^{-1}. \tag{187}$$

The correlation time τ_{I} is defined in Eqs. (135) and (136); ω_λ is defined in Eq. (134). Equation (187) is *a relation between the results of the one- and two-variable theories.* Note that the projected time evolutions are different is the two correlation times; in Eq. (135) the projection is along S_+, whereas in Eqs. (154), (155), and (157) it is along S_+ and $\dot{\mathscr{S}}_+$, and we have indicated this by writing t_{p} in Eq. (135) and $t_{\mathrm{p'}}$ in Eqs. (154) and (155).

The reader satisfied with this outline can go directly on to the next section and a discussion of the physical consequences of the two variable theory and its relation to the one variable theory, since the discussion below is somewhat involved.

In order to prove the above, we will restrict the discussion to $\mathscr{H}_{\mathrm{SL}}^{(\mathrm{sec})}$ in which case we can rewrite $\ddot{\mathfrak{S}}_+(t_{\mathrm{p'}})$ in Eqs. (152) and (155) as

$$\ddot{\mathfrak{S}}_+(t_{\mathrm{p'}}) = -\omega_\lambda[\omega_l(t_{\mathrm{p'}}) + \omega_\Lambda(t_{\mathrm{p'}})], \tag{188}$$

where $\omega_l(t_{\mathrm{p'}})$ and $\omega_\Lambda(t_{\mathrm{p'}})$ are defined as

$$\omega_l(t_{\mathrm{p'}}) = \omega_\lambda^{-1}\{\exp[(1-\P)\, i\mathscr{H}^\times t]\}\,[\mathscr{H}_{\mathrm{L}}, [\mathscr{H}_{\mathrm{SL}}^{(\mathrm{sec})}, S_+]] \tag{189a}$$

and

$$\begin{aligned}\omega_\Lambda(t_{\mathrm{p'}}) = \omega_\lambda^{-1}\{\exp[(1-\P)\, i\mathscr{H}^\times t]\}\,\{\omega_\lambda^2 S_+ - \beta\omega_\lambda[\mathscr{H}_{\mathrm{SL}}^{(\mathrm{sec})}, S_+] \\ - [\mathscr{H}_{\mathrm{SL}}^{(\mathrm{sec})}, [\mathscr{H}_{\mathrm{SL}}^{(\mathrm{sec})}, S_+]]\}\end{aligned} \tag{189b}$$

and ω_λ is defined in Eq. (134). (Remember that $t_{\mathrm{p'}}$ indicates that ¶ in the time evolution operator projects onto both S_+ and $\dot{\mathscr{S}}_+$.) In this notation

$$K''(t) = \langle[\omega_l(t_{\mathrm{p'}}) + \omega_\Lambda(t_{\mathrm{p'}})]\,[\omega_l + \omega_\Lambda]\rangle \tag{190}$$

and $\hat{\mathscr{K}}''$ is the zero-frequency transform of $K''(t)e^{-i\omega_o t}$. We expect $\langle\omega_\Lambda^2\rangle$ to be of the order of ω_λ^2 where ω_λ is of order $\langle|\mathscr{H}_{SL}|^2\rangle$. We can conveniently express the transport coefficient, $\hat{\mathscr{K}}''$, as

$$\hat{\mathscr{K}}'' = \omega_l^2 \tau_{II}' + \omega_l^2 \sum_{n=2} (\omega_\lambda/\omega_l)^n (\tau_{II}^{(n)}) \tag{191}$$

where τ_{II}' and the $\tau_{II}^{(n)}$'s are appropriate correlation times. In particular, τ_{II}' is defined as

$$\tau_{II}' = \int_0^\infty dt \langle \omega_l(t_{p'})\omega_l\rangle e^{-i\omega_o t}/\omega_l^2, \tag{192}$$

where ω_l^2 is defined in Eq. (181). Note the difference between τ_{II}' in Eq. (192) and τ_{II} in Eq. (179). We expect $\omega_l(t_{p'})$ to be a rapidly varying function since it represents the rate of change of lattice motions, i.e. $[\mathscr{H}_L, \mathscr{H}_{SL}]$, but $\omega_\Lambda(t_{p'})$ may be more slowly varying since it represents the lattice motions themselves, i.e. $\mathscr{H}_{SL}$; we will discuss this below. Thus, since $\tau_{II}^{(2)}$ is characteristic of $\omega_\Lambda(t_{p'})$, we expect

$$\tau_{II}' \ll \tau_{II}^{(2)} \tag{193}$$

and

$$[\omega_l \tau_{II}^{(2)}]^2 \gg 1. \tag{194}$$

Therefore, to ensure that Eq. (178) is satisfied, we let ω_Λ get small, i.e. we let $(\omega_\lambda/\omega_l)$ get very small, so that the condition in Eq. (182) applies. Then

$$\hat{\mathscr{K}}'' = \omega_l^2 \tau_{II}' \tag{195}$$

and the condition in Eq. (178) becomes

$$\hat{\mathscr{K}}'' \tau_{II}' \ll 1 \tag{196a}$$

or

$$\omega_l^2 \tau_{II}'^2 \ll 1. \tag{196b}$$

It can readily be shown by an expansion in powers of time that $\langle\omega_l(t_{p'})\omega_l\rangle$ depends upon $\mathscr{H}_{SL}$ through its dependence upon the projection operator; if ω_λ is truly small, this dependence must be minimal and we can replace $\langle\omega_l(t_{p'})\omega_l\rangle$ by $\langle\omega_l(t)\omega_l\rangle$. The replacement of $\omega_l(t_{p'})$ by $\omega_l(t)$ indicates that the projection operator has been neglected, and that $\langle\omega_l(t)\omega_l\rangle$, and hence τ_{II}' defined in Eq. (192), as well as $\hat{\mathscr{K}}''$, are *lattice properties independent of the spins.*[41b] Note that we have replaced ω_L^2 in Eq. (157), i.e. $\langle(\omega_l+\omega_\Lambda)^2\rangle$ in Eq. (181), by ω_l^2 and τ_{II} by τ_{II}'.

[41b] The approximation $\langle\omega_l(t_{p'})\omega_l\rangle \to \langle\omega_l(t)\omega_l\rangle$ is equivalent to assuming

$$\omega_\lambda^2\langle[\dot{\mathscr{S}}_-, \mathscr{H}_L][\mathscr{H}_L, \mathbf{Q}]\rangle \gg \langle\dot{\mathscr{S}}_-\mathbf{Q}\rangle\langle[\dot{\mathscr{S}}_-, \mathscr{H}_L][\mathscr{H}_L, \dot{\mathscr{S}}_+]\rangle$$

where

$$\mathbf{Q} = [\mathscr{H}_L, [\mathscr{H}_L, [\ldots[\mathscr{H}_L, \dot{\mathscr{S}}_+]\ldots]]].$$

In the limit of Eq. (177) we also have

$$\tau'_{\mathrm{II}} = \tau_{\mathrm{II}}. \tag{197}$$

Equations (183)–(187) follow directly.

We have still to show that the condition, $\hat{\mathscr{K}}''\tau_{\mathrm{II}} \ll 1$, implies $(\omega_\lambda/\hat{\mathscr{K}}'')^2 \ll 1$. The correlation times $\tau_{\mathrm{II}}^{(n)}$ in Eq. (191) represent correlation times associated with $\omega_\Lambda(t_{\mathrm{p}'})$, i.e. times associated with $\mathscr{H}_{\mathrm{SL}}(t_{\mathrm{p}'})$. These times should be of the order of τ_{I}, and from Eq. (187) we then see that $\tau_{\mathrm{II}}^{(n)} \approx \gamma_n \hat{\mathscr{K}}''^{-1}$, where γ_n is of the order of unity. In order to guarantee that only the first term in Eq. (191) contributes appreciably we must have

$$\tau'_{\mathrm{II}} \gg (\omega_\lambda/\omega_l)^n \hat{\mathscr{K}}''^{-1}. \tag{198a}$$

This expression can be combined with Eq. (195) and rewritten as

$$1 \gg (\omega_\lambda/\omega_l)^{n-2}(\omega_\lambda/\hat{\mathscr{K}}'')^2. \tag{198b}$$

If we consider $n = 2$, we get Eq. (177). Furthermore, we can show that the expansion through order $(\omega_\lambda/\hat{\mathscr{K}}'')^3$, which we used in obtaining Eqs. (172)–(176), is consistent with the condition in Eq. (177). Equation (177) can be used to rewrite the expression for $\hat{\mathscr{K}}''$ in Eq. (191) to second order in $(\omega_\lambda/\hat{\mathscr{K}}'')$:

$$\hat{\mathscr{K}}'' = \omega_l^2 \tau'_{\mathrm{II}}[1 + \gamma_2(\omega_\lambda/\hat{\mathscr{K}}'')^2], \tag{199}$$

where γ_2 is of the order of unity. In making use of the two-variable theory we have had to drop this second-order term, a procedure which is consistent with the expansions used in obtaining Eqs. (172)–(176). In these expansions we kept terms through order $(\omega_\lambda/\hat{\mathscr{K}}'')^3$. We found that for the sharp line the lowest order contribution was of order $(\omega_\lambda/\hat{\mathscr{K}}'')^2$, and so the γ_2 correction terms in Eq. (199) would only enter in the fourth order. We found that for the broad line the lowest order contribution was of order $(\omega_\lambda/\hat{\mathscr{K}}'')^2$, and the next contribution was of order $(\omega_\lambda/\hat{\mathscr{K}}'')^2$ and could be neglected; consequently, the γ_2 correction to $\hat{\mathscr{K}}''$ could also be neglected.

Some of these results are collected in Table II.

F. Physical Consequences of Extended Theory

In order to understand the results in the last section we shall again examine the special case in which $\mathscr{H}_{\mathrm{SL}}$ is secular, and in particular where it arises from the anisotropy of the g-tensor; from Eq. (53),

$$\mathscr{H}_{\mathrm{SL}}^{(\mathrm{sec})} = \sum_j \mathrm{B}\Gamma_{zz}^{(j)} S_z^{(j)}; \tag{200}$$

$\Gamma_{zz}^{(j)}$ is proportional to the second-order Legendre polynomial which is a function of the Euler angle θ_j and specifies the orientation of the jth

TABLE II

SUMMARY OF CONDITIONS

Theory (variables)	Number of Lorentzians in spectrum	Conditions	Equalities
(a) One-variable (S_+)	1 normal	$\left.\begin{matrix}\tau_I \ll t \\ \lvert\omega-\Omega_0\rvert\tau_I \ll 1\end{matrix}\right\}$	
		$\left.\begin{matrix}\omega_\lambda^2\tau_I^2 \ll 1 \\ \tau_I \ll T_2\end{matrix}\right\}$	$T_2^{-1} = \omega_\lambda^2\tau_I$ $T_2^{-1} = \hat{\mathscr{K}}_{RE}$
(b) Two-variable ($S_+, \dot{\mathscr{S}}_+$)	2 generalized	$\left.\begin{matrix}\tau_{II} \ll t \\ \lvert\omega-\Omega_0\rvert\tau_{II} \ll 1\end{matrix}\right\}$	
		$\left.\begin{matrix}\omega_\lambda^2 \ll \omega_l^2 \\ \omega_l^2\tau_{II}^2 \ll 1 \\ \omega_\lambda^2\tau_{II}^2 \ll 1\end{matrix}\right\}$	$T_-^{-1} = \omega_\lambda^2/\hat{\mathscr{K}}''_{RE}$ $\hat{\mathscr{K}}''_{RE} = \omega_l^2\tau_{II}$ $T_+^{-1} = \hat{\mathscr{K}}''_{RE}$
		$\omega_\lambda^2 \ll (\hat{\mathscr{K}}''_{RE})^2$	
(c) Two-variable in limit of one-variable		Conditions above $\tau_{II} \ll \tau_I$	$\tau_I\tau_{II} = \omega_l^{-2}$ (Hubbard relation) $T_- = T_2$

molecule. We can then readily obtain the following:

$$\omega_\lambda^2 = \langle(\Gamma_{zz})^2\rangle B^2, \tag{201}$$

$$\omega_l^2 = \langle(d\Gamma_{zz}/d\theta)^2\rangle\langle\dot{\theta}^2\rangle/\langle\Gamma_{zz}^2\rangle, \tag{202}$$

where we have dropped the explicit dependence upon j. $\dot{\theta}$ is the angular velocity, and from equilibrium statistical mechanics

$$\langle\dot{\theta}^2\rangle = k_B T/I, \tag{203}$$

where k_B is the Boltzmann constant, T the absolute temperature, and I the molecular moment of inertia for a spherical top. Since Γ_{zz} is proportional to a second-rank Legendre polynomial, Eq. (201) becomes

$$\omega_l^2 = l(l+1)k_B T/I, \tag{204}$$

where $l = 2$. We see that ω_λ, ω_L, and ω_l are equilibrium quantities; ω_λ is the frequency characteristic of the spin-lattice interaction and reflects the strength of this interaction, while ω_l is the mean classical molecular rotational frequency.

We can also obtain physical interpretations for the correlation time τ_I in the simple case discussed above. We see that

$$\tau_I = \int_0^\infty dt\, e^{-i\omega_0 t}\langle[\Gamma_{zz}(t_p)\Gamma_{zz}][S_+(t_p)S_-]\rangle/\langle\Gamma_{zz}^2\rangle\langle S_+S_-\rangle, \tag{205}$$

where the time evolution operator for t_p is $\{\exp[it(1-¶)(\mathscr{H}_S^x+\mathscr{H}_L^x+\mathscr{H}_{SL}^x)]\}$ and ¶ is the projection operator along S_+. We have already seen that for small $\mathscr{H}_{SL}$, we can replace this time evolution operator by $\{\exp[it(\mathscr{H}_S^x+\mathscr{H}_L^x)]\}$ in which case Eq. (205) becomes[29,35,42]

$$\tau_I = \int_0^\infty dt\,\langle \exp[i\mathscr{H}_L^x t]\,\Gamma_{zz}, \Gamma_{zz}\rangle / \langle \Gamma_{zz}^2 \rangle. \tag{206}$$

Thus τ_I is the correlation time for molecular reorientation corresponding to a second-rank Legendre polynomial. The condition under which Eq. (206) holds is small $\mathscr{H}_{SL}$, i.e. small ω_λ, or

$$\tau_I \ll T_2. \tag{207}$$

The correlation time, τ'_{II}, or the related quantity τ_{II}, which enters into the two-variable theory can be considered next [see Eqs. (192 and 197)]:

$$\tau'_{II} = \frac{\int_0^\infty dt\,\langle[\exp i(1-¶)\mathscr{H}^{x}t]\,[S_+\,\theta\, d\Gamma_{zz}/d\theta], [S_-\,\dot{\theta}\, d\Gamma_{zz}/d\theta]\rangle}{\langle (d\Gamma_{zz}/d\theta)^2\rangle\,\langle \dot{\theta}\dot{\theta}\rangle\,\langle S_+ S_-\rangle}\, e^{-i\omega_0 t} \tag{208}$$

where ¶ is the projection operator along both S_+ and $\mathscr{S}_+$. If, furthermore, the angular velocity, $\dot{\theta}$, changes rapidly with respect to both θ and S_+, then

$$\tau'_{II} \approx \int_0^\infty dt\,\langle \dot{\theta}(t)\,\dot{\theta}\rangle / \langle \dot{\theta}^2 \rangle. \tag{209}$$

(We have already shown that $\langle \omega_l(t_{p'})\,\omega_l\rangle$ can be replaced by $\langle \omega_l(t)\,\omega_l\rangle$ in this approximation and τ'_{II} can be replaced by τ_{II}.) Thus τ_{II} is the correlation time for the angular velocity,[29,35,42] and the condition under which Eq. (205) holds can be restated as

$$\tau_{II} \ll \tau_I. \tag{210}$$

[See Eqs. (177) and (187).] The requirement in Eq. (210) is that for rotational diffusion[29,35,42] where the rotational diffusion constant, D_θ, is defined as

$$D_\theta = \int_0^\infty dt\,\langle \dot{\theta}(t)\,\dot{\theta}\rangle. \tag{211}$$

We can then rewrite τ_{II} in Eq. (209) as

$$\tau_{II} = (I/k_B T)\,D_\theta. \tag{212}$$

We can also rewrite Eq. (208) as

$$\tau_I \tau_{II} = I/l(l+1)\,k_B T. \tag{213}$$

[42] D. Kivelson, M. Kivelson, and I. Oppenheim, *J. Chem. Phys.* **52**, 1810 (1970).

This is known as the *Hubbard relation*.[29,31,35] Equations (209) and (208) can be combined to yield

$$\tau_{\rm I}^{-1} = l(l+1)\, D_{\theta}. \tag{214}$$

A summary of many of the results given above is given in Table II. Of particular interest is the fact that within the scope of the two-variable theory, $\tau_{\rm II} \ll \tau_{\rm I}$, that is, rotational diffusion is implied. (A similar conclusion applies to the translational motion if $\mathscr{H}_{\rm SL}$ depends upon molecular translation displacement.) However, if an extension were made to a three-variable theory, this need no longer be the case even in the limit of very small $\omega_{\Lambda}^2 \tau_{\rm I}^2$ where the spectrum reduces to that of a single Lorentzian. We discuss this in the next section.

In the limit that $\omega_{\lambda}^2 \tau_{\rm I}^2$ is very small, the results of the two-variable theory reduce to those of the one-variable theory, and the spectrum, except in the far wings, is well described by a single Lorentzian. In the far wings the two-variable theory always gives a better description. See Table II.

Even in the limit where the two-variable gives the same spectral shape as the one-variable theory, the former yields additional information since it gives a molecular expression for $\tau_{\rm I}$, *i.e. the Hubbard relation, directly within the structure of the theory.*

G. Further Extensions of the Theory

Suppose $\hat{\mathscr{K}}(\omega - \Omega)$ in the two-variable theory depends strongly on ω, i.e. suppose $\mathsf{K}(t)$ does not decay rapidly, then how does one extend the theory? As suggested in previous sections, we could add an additional variable, perhaps $\ddot{S}_+$. To some extent the choice of a third variable would depend upon the cause of the "breakdown" in the two variable theory. There are two physical situations which can give rise to such a breakdown: in the first $\mathscr{H}_{\rm SL}$ gets large, i.e. comparable to the tumbling time, and in the second the correlation time for angular velocity need not be short. The first situation is the well-known slow tumbling problem which arises when magnetic resonance spectra are studied in viscous solutions or if a polymer tumbles in aqueous solution. The second situation arises if the rotational diffusion model no longer applies. Of course, both situations could apply simultaneously.

If the first situation above (the slow tumbling case) applies we need higher powers of $\mathscr{H}_{\rm SL}$ and as a third variable we can choose $[\mathscr{H}_{\rm SL}, \dot{\mathscr{S}}_+]$ or preferably the corresponding variable $\ddot{\mathfrak{S}}_+^{\rm (SL)}$ which is orthogonal to both S_+ and $\dot{\mathscr{S}}_+$:

$$\ddot{\mathfrak{S}}_+^{\rm (SL)} = -[\mathscr{H}_{\rm SL}, [\mathscr{H}_{\rm SL}, S_+] + \omega_{\lambda}^2 S_+ + \beta\omega_{\lambda}[\mathscr{H}_{\rm SL}, S_+] \tag{215}$$

where β is defined in Eq. (150) and ω_λ is Eq. (134). In this formulation we find that the relaxation matrix $\hat{\mathscr{K}}$ has only a single element,

$$\hat{\mathscr{K}}_{33} = \frac{\int_0^\infty dt \langle [\exp i(1-\P)\mathscr{H}^x t](1-\P)\ddot{\mathfrak{S}}_+^{(SL)}(1-\P)\ddot{\mathfrak{S}}_-^{(SL)}\rangle e^{-i\Omega t}}{\langle \ddot{\mathfrak{S}}_+^{(SL)}\ddot{\mathfrak{S}}_-^{(SL)}\rangle}, \qquad (216)$$

where $\P$ projects upon S_+, $\dot{\mathscr{S}}_+$ and $\ddot{\mathfrak{S}}_+^{(SL)}$. The correlation time τ_{III} for the correlation functions in Eq. (216) must be short, i.e.

$$\hat{\mathscr{K}}_{33}{}^3 \tau_{III} \ll 1, \qquad (217a)$$

where τ_{III} is defined as

$$\tau_{III} \equiv \hat{\mathscr{K}}_{33} \langle \ddot{\mathfrak{S}}_+^{(SL)} \ddot{\mathfrak{S}}_-^{(SL)} \rangle / \langle |(1-\P)\dot{\ddot{\mathfrak{S}}}_+^{(SL)}|^2 \rangle. \qquad (217b)$$

We retain the approximation in footnote 41b; this is equivalent to the assumption of rotational diffusion. The spectrum now consists of three generalized Lorentzians. We can keep adding more variables of the form $[\mathscr{H}_{SL}, [\mathscr{H}_{SL}, \ldots [\mathscr{H}_{SL}, S_+] \ldots]]$ in order to obtain a better description of the spectrum for slower rotational diffusion tumbling rates; the addition of each new variable adds an additional generalized Lorentzian to the spectrum and alters the widths, shifts, and dispersive contributions of all the Lorentzians. This type of description of spectra of molecules tumbling slowly by rotational diffusion is discussed, from a rather different point of view, elsewhere.[12,39,40,41a,43–45]

If the tumbling rate is rapid ($\omega_\lambda \tau_I \ll 1$), but the reorientational motion is not well described by rotational diffusion, then we do not seek higher powers of $\mathscr{H}_{SL}$ but of the lattice Hamiltonian, $\mathscr{H}_L$. As a third variable we could then choose

$$\ddot{\mathfrak{S}}_+^{(L)} = -[\mathscr{H}_L, [\mathscr{H}_{SL}, S_+]]. \qquad (218)$$

In this formulation also we find that again the relaxation matrix $\hat{\mathscr{K}}$ has only the single element, $\hat{\mathscr{K}}_{33}$, discussed in Eqs. (216) and (217), with the definition of $\ddot{\mathfrak{S}}_+^{(L)}$ given in Eq. (218) substituted in place of $\ddot{\mathfrak{S}}_+^{(SL)}$. The approximation in footnote 41b no longer holds. τ_{III}, the correlation time for the correlation function in Eq. (216), is now related to the correlation time for the torques, and this time can be short even if the correlation time τ_{II} or τ'_{II}, characteristic of the angular velocities, is long. (This is the case for extended rotational diffusion where the torques are impulsive but

[43] R. G. Gordon and T. Messenger, *in* "Electron Spin Relaxation in Liquids" (L. T. Muus and P. W. Atkins, eds.), Ch. 13, p. 341. Plenum, New York, 1972; see also references cited

[44] H. Sillescu, *J. Chem. Phys.* **54**, 2110 (1971).

[45] R. M. Lynden-Bell, *in* "Electron Spin Relaxation in Liquids" (L. T. Muus and P.W. Atkins, eds.), Ch. 13A, p. 383. Plenum, New York, 1972; R. M. Lynden-Bell, *Mol. Phys.* **22**, 837 (1971).

the time of free rotations long.[35,46,47]) If $\omega_\lambda \tau_I$ is sufficiently small, i.e. if the tumbling is sufficiently rapid, the one-variable theory is valid and the spectrum consists of a single normal Lorentzian; this is the motionally narrowed or Redfield limit. For slightly larger $\omega_\lambda \tau_I$, the two-variable theory is needed; it is restricted to rotational diffusion whereas the three-variable theory is not. The three-variable theory predicts a spectrum consisting of three generalized Lorentzians which reduces to a two- or possibly a one-Lorentzian spectrum in the rotational diffusion limit, $(\tau'_{II}/\tau_I) \ll 1$. A spectrum consisting of three Lorentzians could arise from extended rotational diffusion and small $(\omega_\lambda \tau_I)$ or from slow rotational diffusion, i.e. slightly larger $(\omega_\lambda \tau_I)$; a detailed analysis of the spectrum can, in principle, distinguish between these two situations.

It is interesting to consider further the case where the correlation times for the torques are also long, but the rate of change of torques is rapid. This picture includes as one limiting case the rotational jump model of Ivanov[48] (see Kivelson and Keyes[35] for a detailed discussion). In this case we add a fourth variable, $-[\mathscr{H}_L, [\mathscr{H}_L, [\mathscr{H}_{SL}, S_+]]]$, or rather its orthogonalized analog, and we assume that the correlation time, τ_{IV}, for its projected derivative is short. We then predict a spectrum consisting of four generalized Lorentzian lines, and a careful analysis of the spectrum should yield information concerning the details of the molecular motion.

Of course, if the tumbling is slow and, in addition, the rotational motion nondiffusional, we add as our third variable, $[\mathscr{H}, [\mathscr{H}_{SL}, S_+]]$. In summary, we see that *in the rapid tumbling limit we always observe a single normal Lorentzian near the central part of the spectrum and we cannot distinguish between one kind of molecular motion and another unless we examine the spectral wings. For slower tumbling the spectrum consists of several generalized Lorentzians and one can distinguish between one kind of molecular motion and another without analyzing the far spectral wings; however, this analysis is complicated by the fact that the additional Lorentzian may arise both because of slow tumbling and because of nondiffusional motion*, and only a very careful study can distinguish between these effects.[49]

We have discussed rotational motion but, of course, if $\mathscr{H}_{SL}$ involves translational variables, i.e. if $\mathscr{H}_{SL}$ involves intermolecular interactions, then similar arguments apply to the translational motions.

As a final comment we note that the addition of variables in the Mori approach is equivalent to an expansion of the autocorrelation function, $\langle S_+(t) S_- \rangle$, in terms of exponentials or of the spectrum, $I(\omega)$, in terms of

[46] R. G. Gordon, *J. Chem. Phys.* **44**, 1830 (1966).

[47] W. A. Steele, *J. Chem. Phys.* **38**, 2404, 2411 (1963).

[48] E. N. Ivanov, *Zh. Eksp. Teor. Fiz.* **45**, 1509 (1963) [*Sov. Phys.—JETP* **18**, 1041 (1964)].

[49] S. A. Goldman, G. V. Bruno, C. F. Polnaszek, and J. H. Freed, *J. Chem. Phys.* **56**, 716 (1972).

Lorentzians. Thus

$$\langle S_+(t)S_-\rangle = \sum_n a_n \exp(-\Lambda_n t), \tag{219}$$

where the Λ_n's are the eigenvalues of the transport matrix and $\mathrm{Re}(\Lambda_n) > 0$. There exists a Λ_0 such that $\mathrm{Re}(\Lambda_0) \leqslant \mathrm{Re}(\Lambda_n)$, so that at very large t

$$\langle S_+(t)S_-\rangle \to a_0 \exp(-\Lambda_0 t). \tag{220}$$

The series in Eq. (219) can be truncated at any point since one always has term by term convergence for large t in this exponential series. This expansion in exponentials is also characteristic of various other methods: cumulant expansions,[8,12a] stochastic Liouville approach,[9,12b] the Anderson-Sachs jump theory,[50,51] and numerous other approaches.[39,45] These theories are in contrast to the time-dependent perturbation theories which have been used so successfully in the motionally narrowed limit[6,15-17]; in these perturbation treatments $\langle S_+(t)S_-\rangle$ is expanded in a power series in time,

$$\langle S_+(t)S_-\rangle = \sum_m b_m t^m, \tag{221}$$

where $b_0 = \langle S_- S_+\rangle$. This expression cannot be truncated for large t, but one ordinarily assumes that one can exponentiate the first two terms in this expansion, so that

$$\langle S_+(t)S_-\rangle = \langle S_+ S_-\rangle \exp - [-b_1 t/\langle S_+ S_-\rangle]. \tag{222}$$

This procedure can be justified by the theory of cumulant expansions,[8,12a] but in the various exponential expansions discussed above, the higher order corrections are more readily obtained. It should also be pointed out[51] that *the exponentiation procedure is questionable whenever $\langle S_+(t)S_-\rangle$ cannot be well described by a single exponential;* in this case b_1 is the sum of linear contributions from the various exponentials, and a procedure such as that of Mori discussed in this article should yield more meaningful long time descriptions for $\langle S_+(t)S_-\rangle$. It is only the long time behavior that is of interest.

VI. Nuclear Hyperfine Interaction

We turn now to a commonly confronted situation in both NMR and ESR, i.e. that in which the spectrum is characterized by hyperfine splittings. In NMR these are the so-called "spin-spin splittings" and in ESR they are

[50] P. W. Anderson, *J. Phys. Soc. Jap.* **9**, 316 (1954); P. W. Anderson and P. R. Weiss, *Rev. Mod. Phys.* **25**, 269 (1953).

[51] R. A. Sachs, *Mol. Phys.* **1**, 163 (1958).

"nuclear hyperfine splittings." Both these effects can be described by the same formalism, but we shall introduce a notation used widely among ESR spectroscopists. We shall assume that the correlation times associated with $\mathscr{H}_{SL}$ are short so that we need not introduce the extended Mori theory described in preceeding sections, that is, we will assume that an appropriate collection of spins constitutes a complete set of slow variables and that the time derivatives of the spins need not be included in this set. The isotropic hyperfine interaction resulting from the interaction of a single electron with a single nucleus is described by the term $a\mathbf{I}_j \cdot \mathbf{S}_j$ in the Hamiltonian, where a is the Fermi contact or hyperfine splitting constant, $\mathbf{I}_j$ is the nuclear spin angular momentum, and $\mathbf{S}_j$ is the electron spin angular momentum of the jth molecule. (See Eq. (71).] In Section VI, A we will neglect the nonsecular contributions to $(a\mathbf{I}_j \cdot \mathbf{S}_j)$; in other words we will assume that the Fermi contact interaction can be adequately described by $(aI_{z_j}S_{z_j})$, a situation that pertains in strong applied magnetic fields, and we will study the spin relaxation due to a spin-lattice interaction, $\mathscr{H}_{SL}$. In Section VI, B we will discuss spin relaxation for the case where the isotropic hyperfine interaction contains nonsecular terms.

A. Nuclear Hyperfine Interaction. Secular $\mathscr{H}_S$

In strong applied fields only the secular parts of the pure spin hyperfine Hamiltonian, $\mathscr{H}_S$, need be considered, i.e. Eq. (71) reduces to

$$\mathscr{H}_S = \omega_0 \sum_j S_{z_j} + a \sum_j I_{z_j} S_{z_j}, \tag{223}$$

where the sum runs over all molecules and ω_0 is the Larmor frequency. (The nuclear Zeeman term has been neglected.) The intramolecular spin-lattice interactions, $\mathscr{H}_{SL}$, that affect the relaxation of the hyperfine spectrum are, typically, the anisotropic g-factor and the anisotropic hyperfine interactions. [See Eqs. (66) and (73).] An intermolecular $\mathscr{H}_{SL}$ is discussed in Section VII.

In the formulation of the theory of magnetic spin linewidths by means of the Mori procedures described in Section V, we choose S_+ as our variable. Now we are interested in the S_+-component of the group of molecules which are in a given nuclear state, $|m\rangle$. We choose our set of variables, $\mathbf{A}$, to be

$$\mathbf{A} = \begin{bmatrix} \vdots \\ S_+^{(m)} \\ \vdots \\ S_+^{(m')} \\ \vdots \end{bmatrix} \tag{224}$$

with the elements of **A** defined by

$$S_+^{(m)} = \sum_j S_{+j} P_{mj}, \tag{225a}$$

where P_{mj} is a *nuclear spin projection operator* for the jth molecule, i.e.

$$P_{mj} = |mj\rangle\langle mj|, \tag{225b}$$

where $\langle mj|$ and $|mj\rangle$ are the bra and the ket which describe the molecule j as having the specific quantum number m corresponding to I_z. (Note that the brackets $\langle\,\rangle$ are also used to indicate equilibrium ensemble averages; the meaning implied in a given relationship is always unambiguous.) Nuclear spin energy level separations are small compared to $k_B T$ at room temperature, so we assume that there is an equal number of molecules in each nuclear spin state, i.e. $N/(2I+1)$, since there are $(2I+1)$ nuclear spin states and N spins. The $(2I+1)$ nuclear spin states also mean that **A** is $(2I+1)$-dimensional.

We can take the complete trace over all spin states since the selection of appropriate spin states is automatically performed by the nuclear spin projection operator, P_{mj}. The specification of variables according to a specific nuclear spin state can be made only if the spin Hamiltonian is secular (or is transformed to a secular form as in Section VI, B), because otherwise each state is a mixture of electronic and nuclear spin states.

We now evaluate the $\langle \mathbf{AA}^+ \rangle$ matrix:

$$\langle \mathbf{AA}^+ \rangle = \langle S_+^{(m)} S_-^{(m)} \rangle \, \mathsf{1} = N/[2(2I+1)] \, \mathsf{1}, \tag{226}$$

where $\mathsf{1}$ is the $[(2I+1)\times(2I+1)]$ unit matrix. All off-diagonal terms are zero because they involve the product of spin components from different groups of molecules. The calculation of $\langle \mathbf{AA}^+ \rangle^{-1}$ is straightforward.

Using the Hamiltonian, as given in Eqs. (59) and (223), we find that $\dot{S}_+^{(m)}$ is given by

$$\dot{S}_+^{(m)} = (i\omega_0 + iam) S_+^{(m)} + [\mathscr{H}_{SL}, S_+^{(m)}], \tag{227}$$

where m is the eigenvalue of I_z for a molecule in the mth nuclear spin state. If we again assume that the ensemble average of terms linear in $\mathscr{H}_{SL}$ vanishes, we obtain

$$i\Omega = \begin{bmatrix} i\Omega_1 & & & 0 \\ & i\Omega_2 & & \\ & & \ddots & \\ & & i\Omega_n & \\ 0 & & & \ddots \end{bmatrix}, \tag{228}$$

where we have defined

$$\Omega_m \equiv \omega_0 + am. \tag{229}$$

The variable set, **A**, defined in Eq. (224), satisfies Eq. (123) with $\mathscr{H}_S$ given in Eq. (223) and $i\,\Omega$ given by Eq. (228); so, from our discussion near Eqs. (123) and (124), $\dot{\boldsymbol{\alpha}}$ is given by

$$\dot{\boldsymbol{\alpha}} = i[\mathscr{H}_{SL}, \mathbf{A}], \tag{230a}$$

and from our discussion near Eqs. (139)–(141),

$$\dot{\boldsymbol{\alpha}}(t_p) \approx e^{i(\mathscr{H}_S + \mathscr{H}_L)t}\, i[\mathscr{H}_{SL}, \mathbf{A}]\, e^{-i(\mathscr{H}_S + \mathscr{H}_L)t} \tag{230b}$$

to lowest order in $\mathscr{H}_{SL}$. The memory function matrix, $\mathsf{K}(t)$, is then

$$+ \langle S_{+j}(t_p)(\mathscr{H}_{SL}(t_p))_{m'jm'_j} S_{-j}(\mathscr{H}_{SL})_{m'jm_j}\rangle]\}\, e^{i\Omega_m t}, \tag{231b}$$

($[\mathscr{H}_{SL}(t_p), S_\alpha(t_p)] \equiv \{\exp i[1-¶]\, \mathscr{H}^\times t\}\, [\mathscr{H}_{SL}, S_\alpha]$). If we expand this expression, and keep track of the nuclear spin projection operator, P_{mj}, we obtain

$$\begin{aligned} K_{mm'}(t) = [2/N] \sum_{n_j} \sum_j \{ & [\langle (\mathscr{H}_{SL}(t_p))_{n_j m_j} S_{+j}(t_p)\, S_{-j}(\mathscr{H}_{SL})_{m_j n_j}\rangle \\ & + \langle S_{+j}(t_p)(\mathscr{H}_{SL}(t_p))_{m_j n_j}(\mathscr{H}_{SL})_{n_j m_j}\rangle S_{-j}]\,\delta_{m_j m'_j} \\ & - [\langle (\mathscr{H}_{SL}(t_p))_{m'_j m_j} S_{+j}(t_p)(\mathscr{H}_{SL})_{m_j m_j} S_{-j}\rangle \\ & + \langle S_{+j}(t_p)(\mathscr{H}_{SL}(t_p))_{m'_j m_j} S_{-j}(\mathscr{H}_{SL})_{m'_j m_j}\rangle]\}\, e^{i\Omega_m t}, \end{aligned} \tag{231b}$$

where $(\mathscr{H}_{SL})_{nm}$ is a matrix element evaluated between the $\langle nj|$ and $|mj\rangle$ nuclear spin states. (In this equation the equilibrium averages, $\langle\,\rangle$, are taken over lattice, electron spin states, and over nuclear spins of all but the jth molecule if relevant, but the sums over nuclear spin states for the jth molecule are indicated explicitly; S_{+j} represents $\langle mj|S_{+j}|mj\rangle$ in all four terms.) It can readily be seen that if the only nuclear spin operator in $\mathscr{H}_{SL}$ is I_{zj}, then $(\mathscr{H}_{SL}(t_p))_{nm} = (\mathscr{H}_{SL})_{nn}\delta_{nm}$; in this case, which we shall call nucleo-secular, all four terms in Eq. (231b) contribute to the diagonal $K_{nm}(t)$ element. It can readily be seen that only the first two terms in Eq. (231b), the adiabatic terms, contribute to the diagonal $K_{mm}(t)$ element for those terms in $\mathscr{H}_{SL}$ that depend upon I_{+j} or I_{-j}, terms which we shall call nucleo-nonsecular. Whereas those terms in $\mathscr{H}_{SL}$ that depend upon I_{zj} contribute exclusively to the diagonal $K_{mm}(t)$ terms, the terms in $\mathscr{H}_{SL}$ that depend upon I_{+j} and I_{-j}, the nucleo-nonsecular terms, contribute not only to $K_{mm}(t)$, but also to the off-diagonal $K_{mm'}(t)$ terms with $m \neq m'$. The Mori equations of motion, under appropriate conditions discussed previously, reduce to $(2I+1)$ coupled transport equations. Remembering that $\mathscr{K}(t) = \mathsf{K}(t)\exp(-i\Omega t)$, we then have[27]

$$(d/dt)\,\overline{S_+^{(m)}(t)} = i\Omega_m\,\overline{S_+^{(m)}(t)} - \sum_{m'} \hat{\mathscr{K}}_{mm'}\,\overline{S_+^{(m')}(t)}. \tag{232a}$$

If $a^2 \gg |\hat{\mathscr{K}}|^2$, i.e. if the hyperfine components are well resolved, then the coupling terms are not very important, and the transport equations above reduce to the $(2I+1)$ uncoupled equations:

$$(d/dt)\,\overline{S_+^{(m)}(t)} = (i\omega_m - T_2^{-1}(m))\,\overline{S_+^{(m)}(t)}, \tag{232b}$$

where

$$T_2^{-1}(m) \equiv \frac{2(2I+1)}{N}\,\mathrm{Re}\int_0^\infty dt\,\langle[\exp\{it(\mathscr{H}_S^x + \mathscr{H}_L^x)\}\,\mathscr{H}_{SL}, S_+^{(m)}]\,[S_-^{(m)}, \mathscr{H}_{SL}]\rangle, \tag{232c}$$

$$\omega_m = \Omega_m + \sigma_m, \tag{232d}$$

and the frequency shift, σ_m, is the imaginary part of the right-hand side of Eq. (233); Eqs. (232) and (233) are analogous to Eqs. (129)–(132), with the line now centered at the hyperfine frequency $(\Omega_m + \sigma_m)$ rather than the Larmor frequency $(\omega_0 + \sigma_+)$. $T_2^{-1}(m)$ is the linewidth of each line and is identical in form to Eq. (142), and to results by other authors.[19] Again, we find there may be a nonsecular shift, σ_m, for each line.

The extension of these arguments to molecules with several interacting nuclei is straightforward.[26b, 52–54]

B. Nuclear Hyperfine Interaction. Nonsecular $\mathscr{H}_S$

We now extend the study presented in the last section to include a pure spin Hamiltonian, $\mathscr{H}_S$, which includes nonsecular terms, in particular the off-diagonal nuclear isotropic interaction terms $[\frac{1}{2}a(I_{+j}S_{-j} + I_{-j}S_{+j})]$. [See Eq. (71).] Nonsecular terms in $\mathscr{H}_S$ mix the nuclear spin states; therefore, the nonsecular hyperfine terms, $I_{+j}S_{-j}$ and $I_{-j}S_{+j}$, cannot be directly used in the procedures above.

Wilson and Kivelson[26c] have developed a method of handling nonsecular terms which is readily applicable to the present case. Rather than transform the basis set, they apply a unitary transformation to the spin Hamiltonian itself, a transformation which yields a modified spin Hamiltonian, $\mathscr{H}_S^{(T)}$, which is secular (diagonal) to second order in an ordering parameter, λ. Thus

$$\mathscr{H}_S^{(T)} = e^{i\lambda T}\,\mathscr{H}_S\,e^{-i\lambda T}, \tag{233}$$

[52] J. Freed and G. Fraenkel, *J. Chem. Phys.* **39**, 326 (1963).
[53] D. Kivelson, *J. Chem. Phys.* **41**, 1904 (1964).
[54] A. Carrington and H. C. Lonquet-Higgins, *Mol. Phys.* **5**, 447 (1962).

where T is appropriately constructed as described in Wilson and Kivelson.[26c] The resulting spin Hamiltonian, $\mathscr{H}_S^{(T)}$, through second order in λ, is

$$\mathscr{H}_S^{(T)} = \omega_0 \sum_j S_{z_j} + a \sum_j I_{z_j} S_{z_j}$$

$$+ \tfrac{1}{2}(a^2/\omega_0) \sum_j \{[I(I+1) - I_{z_j}^2] S_{z_j} - I_{z_j}[S(S+1) - S_{z_j}^2]\}. \quad (234)$$

This transformed spin Hamiltonian is secular, and we can use the variable set used earlier in the preceding secion. The unitary transformation described above operates on the complete Hamiltonian, and thus, not only is $\mathscr{H}_S$ transformed, but the spin-lattice Hamiltonian is also transformed;

$$\mathscr{H}_{SL}^{(T)} = e^{i\lambda T} \mathscr{H}_{SL} e^{-i\lambda T}. \quad (235a)$$

To second order in λ, Wilson and Kivelson find that $\mathscr{H}_{SL}^{(T)}$ is given by

$$\mathscr{H}_{SL}^{(T)} = \mathscr{H}_{SL} - \tfrac{1}{2}(a^2/\omega_0) \sum_j [I_{+j} S_{-j} - I_{+j} S_{+j}), \mathscr{H}_{SL}]. \quad (235b)$$

The procedures of the last section can now be readily applied with $\mathscr{H}_S^{(T)}$ and $\mathscr{H}_{SL}^{(T)}$ substituted for $\mathscr{H}_S$ and $\mathscr{H}_{SL}$, respectively. Extensions to higher order in λ could be carried out, i.e. a unitary transformation could be devised to transform $\mathscr{H}_S$ into a spin Hamiltonian which is diagonal to higher order in λ.

VII. Spin Exchange

An intermolecular interaction which can drastically affect spin relaxation is Heisenberg spin exchange. In solution, a collision between two paramagnetic molecules results in the overlap of the electronic wavefunctions of, and hence a very strong electrostratic interaction between, the unpaired electrons. This interaction leads to a "mixing" of the states of the two molecules. Therefore, as a result of the collision, the electrons have a probability of having been exchanged between the molecules; the probability depends upon the exchange integral $(-J_{12})$,[28]

$$-J_{12} = \int \Psi(\mathbf{r}_1, \mathbf{r}_2) \mathscr{H}_{\text{elect}}(\mathbf{r}_1, \mathbf{r}_2) \Psi(\mathbf{r}_2, \mathbf{r}_1)\, d\mathbf{r}_1\, d\mathbf{r}_2, \quad (236)$$

where $\mathscr{H}_{\text{elect}}$ is the electrostatic Hamiltonian of the interacting unpaired electrons which are separated by $\mathbf{r}_{12} = \mathbf{r}_1 - \mathbf{r}_2$; $\Psi(\mathbf{r}_1, \mathbf{r}_2)$ is the wavefunction describing electrons 1 and 2 on molecules A and B, respectively, and $\Psi(\mathbf{r}_2, \mathbf{r}_1)$ is the wavefunction with the electron coordinates reversed. J_{12} is clearly a function of the distance between paramagnetic molecules, and in solutions it is, therefore, a function of time. Van Vleck[28] has demonstrated that the spin exchange Hamiltonian may be written as in Eq. (70). Although spin exchange is in reality an electrostatic coupling, it can be written as if it were just a magnetic coupling given by the scalar product between the electron spin magnetic moments.

The effects of spin exchange can only be seen when the exchanged electron spins end up in a magnetic environment different from their original environment, i.e. when the electron spin energy levels on different molecules are not quite identical. The molecules could have different g-values or chemical shifts and thereby create different magnetic environments for the spins, but we shall consider the case where the different magnetic environments for the electron spins are created by the nuclear spins which can be in different states on different molecules. Therefore, we will study spin exchange in the presence of nuclear hyperfine interactions. Although we could use the transformation described at the end of the last section to convert the nuclear hyperfine interaction into a purely secular form, we shall instead treat the simpler case in which the non-secular hyperfine terms are neglected, i.e. the pure spin Hamiltonian used will be that of Eq. (223).

The problem of spin exchange in solution was first interpreted by Pake and Tuttle.[55] Experimentally, spin exchange effects are seen in the dependence of the ESR spectrum on radical concentration. As the radical concentration is increased, the hyperfine lines start to broaden and shift toward the center of the spectrum, eventually coalescing into one line, which then narrows with increasing radical concentration before reaching a limiting value. Increasing the radical concentration means increasing the radical collision frequency, and its effect on the ESR spectrum reflects the fact that spin exchange is a collision process. The rapidity with which the radicals diffuse through the solution affect the exchange frequency and the spectrum can be used to study the diffusion constant; we shall relate linewidths to diffusion constants.

In the work that follows, we will assume that $\mathcal{H}_{SL}$ is given entirely by the exchange term in Eq. (70), and that the magnetic dipole-dipole interaction between the electron moment and both neighboring nuclear and electronic moments is averaged to zero by the rapid motion of the free radical in solution.

In Section VII, A we study the effect of spin exchange upon the hyperfine spectrum if the correlation time for $\mathcal{H}_{SL}$ is short compared to that for S_+. In Section VII, B, we extend the treatment to the case where this

[55] G. E. Pake and T. R. Tuttle, Jr., *Phys. Rev. Lett.* **3**, 423 (1959).

condition holds less well; this corresponds to the extended Mori treatment presented in preceding sections. In Section VII, C we discuss the diffusion constant obtained from spin exchange measurements. And finally, in Section VII, D we discuss the situation in which the exchange interaction is very large compared to the inverse time of contact of two radicals.

A. Spin Exchange with Hyperfine Lines

We shall use the pure spin Hamiltonian, $\mathscr{H}_S$, given by Eq. (223), i.e. the one that describes an electron spin with Larmor frequency, ω_0, and secular isotropic hyperfine interaction, $a\sum I_{z_j}S_{z_j}$. The spin-lattice interaction, $\mathscr{H}_{SL}$, is taken to be the Heisenberg exchange interaction in Eq. (70);

$$\mathscr{H}_{SL} = \tfrac{1}{2}\sum_{j,k}^{j\neq k} J_{jk}(\mathbf{r}_{jk})\,\mathbf{S}_j\cdot\mathbf{S}_k, \tag{237}$$

where J_{jk} is the exchange interaction which is a function of interradical displacement, $\mathbf{r}_{jk}$. Since J_{jk} depends on $\mathbf{r}_{jk}$, $\mathscr{H}_L$ does not commute with J_{jk}. We use the same "hyperfine spin components", $S_{+j}^{(m)}$, defined in Eqs. (224) and (225), as the elements of the Mori set of variables. **A**. The matrices $\langle \mathbf{AA}^+\rangle$ and $i\Omega$ are given in Eqs. (226) and (228), respectively; and Eqs. (227) and (229)–(230b) are relevant to the present problem. (Remember that $S_+^{(m)}$ is defined in terms of S_{+j} and P_{mj}.) Note that

$$[\mathscr{H}_L, S_+^{(m)}] = [\mathscr{H}_S, S_z^{(m)}] = 0 \tag{238a}$$

and

$$[\mathscr{H}_S, S_+^{(m)}] = i\Omega_m S_+^{(m)}. \tag{238b}$$

Furthermore,

$$[\mathbf{S}_j\cdot\mathbf{S}_k, \sum_l S_{+l}P_{ml}] = (S_{+j}S_{zk} - S_{zj}S_{+k})(P_{mj} - P_{mk}), \tag{239}$$

where $j\neq k$.

In the limit of very small $\mathscr{H}_{SL}$, i.e. the limit in which only the lowest order terms in $\mathscr{H}_{SL}$ are retained, the Mori projection operator and $\mathscr{H}_{SL}$ in the time evolution operator can be dropped as we discussed near Eqs. (139) and (140). Taking $\mathscr{H}_{SL}$ to be the Heisenberg exchange interaction, we write the mth component of $\dot{\boldsymbol{\alpha}}(t_p)$, specified in Eq. (230b), as

$$\dot{\boldsymbol{\alpha}}(t_p)_m = \sum_{m'\neq m}\sum_j\sum_{k\neq j} J_{jk}(t)\,S_{+j}S_{z_k}\mathrm{i}(P_{mj}e^{i\Omega_m t} - P_{mk}e^{i\Omega_{m'}t}), \tag{240}$$

where $J_{jk}(t)$ is defined by

$$J_{jk}(t) = e^{i\mathscr{H}_L t}J_{jk}e^{-i\mathscr{H}_L t}. \tag{241}$$

The time dependence of $J_{jk}(t)$ arises entirely from lattice motions.

We next calculate the effective memory function matrix, $(\mathcal{K}(t))_{mm'} = K_{mm'}(t)\exp - i\Omega_m t$, by means of Eq. (231a) and the results in the precedding paragraph. For most organic radicals, $(\Omega_{m'} - \Omega_m)$, which is proportional to the hyperfine constant, is small enough that the J_{jk} correlation function will decay before terms involving the hyperfine frequency change appreciably. Thus, the exponential factors involving hyperfine interactions can be neglected. We note that

$$\langle P_{mj} P_{m'j} \rangle = \delta_{mm'}/(2I+1) \tag{242a}$$

and

$$\langle P_{mj} P_{m'k} \rangle = 1/(2I+1)^2; \tag{242b}$$

therefore,

$$K_{mm}(t)\, e^{-i\Omega_m t} \cong I(2I+1)^{-1} \sum_k \langle J_{lk} J_{lk}(t) \rangle. \tag{243}$$

and the off-diagonal element is

$$K_{mm'}(t)\, e^{-i\Omega_m t} = -\tfrac{1}{2}(2I+1)^{-1} \sum_k \langle J_{lk} J_{lk}(t) \rangle. \tag{244}$$

The exchange integral depends on overlap of the electronic wavefunctions and falls off rapidly with increasing intermolecular separation; therefore, we expect the correlation function of the J_{lk}'s to fall to zero in a time of the order of molecular collision times. This time scale is much shorter than that for electron spins, so we can assume that the relaxation matrix $\hat{\mathcal{K}}$ in the Mori theory is independent of frequency, i.e.

$$\hat{\mathcal{K}}_{mm} = I(2I+1)^{-1} \int_0^\infty \sum_k \langle J_{lk}(t)\, J_{lk} \rangle\, dt \tag{245}$$

and

$$\hat{\mathcal{K}}_{mm'} = -\tfrac{1}{2}(2I+1)^{-1} \int_0^\infty \sum_k \langle J_{lk}(t)\, J_{lk} \rangle\, dt. \tag{246}$$

It is convenient to define a correlation time τ_J such that

$$\tau_J \equiv \int_0^\infty \langle J_{lk}(t)\, J_{lk} \rangle\, dt / \langle J_{lk}^2 \rangle; \tag{247}$$

this is fundamentally τ_1 defined in Section V. We define *the total exchange*, $\mathcal{J}$, on a given particle as

$$\mathcal{J}^2 \equiv \sum_k \langle J_{lk}^2 \rangle. \tag{248}$$

Then Eqs. (245) and (246) become

$$\hat{\mathcal{K}}_{mm}(0) = I(2I+1)^{-1}\, \mathcal{J}^2 \tau_J, \tag{249}$$

$$\hat{\mathcal{K}}_{mm'}(0) = -\tfrac{1}{2}(2I+1)^{-1}\, \mathcal{J}^2 \tau_J. \tag{250}$$

It is convenient to introduce an exchange frequency, ω_{ex}, defined as

$$\omega_{ex} = \mathscr{J}^2 \tau_J/(2I+1). \tag{251}$$

The assumption that we can use frequency-independent relaxation matrices is equivalent to the condition,

$$\omega_{ex}\tau_J = \mathscr{J}^2 \tau_J^2 (2I+1)^{-1} \ll 1. \tag{252}$$

We have also made use of the approximation

$$a^2 \tau_J^2 \ll 1. \tag{253}$$

Equation (252) also follows from the requirement, stated above, that only lowest order terms in $\mathscr{H}_{SL}$ be retained.

The parameter $\mathscr{J}^2$ is concentration-dependent, but the exact dependence is difficult to specify in terms of the exchange integral, since the exchange integral is a microscopic quantity dependent on intermolecular separation. We can separate out the *major concentration dependence* of $\mathscr{J}^2$ as

$$\mathscr{J}^2 \equiv J^2 x, \tag{254}$$

where x is the mole fraction of the radical in solution, and where we expect J^2 to be only slightly concentration-dependent and of the order of the square of the exchange integral, $\langle J_{12}^2(r_0)\rangle$, where r_0 corresponds to a typical value of the intermolecular separation between two radicals in contact. If one assumes the commonly used model that the exchange interaction is given by[26a]

$$J_{12} = \begin{cases} 0, & r > r_0 \\ J_0, & r < r_0, \end{cases} \tag{255}$$

then we can write

$$\mathscr{J}^2 = J_0^2 \tau_{co}/\tau_{en}, \tag{256}$$

where the contact time, τ_{co}, is the mean time that the interradical separation is within the range $r < r_0$, and the encounter time, τ_{en}, is the mean time between radical–radical encounters. Although this approach allows for the fact that the radical–radical interaction may not be diffusion-controlled, it is model-dependent, whereas Eqs. (247) and (252) are not. The requirement stated above, that only lowest order terms in $\mathscr{H}_{SL}$ are to be retained, implies that the fastest relevant process is the breakup of exchanging pairs, i.e. the time τ_{co}, and $\tau_J \approx \tau_{co}$. Thus the probability of a spin flip during an encounter is of the order of $J_0^2 \tau_J^2$ where

$$J_0^2 \tau_J^2 \ll 1. \tag{257a}$$

Note this is quite different, and more demanding, than the condition in

Eq. (252). What if $\mathscr{H}_{SL}$, i.e. J_0, is very large and

$$J_0^2 \tau_J^2 \gg 1? \tag{257b}$$

Then the expression for $\dot{\boldsymbol{\alpha}}(t_p)$ in Eqs. (240) and (241) is no longer correct; higher order terms in $\mathscr{H}_{SL}$ must be kept and the projection operator must be kept in the time evolution operator. We leave this discussion for the next section.

The coupled linear transport equations, obtained by Mori theory, in the case discussed above, are

$$\frac{d}{dt}\begin{bmatrix} \vdots \\ \overline{S_+^{(m)}(t)} \\ \vdots \\ \overline{S_+^{(m')}(t)} \\ \vdots \end{bmatrix} = -\begin{bmatrix} -i\Omega_1 + I\omega_{ex} & -\frac{1}{2}\omega_{ex} & \cdots \\ -\frac{1}{2}\omega_{ex} & -i\Omega_2 + I\omega_{ex} & \\ \vdots & & \ddots \end{bmatrix}\begin{bmatrix} \vdots \\ \overline{S_+^{(m)}(t)} \\ \vdots \\ \overline{S_+^{(m')}(t)} \\ \vdots \end{bmatrix}, \tag{258}$$

where the bar indicates the instantaneous average.

These linear equations are similar to those obtained by others and have been discussed elsewhere.[56–58] They can be readily solved by numerical techniques, but the most interesting features can be understood by looking at the simple case in which $I = 1/2$; then Eq. (258) reduces to two coupled linear transport equations. In fact, Eq. (258) takes on the form of Eqs. (23) in which the superscripts 1 and 2 now refer to the states $m = -1/2$ and $+1/2$, respectively, and the following identifications can be made:

$$\hat{\mathscr{K}}_{mm'}(0) = -\tfrac{1}{2}\omega_{ex}, \qquad m' \neq m, \tag{259a}$$

$$\hat{\mathscr{K}}_{mm}(0) = \tfrac{1}{2}\omega_{ex}, \tag{259b}$$

where the effective exchange frequency, ω_{ex}, is defined in Eq. (251). The details of this problem have been worked out and discussed in Section II, D; we will not repeat this discussion here but will summarize a few results of special interest.

For slow exchange (Eq. 35),

$$\omega_{ex} = \mathscr{J}^2\tau_J/2 \ll a. \tag{260}$$

This relation states that the effective exchange frequency is much less than

[56] M. P. Eastman, R. G. Kooser, M. R. Das, and J. H. Freed, *J. Chem. Phys.* **51**, 2690 (1969).
[57] J. D. Currin, *Phys. Rev.* **126**, 1995 (1962).
[58] C. S. Johnson, *Mol. Phys.* **12**, 25 (1967).

the hyperfine splitting frequency. Under these conditions the spectrum, $I(\omega)$, is given by Eqs. (31)–(39) or, in the present notation, by

$$I(\omega) = \frac{\omega_{\text{ex}}}{2}\left\{\frac{1-2a^{-1}(\omega-\Omega_1-\omega_{\text{ex}}^2/a)}{\omega_{\text{ex}}^2+(\omega-\Omega_1-\omega_{\text{ex}}^2/a)}+\frac{1+2a^{-1}(\omega-\Omega_2+\omega_{\text{ex}}^2/a)}{\omega_{\text{ex}}^2+(\omega-\Omega_2+\omega_{\text{ex}}^2/a)}\right\}. \tag{261}$$

In this slow exchange limit, the spectrum is seen to consist of two nearly Lorentzian lines centered near the hyperfine frequencies, Ω_1 and Ω_2. The linewidth of the Lorentzian contribution is $\omega_{\text{ex}} = \frac{1}{2}\mathscr{J}^2\tau_J$, which increases with increasing radical concentration. The shift of the lines toward the center is second order, but each line also has a dispersion-like contribution of magnitude $(\mathscr{J}^2\tau_J)/(2a)$. In the limit of $\mathscr{J}^2\tau_J^2 \ll 1$, Eastman *et al.*[56] and Jones[59] have confirmed the expected linear dependence of the linewidth on concentration. [See Eq. (256).]

In the fast exchange limit, [Eq. (40)],

$$\omega_{\text{ex}}^2 = (\mathscr{J}^2\tau_J)^2/4 \gg a^2, \tag{262}$$

and the spectrum, $I(\omega)$, is given by Eqs. (31)–(34) and (40)–(42); in the present notation

$$I(\omega) = \frac{2|a^2/4\omega_{\text{ex}}|}{(a^2/4\omega_{\text{ex}})^2+(\omega-\omega_0)^2} - \frac{(a^2/2\omega_{\text{ex}}^2)\,\omega_{\text{ex}}}{(\omega_{\text{ex}})^2+(\omega-\omega_0)^2}. \tag{263}$$

In this limit, the spectrum consists of a broad and a narrow Lorentzian, both centered at the Larmor frequency, ω_0, which is the mean of the original hyperfine frequencies. The narrow Lorentzian dominates since it has greater integrated intensity and is concentrated near ω_0; its width, $\frac{1}{8}a^2/\mathscr{J}^2\tau_J$, narrows with increasing radical concentration in agreement with experimental observation. The broad Lorentzian has negative intensity, $2(a/\mathscr{J}^2\tau_J)^2$, which pulls the wings of the total spectrum closer to the base line; the intensity decreases with increasing radical concentration and the width, $\mathscr{J}^2\tau_J/2$, increases. These results are identical with the results of Johnson[58] in this limit.

By using the techniques given in Section VI, B we could develop the theory of exchange for nonsecular $\mathscr{H}_S$.

B. Extended Theory of Spin Exchange

The theory we developed in the last section predicts the effect of spin exchange on the ESR spectrum in the limit of $\mathscr{J}^2\tau_J^2 \ll 1$, but it does not exhibit an explicit dependence of τ_J upon the translational diffusion

[59] M. T. Jones, *J. Chem. Phys.* **38**, 2892 (1963); see also, N. Edelstein, N. Kwok, and A. H. Maki, *J. Chem. Phys.* **41**, 3473 (1964); W. Z. Plachy and D. Kivelson, *J. Chem. Phys.* **47**, 3312 (1967).

constant. By using an increased variable set, we can obtain a more general result and can also obtain an explicit relation between τ_J and the diffusion constant. In analogy with the treatment in Section V, C, the theory of spin relaxation can be improved at shorter times by expanding the variable set, **A**, to include the variables, $\dot{S}_+^{(m)}$, in addition to our original set, $S_+^{(m)}$. The variable $\overline{\dot{S}_+^{(m)}(t)}$ is likely to vary more rapidly than $\overline{S_+^{(m)}(t)}$ but, hopefully, more slowly than $\overline{\ddot{S}_+^{(m)}(t)}$; therefore, we neglect higher derivatives as variables.

We use the same Hamiltonian as we did in the last subsection, i.e. that which contains a Zeeman term, the secular part of the isotropic hyperfine interaction, the exchange interaction, and $\mathscr{H}_L$. As in Section V, C, we do not use $\dot{S}_+^{(m)}$ as a variable, but rather $\mathscr{S}_+^{(m)}$, the component of $\dot{S}_+^{(m)}$ that is orthogonal to $S_+^{(m)}$:

$$\dot{\mathscr{S}}_+^{(m)} \equiv i[\mathscr{H}_{SL}, S_+^{(m)}]. \tag{264}$$

Since $\mathscr{H}_{SL}$ commutes with the total S_+, the set of variables, $\dot{\mathscr{S}}_+^{(m)}$, are not linearly independent, and lead to a singular $\langle \mathbf{A}^+\mathbf{A}\rangle$ matrix. This singularity is removed by including only $2I$ of the $(2I+1)$ derivatives, $\dot{\mathscr{S}}_+^{(n)}$. Thus we use $\dot{\mathscr{S}}_+^{(m)}$'s where m takes on $2I$ of the possible $2I+1$ values. Our variable set is then

$$\tilde{\mathbf{A}} = (S_+^{(1)}, \dot{\mathscr{S}}_+^{(1)}, \ldots S_+^{(n)}, \dot{\mathscr{S}}_+^{(n)}, \ldots S_+^{(2I+1)}), \tag{265a}$$

where the tilde indicates the transpose.

The essence of the problem can be understood by once again studying the simple case where $I = 1/2$. Equation (265a) is now

$$\tilde{\mathbf{A}} = (S^{(1/2)}, \dot{\mathscr{S}}_+^{(1/2)}, S_+^{(-1/2)}), \tag{265b}$$

where the superscripts represent the group of molecules with nuclear spin $m_I = +1/2$ or $-1/2$, respectively. With the aid of Eq. (239) we see that $\dot{\mathscr{S}}_+^{(1/2)}$ is

$$\dot{\mathscr{S}}_+^{(1/2)} = i\sum_l \sum_k J_{lk}(S_{+l}S_{z_k} - S_{z_l}S_{+k})\, P_{(1/2)l}\, P_{(-1/2)k} \tag{266}$$

and $\dot{\mathscr{S}}^{(-1/2)} = -\dot{\mathscr{S}}_+^{(1/2)}$. Clearly, as indicated above, $\dot{\mathscr{S}}_+^{(-1/2)}$ is not independent of $\dot{\mathscr{S}}_+^{(1/2)}$ and so only one of them has been used as an independent variable. Since

$$\langle \dot{\mathscr{S}}_+^{(1/2)} \dot{\mathscr{S}}_-^{(1/2)} \rangle = \tfrac{1}{8} N \mathscr{J}^2, \tag{267}$$

where $\mathscr{J}^2$ is defined in Eq. (254), it follows that $(\mathbf{AA}^+\rangle$ is

$$\langle \mathbf{AA}^+ \rangle = \tfrac{1}{4}N \begin{bmatrix} 1 & 0 & 0 \\ 0 & \tfrac{1}{2}\mathscr{J}^2 & 0 \\ 0 & 0 & 1 \end{bmatrix}. \tag{268}$$

The calculation of the frequency matrix, $i\Omega$, requires the evaluation of the time derivative of $\dot{\mathscr{S}}_{+}^{(1/2)}$,

$$(d/dt)\,\dot{\mathscr{S}}_{+}^{(1/2)} = i[\mathscr{H}_{\mathrm{L}}, \dot{\mathscr{S}}_{+}^{(1/2)}] + i[\mathscr{H}_{\mathrm{S}}, \dot{\mathscr{S}}_{+}^{(1/2)}] + i[\mathscr{H}_{\mathrm{SL}}, \dot{\mathscr{S}}_{+}^{(1/2)}], \quad (269)$$

where

$$i[\mathscr{H}_{\mathrm{L}}, \dot{\mathscr{S}}_{+}^{(1/2)}] = \sum_{l}\sum_{k} \dot{J}_{lk}(S_{+l}S_{z_k} - S_{z_l}S_{+k})\,P_{(1/2)l}P_{(-1/2)k}. \quad (270)$$

The only element in $i\Omega$ which involves combinations not seen in previous sections is

$$\langle \ddot{\mathscr{S}}^{(1/2)}\dot{\mathscr{S}}_{+}^{(1/2)} \rangle = \tfrac{1}{4}N(i\omega_0 \tfrac{1}{2}\mathscr{J}^2) + \tfrac{1}{4}\sum_{l}\sum_{k} \langle J_{lk}\dot{J}_{lk}\rangle \langle P_{(1/2)l}P_{(-1/2)k}\rangle. \quad (271)$$

If we make the reasonable assumption that the average in the second term of Eq. (271) is zero, then the frequency matrix is

$$i\Omega = \begin{bmatrix} i\Omega_{1/2} & 1 & 0 \\ -\tfrac{1}{2}\mathscr{J}^2 & i\omega_0 & \tfrac{1}{2}\mathscr{J}^2 \\ 0 & -1 & i\Omega_{-1/2} \end{bmatrix}, \quad (272)$$

where $\Omega_{1/2} = \omega_0 + \tfrac{1}{2}a$ and $\Omega_{-1/2} = \omega_0 - \tfrac{1}{2}a$.

Next we must calculate $\dot{\boldsymbol{\alpha}}$ in order to evaluate the memory matrix, $\mathsf{K}(t)$. $\dot{\boldsymbol{\alpha}}$ in the present case is $\dot{\mathbf{A}}$ projected perpendicularly to $S_{+}^{(1/2)}$, $\dot{\mathscr{S}}_{+}^{(1/2)}$, and $S_{+}^{(-1/2)}$:

$$\dot{\boldsymbol{\alpha}} = \begin{bmatrix} 0 \\ \ddot{\mathfrak{S}}_{+} \\ 0 \end{bmatrix} \quad (273)$$

where $\ddot{\mathfrak{S}}_{+}$ is

$$\ddot{\mathfrak{S}}_{+} = \tfrac{1}{2}\mathscr{J}^2(S_{+}^{(1/2)} - S_{+}^{(-1/2)} + i[a\sum_{j} I_{z_j}S_{z_j}, i[\mathscr{H}_{\mathrm{SL}}, S_{+}^{(1/2)}]]$$
$$+ i[\mathscr{H}_{\mathrm{SL}}, i[\mathscr{H}_{\mathrm{SL}}, S_{+}^{(1/2)}]] + i[\mathscr{H}_{\mathrm{L}}, i[\mathscr{H}_{\mathrm{SL}}, S_{+}^{(1/2)}]]. \quad (274)$$

The memory function, $\mathsf{K}(t)$, takes the form

$$\mathsf{K}(t) = \frac{8\langle \ddot{\mathfrak{S}}_{+}(t_p)\,\ddot{\mathfrak{S}}_{-}\rangle}{N\mathscr{J}^2} \begin{bmatrix} 0 & 0 & 0 \\ 0 & 1 & 0 \\ 0 & 0 & 0 \end{bmatrix}, \quad (275)$$

where

$$\ddot{\mathfrak{S}}_{+}(t_p) \equiv e^{t(1-\P)i\mathscr{H}^{x}}\ddot{\mathfrak{S}}_{+}. \quad (276)$$

Our experience in Section V, C with the extended Mori theory applied to spin relaxation indicates that the present formulation of the spin exchange problem is valid only under certain restrictive conditions. The conditions arise from the fact that the transport equations, to be useful, must have frequency-independent coefficients, $\hat{\mathscr{K}}$; this condition applies only *if* $\hat{\mathscr{K}}(t)$, and hence $\langle \ddot{\mathfrak{S}}_+(t)\, \ddot{\mathfrak{S}}_- \rangle$, vary rapidly compared to $\overline{S_+^{(m)}(t)\, e^{-i\Omega_m t}}$ and $\overline{\mathscr{S}_+^{(m)}(t)\, e^{-i\Omega_m t}}$. Under these conditions, the relaxation matrix, $\hat{\mathscr{K}}$, is

$$\hat{\mathscr{K}} = \hat{\mathscr{K}}'' \begin{bmatrix} 0 & 0 & 0 \\ 0 & 1 & 0 \\ 0 & 0 & 0 \end{bmatrix}, \tag{277}$$

where

$$\hat{\mathscr{K}}'' = \int_0^\infty dt\, K''(t)\, e^{i\omega_0 t}, \tag{278}$$

$$\tau_{\mathrm{II}} = \hat{\mathscr{K}}''_{\mathrm{RE}}/\omega_{\mathrm{L}}^2, \tag{279}$$

and

$$\omega_{\mathrm{L}}^2 = 8\langle \ddot{\mathfrak{S}}_+ \ddot{\mathfrak{S}}_- \rangle / N\mathscr{J}^2; \tag{280}$$

ω_{L} is defined as in Eq. (158); τ_{II} is the correlation time analogous to that defined in Eq. (157), and the subscript II indicates the correlation time associated with the second derivative of S_+, in this case that associated with $\dot{J}$. The requirement on our solutions is again given by Eq. (159),

$$\omega_{\mathrm{L}}^2 \tau_{\mathrm{II}}^2 \ll 1. \tag{281}$$

The same arguments as those discussed in detail in Section V, D hold here, and it follows that Eq. (281) holds provided

$$(\mathscr{J}^2)^2 \ll \langle |[\mathscr{H}_{\mathrm{L}}, \mathscr{H}_{\mathrm{SL}}]|^2 \rangle, \tag{282a}$$

which is equivalent to Eq. (187a), and if

$$1 \gg (\omega_{\mathrm{L}}/\hat{\mathscr{K}}''_{\mathrm{RE}})^2, \tag{282b}$$

which is equivalent to Eq. (198). Equation (282b) allows us to approximate ω_{L} by ω_J where

$$\omega_{\mathrm{J}}^2 \equiv \sum_k \langle \dot{J}_{1k}^2 \rangle / \sum_j \langle J_{1j}^2 \rangle; \tag{283}$$

the sums run over all particles except particle 1. In keeping with these approximations, as explained in Section V, E, we can also ignore the projected time dependence in $K''(t)$, and we can approximate τ_{II} by τ'_{II} where

$$\tau'_{\mathrm{II}} = \left\{ \sum_{k,j,l} \int_0^\infty dt\, \langle (\exp(i\mathscr{H}_{\mathrm{L}}^{\times} t)\, \dot{J}_{jl})\, \dot{J}_{1k} \rangle \right\} \Big/ \omega_{\mathrm{J}}^2 \mathscr{J}^2, \tag{284}$$

We can now summarize the conditions in Eqs. (282ab) in the more useful form:

$$\omega_J^2(\tau_{II}')^2 \ll 1, \tag{285}$$

$$\mathscr{J}^2 \ll \omega_J^2, \tag{286}$$

and we also require the additional condition

$$a^2 \ll \omega_J^2. \tag{287}$$

The time dependence in Eq. (284) is valid only if $\mathscr{H}_{SL}$ is small, and only lowest order terms in $\mathscr{H}_{SL}$ are retained.

The three coupled transport equations which are relevant to the present problem can be written as

$$\frac{d}{dt}\begin{bmatrix} \overline{S_+^{(1/2)}(t)} \\ \overline{\dot{\mathscr{S}}_+^{(1/2)}(t)} \\ \overline{S_+^{(-1/2)}(t)} \end{bmatrix} = -\begin{bmatrix} -i\Omega_{1/2} & -1 & 0 \\ +\frac{1}{2}\mathscr{J}^2 & -i\omega_0 + \omega_J^2\tau_k & -\frac{1}{2}\mathscr{J}^2 \\ 0 & +1 & -i\Omega_{-1/2} \end{bmatrix} \begin{bmatrix} \overline{S_+^{(1/2)}(t)} \\ \overline{\dot{\mathscr{S}}_+^{(1/2)}(t)} \\ \overline{S_+^{(-1/2)}(t)} \end{bmatrix}, \tag{288}$$

where the bar indicates the instantaneous average. The solution to Eq. (288) is the sum of three time exponentials with characteristic frequencies equal to the eigenvalues of the transport matrix. These eigenvalues are the solutions of a cubic characteristic equation, which cannot be solved exactly; however, we shall present approximate solutions in the limits of slow and fast exchange (details of these calculations are given in Ogan[60]).

In order to connect the present extended three-variable theory with the two-variable theory in the last section, we identify τ_J as

$$\tau_J = (\omega_J^2\tau'_{II})^{-1}; \tag{289a}$$

consequently, the effective exchange frequency, ω_{ex}, defined in Eq. (251) becomes

$$\omega_{ex} = \mathscr{J}^2/\omega_J^2\tau'_{II}. \tag{289b}$$

We shall see below that these identifications are meaningful and readily satisfied.

In the slow exchange limit

$$\omega_{ex}^2 = (\mathscr{J}^2/\omega_J^2\tau'_{II}) \ll a^2, \tag{290}$$

and the eigenvalues $(\lambda_1, \lambda_2, \lambda_3)$ of the transport matrix in Eq. (288) are

[60] K. Ogan, Ph.D. Thesis, Univ. of California, Los Angeles, 1972.

given by the approximate relations

$$\lambda_1 = i(\Omega_{1/2} + \omega_{\mathrm{ex}}^2/a) - \omega_{\mathrm{ex}}, \tag{291a}$$

$$\lambda_2 = i(\Omega_{-1/2} - \omega_{\mathrm{ex}}^2/a) - \omega_{\mathrm{ex}}, \tag{291b}$$

$$\lambda_3 = i\omega_0 - \tau_{\mathrm{J}}^{-1}. \tag{291c}$$

These solutions correspond to two lines near the hyperfine frequencies, $\Omega_{1/2}$, $\Omega_{-1/2}$, each of width ω_{ex}, and a broad line centered at the Larmor frequency, ω_0, with width, τ_J^{-1}. Each of the two sharp lines are shifted by an amount $4\omega_{\mathrm{ex}}^2/a$ towards the central frequency, ω_0. The effect of the broad line would be noticeable only in the far wings, and this three-variable theory reduces to the two-variable theory except in the spectral wings; in the wings, as discussed before, this three-variable theory is better than the two-variable theory. It can also be seen that the identification in Eqs. (289ab) is consistent.

In the fast exchange limit

$$\omega_{\mathrm{ex}}^2 = (\mathscr{J}^2/\omega_{\mathrm{J}}^2 \tau_{\mathrm{II}}')^2 \gg a^2, \tag{292}$$

and the eigenvalues of the transport matrix in Eq. (288) are

$$\lambda_1 = i\omega_0 - \tfrac{1}{8}a^2/\omega_{\mathrm{ex}}, \tag{293a}$$

$$\lambda_2 = i\omega_0 - 2\omega_{\mathrm{ex}}, \tag{293b}$$

$$\lambda_3 = i\omega_0 - \tau_{\mathrm{J}}^{-1}. \tag{293c}$$

These solutions correspond to three lines at the Larmor frequency; a very broad line with width τ_J^{-1}, a second broad line with width $2\omega_{\mathrm{ex}}$ (which corresponds to the broad line in Section VII, A), and a narrow line with width $a^2/8\omega_{\mathrm{ex}}$ (which corresponds to the narrow line in Section VII, A). The very broad line with width τ_J^{-1} is probably not observable, and the three-variable theory reduces properly to the two-variable theory presented in the last section. The identifications in Eqs. (289ab) are still valid.

One could solve for intermediate cases where $\omega_{\mathrm{ex}} \approx a$, but we shall not do so here. We could also account for the effect of nonsecular isotropic hyperfine interactions by the procedures of Section VI, B.

C. Physical Significance of Extended Theory

The extended theory described in the last section is valid only if the conditions in Eqs. (285)–(287) apply and $\mathscr{H}_{\mathrm{SL}}$ is small. We wish to discuss the physical implications of these conditions and then study the way in which the results can be used to determine diffusion constants.

The frequency ω_J *is an equilibrium property which measures the rate of change of the exchange integral.* If we assume that $J_{lm}(r_{lm})$ is isotropic, then

$$\dot{J}_{lm} = (\partial J_{lm}/\partial r_{lm})\,\dot{r}_{lm}. \tag{294}$$

We can now write ω_J^2 in the suggestive form

$$\omega_J^2 = Q\langle \dot{r}_{12}^2\rangle r_0^{-2}, \tag{295}$$

where the dimensionless number, Q, is

$$Q = r_0^2\langle(\partial J_{12}/\partial r_{12})^2\rangle/\langle J_{12}^2\rangle \tag{296}$$

and r_0 is an arbitrary distance chosen to be the "contact" distance between two "colliding" free radicals. Since J_{12} varies strongly with r_{12}, we expect Q to be a rather large number of the order of 40. r_0 might typically be about 4Å and $\langle \dot{r}_{12}^2\rangle$ can be evaluated by classical statistical mechanics. With these results we obtain the approximate value $\omega_J \approx 3\times 10^{12}\,\text{sec}^{-1}$. τ'_{II} is the correlation time for the rate of change of the exchange integral and, as discussed below, it is probably very short, perhaps of the order of 10^{-14} to 10^{-13} sec. It thus seems very likely that $\omega_J^2(\tau'_{\text{II}})^2 \ll 1$ as required by Eq. (285). At the same time, for free radicals such as the nitroxides, the hyperfine constant $a \approx 3\times 10^8\,\text{sec}^{-1}$, and it therefore appears that $a^2 \ll \omega_J^2$ as required by Eq. (287). Exchange integrals, J_0, of radicals in contact may well be in the range 10^{10}–$10^{13}\,\text{sec}^{-1}$, which implies that at least for dilute solutions and, in most cases, even for concentrated solutions, $\mathscr{J}^2 \ll \omega_J^2$ as required by Eq. (286). [See Eqs. (254) and (256) for definitions of $\mathscr{J}^2$ and J_0^2.]

As we saw in the last section, the "fast" and "slow" exchange regimes are those for which ω_{ex}^2 is large and small, respectively, compared to the hyperfine frequency, a, where

$$\omega_{\text{ex}} = (\mathscr{J}^2/2\omega_J^2\tau'_{\text{II}}). \tag{297}$$

In all cases the exchange frequency, ω_{ex}, can be determined from the line widths. We can define an "exchange diffusion constant," D_J, in terms of ω_{ex}:

$$D_J \equiv (\mathscr{J}^2)^2[2\omega_{\text{ex}}\langle(\partial J_{12}/\partial r_{12})^2\rangle]^{-1}. \tag{298}$$

Under appropriate conditions, D_J reduces to the usual translational diffusion constant, D. To see this, we rewrite Eq. (284) making use of Eqs. (294) and (283):

$$\tau'_{\text{II}} = \sum_{i,j,l}\int_0^\infty dt\left\langle\left[\frac{dJ_{1i}(t)}{dr_{1i}}\frac{dJ_{jl}}{dr_{jl}}\right][\dot{r}_{1i}(t)\dot{r}_{jl}]\right\rangle\Big/\sum_k\left\langle\left(\frac{dJ_{1k}}{dr_{1k}}\right)^2\right\rangle\langle\dot{r}_{1k}^2\rangle \tag{299}$$

We can neglect three- and four-body interactions, since the dependence upon spins, which has already been factored out, insures that such correlations vanish. If dJ/dr varies slowly compared to $\dot{r}$, then

$$\tau'_{\text{II}} \approx \int_0^\infty dt \langle \dot{r}_{12}\dot{r}_{12}(t)\rangle / \langle \dot{r}_{12}^2 \rangle. \tag{300}$$

We can define a relative diffusion constant, D_{rel},

$$D_{\text{rel}} = \int_0^\infty dt \langle \dot{x}_{12}\dot{x}_{12}(t)\rangle \tag{301}$$

in which case

$$\tau'_{\text{II}} = D_{\text{rel}} M/(k_{\text{B}} T), \tag{302}$$

where M is the mass of the paramagnetic particle, k_{B} is the Boltzmann constant, and T the temperature. If Eq. (302) is inserted back into Eqs. (297) and (298), then

$$D_J \approx D_{\text{rel}}. \tag{303}$$

D_{rel} is closely related to the self- or particle diffusion constant.[61]

If dJ/dr varies rapidly compared to $\dot{r}$, then Eq. (299) reduces to

$$\tau'_{\text{II}} = \int_0^\infty dt \left\langle \frac{dJ_{12}(t)}{dr_{12}} \frac{dJ_{12}}{dr_{12}} \right\rangle \Big/ \left\langle \left(\frac{dJ_{12}}{dr_{12}} \right)^2 \right\rangle. \tag{304}$$

In this case

$$D_J \approx \langle \dot{r}_{12}^2 \rangle \tau'_{\text{II}} \tag{305}$$

and the "exchange diffusion" constant has little relation to the usual translational diffusion constant, D; in fact, under these conditions, $D \ll D_J$.

Which conditions do we expect to hold? If $J(r)$ varies as r^{-n} with $n \geqslant 9$, then dJ/dr should vary rapidly compared to $J(r)$, i.e. $\tau'_{\text{II}} \ll \tau_{\text{I}}$. The contact time, $\tau_{\text{co}} \approx \tau_J$, is probably less than 10^{-10} sec. Furthermore, under these conditions $(dJ(r)/dr)$ behaves similarly to the intermolecular forces, and τ'_{II} should be comparable to "shear times" or correlation times associated with microscopic stress tensors; these times are probably 10^{-13} sec or less. On the other hand, correlation times for $\dot{r}$, as determined from diffusion constants, are often even shorter, but this is probably because the velocity correlation function has negative contributions[62]; the frequency of fluctuation of velocities may well be considerably less than that for forces or for $dJ(r)/dr$. It seems more than likely that the expressions in Eqs. (304) and (305) are more applicable than that of Eq. (303), although in many cases intermediate situations probably apply.

[61] M.-K. Ahn, S. J. K. Jensen, and D. Kivelson, *J. Chem. Phys.* **57**, 2940 (1972).
[62] D. Kivelson, submitted (1974).

In conclusion, the association of D_J with the usually encountered translational diffusion constants is probably *never* justified.

Note that the extended Mori theory is required to give information concerning the diffusion constant, D_J.

D. Further Extension of the Spin Exchange Problem

Throughout Sections VII, A–VII, C we have referred to the requirement that $\mathscr{H}_{\mathrm{SL}}$ be small. This, as indicated in Eq. (257a), is equivalent to $J_0^2\tau_J^2 \ll 1$ where τ_J is roughly the encounter time and J_0 is the exchange integral at encounter. If $\mathscr{H}_{\mathrm{SL}}$ is large, $J_0^2\tau_J^2 \gg 1$ and the probability of a spin flip is very large because the spin flip period, J_0^{-1}, is small compared to a single encounter time, τ_J. When $\mathscr{H}_{\mathrm{SL}}$ is large, it can be approximated by $J_0\mathbf{S}_1\cdot\mathbf{S}_2$ during an encounter, and plays an important role in the time dependence of $\dot{\alpha}(t_p)$ in Eq. (240). If we tentatively ignore the projection operator in the propogator for $\dot{\alpha}(t_p)$, and assume that

$$a^2 \ll J_0^2, \tag{306}$$

then we can approximate $\dot{\alpha}(t_p)_m$, not by the expression in Eq. (240), but by

$$\dot{\alpha}(t_p)_m = \tfrac{1}{2}\sum_l \sum_{k=1} J_{lk}(t)\,J_{lk}\cdot[S_{+l}^{\#}(t)\,S_{zl}^{\#}(t) - S_{+k}^{\#}(t)]\,[P_{mj}-P_{ml}], \tag{307}$$

where $J_{lk}(t)$ is given by Eq. (241) and

$$S_{\alpha_j}^{\#}(t) \approx \exp(iJ_0\sum_m \mathbf{S}_j\cdot\mathbf{S}_m t)S_{\alpha_j}\exp(-iJ_0\sum_m \mathbf{S}_j\cdot\mathbf{S}_m t). \tag{308}$$

We have approximated $\mathscr{H}_{\mathrm{SL}}$ by $\frac{1}{2}J_0\sum_{i,j}\mathbf{S}_i\cdot\mathbf{S}_j$ where J_0 is a constant; this approximation is quite good since this is its value when the free radicals are in contact, and if the free radicals are not in contact, it makes no difference since then $J_{lk} = 0$ in Eq. (307). We have also made the approximation that $\mathscr{H}_{\mathrm{S}}$, $\mathscr{H}_{\mathrm{SL}}$, and $\mathscr{H}_{\mathrm{L}}$ all commute with each other in the time propogating exponentials. Note that whereas $\overline{S_{+l}(t)}$ may be slowly varying, $S_{+l}(t)$ need not be. A straight-forward calculation, as discussed in Appendix C, yields

$$K_{mm}(t)\,e^{-i\Omega_m t} = I(2I+1)^{-1}\sum_k \langle J_{lk}(t)\,J_{lk}\rangle\,e^{iJ_0 t}. \tag{309}$$

If we assume that $\langle J_{lk}(t)\,J_{lk}\rangle = \langle J_{lk}^2\rangle\,e^{-t/\tau_J}$, then upon integration of Eq. (309),

[63] See Ref. 17, p. 113.

[64] H. Y. Carr and E. M. Purcell, *Phys. Rev.* **94**, 63 (1954).

we get

$$(\hat{\mathscr{K}}_{\mathrm{RE}})_{mm} = [I/(2I+1)]\,\mathscr{J}^2\tau_J/(1+J_0^2\tau_J^2), \tag{310}$$

$$(\hat{\mathscr{K}}_{\mathrm{IM}})_{mm} = [1/(2I+1)]\,\mathscr{J}^2\tau_J^2 J_0/(1+J_0^2\tau_J^2). \tag{311}$$

It is useful at this point to redefine the exchange frequency as

$$\omega_{\mathrm{ex}} = \tfrac{1}{2}\mathscr{J}^2\tau_J/(1+J_0^2\tau_J^2). \tag{312}$$

In terms of the model in Eq. (256), this relation becomes

$$\omega_{\mathrm{ex}} = \tfrac{1}{2}[J_0^2\tau_J^2/(1+J_0^2\tau_J^2)]\,\tau_{\mathrm{en}}^{-1} \tag{313}$$

where τ_{en} is the time between radical encounters and τ_J is approximately the time of contact. In the limit $J_0^2\tau_J^2 \ll 1$, studied in previous sections, i.e. the limit in which $\mathscr{H}_{\mathrm{SL}}$ is very small,

$$\omega_{\mathrm{ex}} = J_0^2\tau_J^2/2\tau_{\mathrm{en}}. \tag{314}$$

On the other hand, in the limit $J_0^2\tau_J^2 \gg 1$, i.e. the limit in which $\mathscr{H}_{\mathrm{SL}}$ is very large,

$$\omega_{\mathrm{ex}} = \tfrac{1}{2}\tau_{\mathrm{en}}^{-1}. \tag{315}$$

The entire treatment is valid only if $\mathscr{J}^2\tau_J^2 \ll 1$, or in terms of the model used, if

$$J_0^2\tau_J^3/\tau_{\mathrm{en}} \ll 1; \tag{316}$$

if this were not so, the relaxation matrix would be frequency dependent. It also follows from Eqs. (311) and (312) that $(\hat{\mathscr{K}}_{\mathrm{IM}})_{mm}\tau_J \ll 1$, and $(\hat{\mathscr{K}}_{\mathrm{RE}})_{mm}\tau_J \ll 1$.

If we neglect the imaginary part $(\hat{\mathscr{K}}_{\mathrm{IM}})_{mm}$ in Eq. (310), the problem in the limit $J_0^2\tau_J^2 \gg 1$ is the same as that for $J_0^2\tau_J^2 \ll 1$ except for the difference in the definitions of ω_{ex}. [See Eqs. (313) and (314).] The expression used for ω_{ex} in Eq. (312) has been proposed by several authors.[56,58] Is it valid? As we have shown, it appears to be valid in both extreme limits, but it is probably not correct in the intermediate region because in the region $J_0\tau_J \approx 1$, i.e. the region of intermediate values of $\mathscr{H}_{\mathrm{SL}}$, we cannot replace $\mathscr{H}_{\mathrm{SL}}$ by $\sum_{l,m} J_0 \mathbf{S}_l \cdot \mathbf{S}_m$, and we cannot assume $\mathscr{H}_{\mathrm{SL}}$, $\mathscr{H}_{\mathrm{S}}$, and $\mathscr{H}_{\mathrm{L}}$ commute in the exponential time propagator.

If Eq. (315) holds, then ω_{ex} depends directly on τ_{en}^{-1} which, in turn, can be related to the particle diffusion constant.

VIII. Diffusion Constants

In Section V, F we discussed the rotational diffusion constant and the way in which it enters into magnetic spin relaxation, while in Sections VII, C, and VII, D we did the same for the translational diffusion constant.

In both cases the extended Mori theory was required to obtain information concerning the diffusion constant even if, as is usually the case, we are interested in the extreme situation of slow spin relaxation where the un-extended theory is sufficient to describe the magnetic resonance spectrum. In Section VIII, A, we discuss the more direct relationship between translational diffusion constants and spin relaxation in an inhomogeneous magnetic field; this is the classic relationship[63] which is commonly used to determine diffusion constants from nuclear spin echo experiments. We include this discussion for completeness since it has already been discussed elsewhere in terms of the Mori theory[61] although we have made some changes in the present treatment. In Section VIII, B we make a few additional comments concerning diffusion constants.

A. Diffusion in an Inhomogeneous Field

In this section we consider spin relaxation due to molecular diffusion in an inhomogeneous magnetic field, a phenomenon that enables us to determine particle[61] or self-diffusion constants from spin-echo measurements.[63,64] The analysis of this problem by means of the Mori approach has already been presented,[61] but here we consider the problem in more detail and somewhat more carefully. Also, at the end of this section, we investigate briefly the "rapid diffusion" limit where the usual relationships must be modified.

In an inhomogeneous dc field applied along the laboratory Z-axis, the pure spin Hamiltonian takes the form

$$\mathscr{H}_S = \omega_0 \sum_j [1 + Gz_j]\, S_{z_j}, \tag{317}$$

where z_j is the z-position of the jth spin, ω_0 is the Larmor frequency for a spin at $z_j = 0$,

$$G \equiv (1/\mathrm{B}_0)\, dB/dZ, \tag{318}$$

B_0 is the magnetic field at $Z = 0$, and the field gradient, dB/dZ, is assumed to be constant. Z is the Z-component of a space point. The total Hamiltonian is once again $(\mathscr{H}_S + \mathscr{H}_{SL} + \mathscr{H}_L)$. The slow collective variable, A, is chosen to be the *local spin density*, i.e. the number of spins per unit volume at a given spatial point, $\mathbf{R}$. Thus

$$\mathbf{A}(\mathbf{R}) = \sum_j S_{+j}\, \delta(\mathbf{R} - \mathbf{r}_j), \tag{319}$$

where S_{+j} and $\mathbf{r}_j$ are the spin and position, respectively, of the jth molecule.

The quantity, $\dot{\mathbf{A}}(\mathbf{R})$, is

$$\dot{\mathbf{A}}(\mathbf{R}) = -\sum_j \dot{\mathbf{r}}_j \cdot \nabla_{\mathbf{R}} \delta(\mathbf{R}-\mathbf{r}_j) S_{+_j}$$

$$+ i \sum_j \delta(\mathbf{R}-\mathbf{r}_j) \{\omega_0 S_{+_j} + \omega_0 G z_j S_{+_j} + [\mathscr{H}_{\mathrm{SL}}, S_{+_j}]\}. \qquad (320)$$

The variable $\mathbf{A}(\mathbf{R})$ is in reality an infinite array of variables, one for each $\mathbf{R}$, and these variables are all coupled in the Mori theory through the corresponding collection of coupled transport equations. However, if we make use of the Fourier transform variable, $\mathbf{A}(\mathbf{k})$, where

$$\mathbf{A}(\mathbf{k}) = \int d\mathbf{R}\, e^{i\mathbf{k}\cdot\mathbf{R}} \mathbf{A}(\mathbf{R}), \qquad (321)$$

then, as shown by Mori,[1] the resulting transport equations for each $\mathbf{A}(\mathbf{k})$ are uncoupled. ($d\mathbf{R} = dX\,dY\,dZ$, and $\mathbf{A}(\mathbf{k})$ is distinguished from its Fourier transform, $\mathbf{A}(\mathbf{R})$, by the explicitly indicated variable.) Thus, we write

$$(d/dt)\overline{\mathbf{A}(\mathbf{k},t)} = i\Omega(\mathbf{k})\overline{\mathbf{A}(\mathbf{k},t)} - \int_0^\infty d\tau\, \mathsf{K}(\mathbf{k},\tau)\overline{\mathbf{A}(\mathbf{k},t-\tau)}, \qquad (322)$$

where $\Omega(\mathbf{k})$ and $\mathsf{K}(\mathbf{k},\tau)$ are the spatial Fourier transforms of Eqs. (85)–(88). As we have discussed frequently above, under the appropriate conditions, i.e. that $\mathsf{K}(\mathbf{k},t)\, e^{-i\Omega(\mathbf{k})t}$ relaxes rapidly compared to $e^{-i\Omega(\mathbf{k})t}\, \mathbf{A}(\mathbf{k},t)$, Eq. (322) reduces to

$$(d/dt)\overline{\mathbf{A}(\mathbf{k},t)} = [i\Omega(\mathbf{k}) - \hat{\mathscr{K}}(\mathbf{k})]\overline{\mathbf{A}(\mathbf{k},t)}, \qquad (323)$$

where

$$\hat{\mathscr{K}}(\mathbf{k}) = \int_0^\infty \mathsf{K}(\mathbf{k},t)\, e^{-i\Omega(\mathbf{k})t}\, dt. \qquad (324)$$

In most cases of interest the equilibrium distribution of molecular positions, $\mathbf{r}_j$, can be adequately represented by $\delta(\mathbf{r}_j - \mathbf{R})$. This means that the particle does not move appreciably in the times relevant to the problem, that is, in times of the order of T_2. Since typical molecular diffusion constants are of order $10^{-5}\,\mathrm{cm}^2/\mathrm{sec}$, it takes about 10 sec for a molecule to diffuse 10^{-2} cm. For most NMR spin-echo experiments, this ensures a small fractional change in positional coordinate in a times of the order of T_2. We shall assume the distribution $\delta(\mathbf{r}_j - \mathbf{R})$, but we shall later discuss its deficiencies.

From Eqs. (319)–(321) it can readily be seen that

$$\mathbf{A}(\mathbf{k}) = \sum_j S_{+_j} e^{i\mathbf{k}\cdot\mathbf{r}_j} \qquad (325)$$

and

$$\dot{\mathbf{A}}(\mathbf{k}) = i\sum_j S_{+_j} e^{i\mathbf{k}\cdot\mathbf{r}_j}[\mathbf{k}\cdot\dot{\mathbf{r}}_j + \omega_0 + \omega_0 G z_j] + i\sum_j e^{i\mathbf{k}\cdot\mathbf{r}_j}[\mathscr{H}_{\mathrm{SL}}, S_{+_j}]. \quad (326)$$

We note that when averaged over the ensemble, terms linear in $\mathscr{H}_{\mathrm{SL}}$ vanish. If we choose $\mathscr{H}_{\mathrm{SL}}$ to be intramolecular, then $\langle S_{+_j} S_{-_{j'}}\rangle = \langle S_{+_j} S_{-_j}\rangle \delta_{jj'}$. With the aid of Eq. (85) we see that

$$i\Omega = i\omega_0 + i\omega_0 G\langle z_j\rangle, \quad (327)$$

and since the equilibrium distribution is $\delta(\mathbf{r}_j - \mathbf{R})$,

$$i\Omega = i\omega_0 + i\omega_0 GZ \quad (328)$$

where Z is the Z-component of the spatial position.

Next we evaluate $\mathsf{K}(\mathbf{k}, t) = \langle (e^{i(1-\P)\mathscr{H}^{\times} t}\dot{\alpha}(\mathbf{k}))\dot{\alpha}(\mathbf{k})^{+}\rangle$, where the quantity, $\dot{\alpha}(\mathbf{k})$, is given by

$$\dot{\alpha}(\mathbf{k}) = (1-\P)\dot{\mathbf{A}}(\mathbf{k}) = \dot{\mathbf{A}}(\mathbf{k}) - i\Omega\mathbf{A}(\mathbf{k}) \quad (329)$$

and, consequently,

$$\dot{\alpha}(\mathbf{k}) = i\sum_j S_{+_j} e^{i\mathbf{k}\cdot\mathbf{r}_j}[\mathbf{k}\cdot\dot{\mathbf{r}}_j + \omega_0 G(z_j - Z)] + i\sum_j e^{i\mathbf{k}\cdot\mathbf{r}_j}[\mathscr{H}_{\mathrm{SL}}, S_{+_j}]. \quad (330)$$

If we again consider the distribution, $\delta(\mathbf{r}_j - \mathbf{R})$, the $\omega_0 G(z_j - Z)$ term in Eq. (330) vanishes. We then obtain

$$\hat{\mathscr{K}}(\mathbf{k}) = T_{2G}^{-1} + D(\mathbf{k})k^2 \quad (331)$$

where

$$T_{2G}^{-1}(\mathbf{k}) = \sum_j \int_0^\infty dt \langle \{\exp[i(1-\P)\mathscr{H}^{\times} t]\, e^{i\mathbf{k}\cdot\mathbf{r}_j}[\mathscr{H}_{\mathrm{SL}}, S_{+_j}]\}\, e^{-i\mathbf{k}\cdot\mathbf{r}_j}[S_{-_j}, \mathscr{H}_{\mathrm{SL}}]\rangle$$

$$\times\, e^{-i\Omega t}\,[\sum_j \langle S_{-_j} S_{+_j}\rangle]^{-1}, \quad (332)$$

$$D(\mathbf{k}) = \sum_j \int_0^\infty dt \langle \{\exp[i(1-\P)\mathscr{H}^{\times} t]\, e^{i\mathbf{k}\cdot\mathbf{r}_j}\dot{x}_j S_{+_j}\}\, e^{-i\mathbf{k}\cdot\mathbf{r}_j} S_{-_j}\dot{x}_j\rangle$$

$$\times\, e^{-i\Omega t}[\sum_j \langle S_{+_j} S_{-_j}\rangle]^{-1}. \quad (333)$$

In obtaining this result we again note that for intramolecular $\mathscr{H}_{\mathrm{SL}}$, we have $\langle S_{+_{j'}}(t) S_{-_j}\rangle = \langle S_{+_j}(t) S_{-_j}\rangle \delta_{jj'}$, and that in Eq. (333) we could equally well substitute $\dot{y}_j$ or $\dot{z}_j$ for $\dot{x}_j$. Furthermore, we have neglected cross terms which contain $\dot{\mathbf{r}}_j$ and $\mathscr{H}_{\mathrm{SL}}$; we have assumed that the autocorrelation time for $\dot{\mathbf{r}}_j$ is short compared to that for $\mathscr{H}_{\mathrm{SL}}$ and, consequently, that these two quantities are uncorrelated.

For *small* k and for *small* $\mathscr{H}_{SL}$, i.e. small gradients and small ω_λ, the relaxation time T_{2G} given above reduces to

$$T_{2G}^{-1} = \sum_j \int_0^\infty dt \langle [\mathscr{H}_{SL}(t), S_{+j}] [S_{-j}, \mathscr{H}_{SL}] \rangle \cdot [\sum_j \langle S_{+j} S_{-j} \rangle]^{-1}, \quad (334)$$

where the precessional terms in $\mathscr{H}_{SL}(t)$ have frequency $\omega_0(1+GZ)$ rather than simply ω_0 as in Eq. (143). See the discussion in Section V, B. In the same limit, provided $\dot{x}_j$ relaxes rapidly compared to $\mathbf{r}_j$ and S_{+j}, the diffusion constant becomes

$$D = \int_0^\infty dt \langle \dot{x}_j(t) \dot{x}_j \rangle, \quad (335)$$

the usual particle or self-diffusion constant.[61] In these limits if follows that

$$d\mathbf{A}(\mathbf{k}, t)/dt = i\omega_0(1+GZ)\mathbf{A}(\mathbf{k}, t) - Dk^2\mathbf{A}(\mathbf{k}, t) - T_{2G}^{-1}\overline{\mathbf{A}(\mathbf{k}, t)}. \quad (336)$$

If we Fourier transform back to **R**-space.

$$\overline{d\mathbf{A}(\mathbf{R}, t)/dt} = i\omega_0(1+GZ)\overline{\mathbf{A}(\mathbf{R}, t)} - T_{2G}^{-1}\overline{\mathbf{A}(\mathbf{R}, t)} + D\nabla_{\mathbf{R}}^2\overline{\mathbf{A}(\mathbf{R}, t)}. \quad (337)$$

This is a well-known expression.[63]

In summary, the approximations used in obtaining Eq. (337) are small ω_λ or $\mathscr{H}_{SL}$, small or hydrodynamic k, rapidly fluctuating $\mathscr{H}_{SL}$, and still more rapidly fluctuating $\dot{x}_j$, no cross-correlation between $\dot{x}_j$ and $\mathscr{H}_{SL}$, and the equilibrium distribution $\delta(\mathbf{r}_j - \mathbf{R})$. Suppose the spin relaxation T_{2G} is very long, the diffusion very fast and the sample very small, then the equilibrium distribution $\delta(\mathbf{r}_j - \mathbf{R})$ may not be applicable and should be replaced by a constant, V^{-1}, where V is the sample volume. In this case $\langle z_j \rangle = 0$ in Eq. (327), and the term linear in Z does not appear in Eqs. (328) and (330); if we also assume no cross-correlations between z_j and $\dot{\mathbf{r}}_j$ and $\mathscr{H}_{SL}$, we have, instead of Eq. (336),

$$\begin{aligned}\overline{d\mathbf{A}(\mathbf{k}, t)/dt} &= i\omega_0\overline{\mathbf{A}(\mathbf{k}, t)} - Dk^2\overline{\mathbf{A}(\mathbf{k}, t)} - T_2^{-1}\overline{\mathbf{A}(\mathbf{k}, t)} \\ &\quad - \omega_0^2 G^2 \int_0^t \sum_j d\tau \langle S_{+j}(\tau) S_{-j} e^{i\mathbf{k}\cdot(\mathbf{r}_j(\tau) - \mathbf{r}_j)} z_j(\tau) z_j \rangle \\ &\quad \times \left[\sum_j \langle S_{+j} S_{-j} \rangle\right]^{-1} \overline{\mathbf{A}(\mathbf{k}, t-\tau)},\end{aligned} \quad (338)$$

where in the last term we have neglected the small effects of $\mathscr{H}_{SL}$ on $\mathbf{r}_j(\tau)$ and $z_j(\tau)$ and have only kept lowest order terms in G. The last term is difficult to handle because the correlation time of the quantity in brackets $\langle\,\rangle$ may be comparable to that of $\overline{\mathbf{A}(\mathbf{k}, t)}$; therefore, we cannot make the usual approximations of a short memory function. (Of course, the time dependences evolve from a projected propagator.) However, if the diffusion is very rapid and the spin relaxation time very long, the last term in Eq. (338) becomes

$$-\omega_0^2 G^2 \left[\int_0^{\infty} d\tau \langle z_j(\tau) z_j \rangle \right] \overline{\mathbf{A}(\mathbf{k}, t)}, \tag{339}$$

provided this correlation function is meaningful. In this case, instead of Eq. (337), we have

$$\overline{d\mathbf{A}(\mathbf{R}, t)/dt} = [i\omega_0 - T_2^{-1} - T_G^{-1} + D\nabla_{\mathbf{R}}^2]\, \overline{\mathbf{A}(\mathbf{R}, t)}, \tag{340}$$

where

$$T_G^{-1} = \omega_0^2 G^2 \int dt \langle z_j(t) z_j \rangle. \tag{341}$$

B. Additional Comments on Diffusion

In Section V, F we discussed the relationship of rotational diffusion constants to magnetic resonance experiments; this connection arises when the relaxation interaction, $\mathcal{H}_{SL}$, is intramolecular and is modulated by molecular tumbling. Since spins on neighboring molecules, except in the presence of exchange interactions, are almost completed uncorrelated, the rotational diffusion constants or correlation times obtained from magnetic resonance are strictly single particle properties in contrast to those obtained from dielectric relaxation and light scattering.[65] In our treatment we only considered isotropic rotational diffusion, but the method can be readily applied to anisotropic rotational diffusion.[49,66] If we choose S_+ and $\dot{\mathcal{S}}_+$ as our variables, we find terms in $\ddot{\mathfrak{S}}_+$ of the form $[\mathcal{H}_L, \mathcal{H}_{SL}]$; $\mathcal{H}_{SL}$ in many cases is a linear combination of Wigner rotation functions and $[\mathcal{H}_L, \mathcal{H}_{SL}]$ is then a linear combination of the components, relative to molecular axes of the molecular angular velocity. Consequently, the quantity $\langle \ddot{\mathfrak{S}}_+(t_{p'}) \ddot{\mathfrak{S}}_- \rangle$ is bilinear in the molecular angular velocity, i.e. it is a linear combination of angular velocity correlation functions. The relaxation matrix, $\hat{\mathcal{K}}$, is the integral over $\langle \ddot{\mathfrak{S}}_+(t_{p'}) \ddot{\mathfrak{S}}_- \rangle$, and since the components of the rotational diffusion tensor are integrals over angular velocity correlation functions, $\hat{\mathcal{K}}$ is a linear combination of the components of the rotational diffusion tensor. We will not pursue this problem here since the results have been adequately studied by different approaches.[61]

If $\mathcal{H}_{SL}$ is the spin rotational interaction, i.e. if

$$\mathcal{H}_{SL} = \sum_j \mathbf{S}_j \mathbf{C}_j \mathbf{J}_j \tag{342}$$

[65] T. Keyes, *Mol. Phys.* **23**, 699 (1972).

[66] L. D. Favro, *in* "Fluctuation Phenomena in Solids" (R. E. Burgess, ed.), p. 79. Academic Press, New York, 1965; J. H. Freed, *J. Chem. Phys.* **41**, 2077 (1964) and references cited therein.

where $\mathbf{J}_j$ is the rotational angular momentum of the jth molecule and C_j is the spin rotational coupling matrix of the jth molecule, then information concerning autocorrelation functions of $\mathbf{J}$ can be obtained. (Elsewhere we have used J to indicate exchange integrals.) If $\mathbf{S}_j$ is quantized along the laboratory Z-axis and $\mathbf{J}_j$ along the molecular axes, C_j depends upon the Eulerian angles which determine the orientation of the molecule. Once again $\ddot{\mathfrak{S}}_+$ depends upon $[\mathscr{H}_{\mathrm{L}}, \mathscr{H}_{\mathrm{SL}}]$; $[\mathscr{H}_{\mathrm{L}}, \mathscr{H}_{\mathrm{SL}}]$ is a function of spin, Eulerian angles, angular velocity or momentum, and the rate of change of angular momentum, that is, intermolecular torques. It is often the case that the fluctuation or relaxation times of these quantities differ greatly from each other: spin times are of the order 10^{-1} to 10^{-7} sec, Eulerian angles fluctuate in times of order 10^{-11} sec, angular velocity correlation times may be considerably shorter, and we are not sure whether or not the torques vary more rapidly than the angular momentum. We can neglect the time dependence of all quantities other than torques and angular velocities in $\langle \ddot{\mathfrak{S}}_+(t_{p'})\ddot{\mathfrak{S}}\rangle$, but if the characteristic fluctuating times for torques and angular velocities are comparable, we would probably have to extend our treatment and introduce additional, faster variables such as $\ddot{\mathscr{S}}_+, \dddot{S}_+, \ddddot{S}_+$, etc. to describe the situation adequately. This extended procedure might not work well since it is unlikely that higher derivatives fluctuate appreciably faster than do the angular velocities and torques. Evidence that the fluctuating frequencies of torques and angular velocities are comparable in many liquids is available: the angular velocity correlation times, which are the integrals over the normalized correlation functions obtained from ESR experiments, are often as small as 10^{-16} sec, which indicates that *the correlation function for angular velocity must be negative over part of the time span;* this suggests that the torques do not fluctuate rapidly compared to the fluctuations in angular velocity[26b, 30, 31] since the angular velocity correlation function does not decay exponentially but gives evidence of negative correlations after some elapsed time, presumably the result of a reversal of angular momentum upon intermolecular impact.

Finally, we note that if $\mathscr{H}_{\mathrm{SL}}$ is intermolecular, the effective correlation times depend both upon the molecular angular velocity and the intermolecular separation velocities. Thus we obtain diffusion constants which have both rotational and translational contributions. This has been discussed elsewhere.[61]

Appendix A. Derivation of Mori Relations

A simple derivation of the Mori relations follows (the derivation below is based in large part on that in Hynes and Deutch[4]): Let $\mathbf{A}$ be the set of variables and, as in the text, let ¶ be the projection operator, $\mathscr{H}^{x}$ the

Hamiltonian superoperator, and Ω a frequency matrix. Thus, if $\mathbf{B}$ is an arbitrary set of variables with the same dimensionality as $\mathbf{A}$,

$$¶\mathbf{B} = \langle \mathbf{B}\mathbf{A}^{+}\rangle\langle \mathbf{A}\mathbf{A}^{+}\rangle^{-1}\mathbf{A}, \tag{A.1}$$

$$i\,\Omega = \langle \dot{\mathbf{A}}\mathbf{A}^{+}\rangle\langle \mathbf{A}\mathbf{A}^{+}\rangle^{-1}, \tag{A.2}$$

$$\dot{\mathbf{A}} = i\mathscr{H}^{x}\mathbf{A}, \tag{A.3}$$

$$\mathbf{A}(t) = e^{i\mathscr{H}^{x}t}\mathbf{A}. \tag{A.4}$$

In these expressions $\mathbf{A} = \mathbf{A}(0)$ and in a scalar product, $\langle \mathbf{B}\mathbf{C}^{+}\rangle$, $\mathbf{C}^{+}$ is the adjoint of $\mathbf{C}$. We also define the normalized correlation function, $\mathsf{G}(t)$,

$$\mathsf{G}(t) = \langle \mathbf{A}(t)\,\mathbf{A}^{+}\rangle\langle \mathbf{A}\mathbf{A}^{+}\rangle^{-1}. \tag{A.5}$$

Consider the quantity $\mathbf{F}(t)$:

$$\mathbf{F}(t) = e^{-i\mathscr{H}^{x}t}\,e^{i(1-¶)\mathscr{H}^{x}t}(1-¶)\,\dot{\mathbf{A}}. \tag{A.6}$$

Differentiate Eq. (A.6) with respect to time:

$$\partial\mathbf{F}(t)/\partial t = e^{-i\mathscr{H}^{x}t}[-i¶\mathscr{H}^{x}]\,e^{i(1-¶)\mathscr{H}^{x}t}(1-¶)\,\dot{\mathbf{A}}. \tag{A.7}$$

Integrate Eq. (A.7) from 0 to t and note that $\mathbf{F}(0) = (1-¶)\,\dot{\mathbf{A}}$:

$$\mathbf{F}(t) = (1-¶)\,\dot{\mathbf{A}} + \int_0^t d\tau\, e^{-i\mathscr{H}^{x}\tau}[-i¶\mathscr{H}^{x}]\,e^{i(1-¶)\mathscr{H}^{x}\tau}(1-¶)\,\dot{\mathbf{A}}. \tag{A.8}$$

Premultiply Eq. (A.8) by $\exp i\mathscr{H}^{x}t$ and make use of Eq. (A.1) and (A.6):

$$e^{i(1-¶)\mathscr{H}^{x}t}(1-¶)\,\dot{\mathbf{A}} = [e^{i\mathscr{H}^{x}t}\dot{\mathbf{A}} - e^{i\mathscr{H}^{x}t}\langle \dot{\mathbf{A}}\mathbf{A}^{+}\rangle\langle \mathbf{A}\mathbf{A}^{+}\rangle^{-1}\mathbf{A}]$$
$$- \int_0^t d\tau\, e^{i\mathscr{H}^{x}(t-\tau)}\langle i\mathscr{H}^{x}e^{i(1-¶)\mathscr{H}^{x}\tau}(1-¶)\,\dot{\mathbf{A}}, \mathbf{A}^{+}\rangle\langle \mathbf{A}\mathbf{A}^{+}\rangle^{-1}\mathbf{A}. \tag{A.9}$$

Note that $\exp i\mathscr{H}^{x}t$ commutes with an ensemble average, and making use of Eqs. (A.2) and (A.4), and rearranging Eq. (A.9) we have

$$e^{i\mathscr{H}^{x}t}\dot{\mathbf{A}} = i\Omega\,\mathbf{A}(t) + \int_0^t d\tau\,\langle i\mathscr{H}^{x}\dot{\boldsymbol{\alpha}}(\tau_{\mathrm{p}}), \mathbf{A}^{+}\rangle\langle \mathbf{A}\mathbf{A}^{+}\rangle^{-1}\,\mathbf{A}(t-\tau) + \dot{\boldsymbol{\alpha}}(t_{\mathrm{p}}), \tag{A.10}$$

where

$$\dot{\boldsymbol{\alpha}}(t_{\mathrm{p}}) = e^{i(1-¶)\mathscr{H}^{x}t}(1-¶)\,\dot{\mathbf{A}}. \tag{A.11}$$

Since

$$\langle [i\mathscr{H}^{x}\dot{\boldsymbol{\alpha}}(t_{\mathrm{p}})]\,\mathbf{A}^{+}\rangle = -\langle \dot{\boldsymbol{\alpha}}(t_{\mathrm{p}})\,\dot{\mathbf{A}}^{+}\rangle \tag{A.12}$$

we can rewrite Eq. (A.10) as

$$d\mathbf{A}(t)/dt = i\Omega\mathbf{A}(t) - \int_0^t d\tau\,\mathsf{K}(\tau)\,\mathbf{A}(t-\tau) + \dot{\boldsymbol{\alpha}}(t_{\mathrm{p}}), \tag{A.13}$$

where

$$\mathsf{K}(t) = \langle \dot{\boldsymbol{\alpha}}(t_p)\dot{\mathbf{A}}^+\rangle \ \langle \mathbf{A}\mathbf{A}^+\rangle^{-1}. \tag{A.14}$$

Since it can readily be seen, by direct expansion, that

$$\langle \dot{\boldsymbol{\alpha}}(t_p)(\P\dot{\mathbf{A}})^+\rangle = i\Omega\langle \dot{\boldsymbol{\alpha}}(t_p)\mathbf{A}^+\rangle = 0, \tag{A.15}$$

$\mathsf{K}(t)$ is equally well given as $\langle \dot{\boldsymbol{\alpha}}(t_p)\dot{\boldsymbol{\alpha}}^+\rangle$. Equation (A.13) is the Mori equation, Eq. (81). If we average over Eq. (A.13) and assume a linear response,[1] as we shall see below, $\overline{\dot{\boldsymbol{\alpha}}(t_p)}$ vanishes, and we obtain Eq. (90).

Now post-multiply Eq. (13) by $\mathbf{A}^+\langle \mathbf{A}\mathbf{A}^+\rangle^{-1}$ and average over the ensemble. Since

$$\langle \dot{\boldsymbol{\alpha}}(t_p)\mathbf{A}^+\rangle = 0, \tag{A.16}$$

a result which is essentially equivalent to that in Eq. (A.15), we obtain Eq. (103a). The correspondence between Eqs. (103a) and (90) is equivalent to Onsager's assumption that the time dependence of fluctuations and dissipations are the same.[19]

Finally we wish to show that in the linear response regime, $\overline{\dot{\boldsymbol{\alpha}}(t_p)} = 0$. If $\mathscr{H}$ is the Hamiltonian governing the system, let us assume that in the presence of a small perturbing force $\mathscr{F}$ which couples to the system via the variable $\mathbf{A}$, the interaction energy is $-\mathbf{A}^+\cdot\mathscr{F}$. Thus the total Hamiltonian is $\mathscr{H}-\mathbf{A}^+\cdot\mathscr{F}$, and we assume that only the term linear in $\mathscr{F}$ need be considered. Thus

$$\overline{\dot{\boldsymbol{\alpha}}(t_p)} = \mathrm{Tr}\,\dot{\boldsymbol{\alpha}}(t_p)e^{-\beta(\mathscr{H}-\mathbf{A}^+\cdot\mathscr{F})}/\mathrm{Tr}\,e^{-\beta(\mathscr{H}-\mathbf{A}^+\cdot\mathscr{F})}, \tag{A.17}$$

where Tr is the quantum mechanical trace over all states of the system or the classical integral over all variables. It can be readily shown that[67]

$$e^{-\beta(\mathscr{H}-\mathbf{A}^+\cdot\mathscr{F})} = e^{-\beta\mathscr{H}}\left[1-\int_0^\beta d\lambda e^{\lambda\mathscr{H}}A^+\cdot\mathscr{F}e^{-\lambda\mathscr{H}}+0(\mathscr{F}^2)\right] \tag{A.18}$$

Note that the trace does not affect the externally applied force, $\mathscr{F}$; since

$$\langle A\rangle = \mathrm{Tr}\,Ae^{-\beta\mathscr{H}}/\mathrm{Tr}\,e^{-\beta\mathscr{H}} = 0, \tag{A.19}$$

to first order in $\mathscr{F}$, Eq. (A.17) becomes

$$\dot{\boldsymbol{\alpha}}(t_p) = -\langle \dot{\boldsymbol{\alpha}}(t_p)\mathbf{A}^+\rangle\mathscr{F}. \tag{A.20}$$

From Eq. (A.16), we see that this quantity vanishes.

Appendix B

We wish to justify the approximation in Eq. (162). Consider the effect of the off-diagonal terms in Ω on the value of $[\mathsf{K}(t)\exp\{-i\Omega t\}]$ in

[67] R. Kubo and K. Tomita, *J. Phys. Soc. Jap.* **9**, 888 (1954).

Eq. (92). We write $\Omega = \Omega^{(D)} + \Omega^{(OF)}$ where $\Omega^{(D)}$ and $\Omega^{(OF)}$ are the diagonal and off-diagonal components, respectively, of Ω. Consider the special case where $\mathscr{H}_{SL}$ is secular, i.e. $b = 1$ and $\beta = 0$; then

$$[\Omega^{(D)}, \Omega^{(OF)}] = 0. \tag{B.1}$$

This leads to the relation

$$e^{+i\Omega t} = e^{+i\Omega^{(D)}t} e^{+i\Omega^{(OF)}t}. \tag{B.2}$$

With the aid of Eqs. (153) and (148), we then find that

$$\mathsf{K}(t)\, e^{-i\Omega t} = K''(t) \begin{bmatrix} 0 & 0 \\ +\omega_\lambda \sin \omega_\lambda t & \cos \omega_\lambda t \end{bmatrix} e^{i\omega_0 t}. \tag{B.3}$$

If the decay or correlation time corresponding to $K(t)$ is very rapid, i.e. if

$$\tau_{\text{II}}^2 \omega_\lambda^2 \ll 1, \tag{B.4}$$

then

$$\hat{\mathscr{K}} \approx \hat{\mathscr{K}}'' \begin{bmatrix} 0 & 0 \\ 0 & 1 \end{bmatrix}, \tag{B.5}$$

where

$$\hat{\mathscr{K}}'' = \int_0^\infty K(t)\, e^{i\omega_0 t}\, e^{+i\beta\omega_\lambda t}\; dt, \tag{B.6}$$

and $\Omega^{(OF)}$ does not contribute to $\hat{\mathscr{K}}$. This is what we have already assumed. The condition in Eq. (B.4) is thus an additional requirement for a tractable two variable $(S_+, \dot{\mathscr{S}}_+)$ theory. We also assume that $\beta\omega_\lambda \ll \omega_0$.

Appendix C

In Eq. (308), $S_{\alpha_j}^{\#}(t)$ is defined as

$$S_{\beta_l}^{\#}(t) = \exp\left(iJ_0 t \sum_m \mathbf{S}_l \cdot \mathbf{S}_m\right) S_{\beta_l} \exp - \left(iJ_0 \sum_m \mathbf{S}_l \mathbf{S}_m t\right) \tag{C.1}$$

Since, in Eq. (307), $S_{\beta_l}^{\#}(t)$ is premultiplied by J_{lk}, and J_{lk} vanishes unless the lth and kth molecules are in contact, we can approximate Eq. (C.1) by

$$S_{\beta_l}^{\#}(t) = \exp(iJ_0 \mathbf{S}_l \cdot \mathbf{S}_k t)\, S_{\beta_l} \exp(-iJ_0 \mathbf{S}_l \cdot \mathbf{S}_k t). \tag{C.2}$$

We wish to find a representation in which $J_0 \mathbf{S}_l \cdot \mathbf{S}_k$ is diagonal. If $S = \frac{1}{2}$, the appropriate representation is the triplet-singlet representation, i.e. $|SM_s)$ where $(S = 1, M_s = 0, \pm 1)$ and $(S = 0, M_s = 0)$. In this representation

$(1M_s|J_0\mathbf{S}_l\cdot\mathbf{S}_k|1M_s) = J_0/4$ and $(00|J_0\mathbf{S}_l\cdot\mathbf{S}_k|00) = -3J_0/4$. Furthermore, the nonvanishing matrix elements of interest are

$$(S\pm1, M_s+1|S^{\#}_{+k}(t)|S, M_s) = \pm(1/\sqrt{2})\,e^{\pm iJ_0t}(-1)^k, \tag{C.3}$$

$$(S, M_s+1|S^{\#}_{+k}(t)|S, M_s) = 1/\sqrt{2}, \tag{C.4}$$

$$(S, M_s|S^{\#}_{z_k}(t)|S, M_s) = \tfrac{1}{2}M_s, \tag{C.5}$$

$$(S\pm1, 0|S^{\#}_{z_k}(t)|S, 0) = \tfrac{1}{2}e^{\pm iJ_0t}(-1)^{k+1}, \tag{C.6}$$

$$(S, M_s-1|S_{-k}|S, M_s) = 1/\sqrt{2}, \tag{C.7}$$

$$(S\pm1, M_s-1|S_{-k}|SmM_s) = \mp(1/\sqrt{2})(-1)^k. \tag{C.8}$$

It is then an easy matter to obtain Eq. (309) from Eq. (307).

Acknowledgments

One of the authors (DK) would like to thank Professor John Deutch and Dr. Margaret Kivelson for many interesting and helpful discussions on a number of topics treated in this article. Professors Jack Freed and Peter Atkins have also made useful contributions. We are grateful to the National Science Foundation for its support. We would also like to recognize the patient, rapid and proficient work of Miss Gail Wilbur in typing the manuscript in its many revisions.

Chemically Induced Dynamic Nuclear Polarization

G. L. CLOSS

DEPARTMENT OF CHEMISTRY, THE UNIVERSITY OF CHICAGO, CHICAGO, ILLINOIS

I. Introduction

In 1967 two independent research groups reported the observation of strong nuclear polarization in diamagnetic products obtained from reactions proceeding through radical intermediates.[1,2] Bargon and Fischer observed NMR emission of the protons of benzene resulting from the thermal decomposition of benzoyl peroxide, and Ward and Lawler reported enhanced absorption and emission lines in the proton NMR spectra recorded on a reaction mixture of alkyllithium with alkyl halides. Both groups suggested that a special type of Overhauser effect might be responsible for the nuclear polarization and the effect became known as Chemically Induced Dynamic Nuclear Polarization (CIDNP). Following the initial reports a number of publications appeared describing similar observations in widely

[1] J. Bargon, H. Fischer, and U. Johnsen, *Z. Naturforsch. A* **22**, 1551 (1967); J. Bargon and H. Fischer, *Z. Naturforsch. A* **22**, 1556 (1967); J. Bargon and H. Fischer, *Z. Naturforsch. A* **23**, 2109 (1968).

[2] H. R. Ward, *J. Amer. Chem. Soc.* **89**, 5517 (1967); H. R. Ward and R. G. Lawler, *J. Amer. Chem. Soc.* **89**, 5518 (1967); R. G. Lawler, *J. Amer. Chem.* 5519 *Soc.* **89**, (1967).

different reactions, having in common only the apparent paramagnetism of reaction intermediates.[3–6]

As more experimental data accumulated it became clear that the initial explanation of the effect could no longer be upheld since it failed to account for a frequently observed polarization pattern which subsequently became known as the "multiplet effect." In essence, this effect is the presence of both enhanced absorption and emission in the multiplet of lines originating from transitions of a nucleus or a group of identical nuclei coupled to other nuclei via indirect nuclear spin coupling. The basis of the early theory may be summarized by stating that it assumed the formation of free radicals from diamagnetic precursors with infinite electron spin temperature followed by subsequent electron relaxation. Dipolar and scalar interactions among the electrons and the nuclei were thought to lead to electron-nuclear cross relaxation, resulting in polarization of the nuclear spin states. Subsequent annihilation of the radicals by combination and disproportionation or radical transfer would lead to a carry-over of this polarization to the diamagnetic products in which the polarization was detected by NMR. A detailed consideration of this mechanism shows that it can only lead to polarizations in which all lines associated with transitions of any given nucleus are either in enhanced absorption or emission, but not both as is the case in spectra showing the multiplet effect.

Additional evidence against the mechanism was discovered in our laboratory in 1968 when we found that enhancement factors of individual lines could exceed the "Overhauser limit" of 658 for proton spectra.[7] Also, products derived from benzhydryl radicals which had been generated from diphenylmethylene via hydrogen abstraction showed strong polarization, although comparison of the rate of hydrogen abstraction with the estimated relaxation time showed that the electron spin state distribution in the triplet state of diphenylmethylene must have been in equilibrium with the lattice.

In 1969 we, and independently Kaptein and Oosterhoff, proposed a different mechanism which subsequently became known as the radical pair theory of chemically induced nuclear polarization.[8,9] This mechanism, together with a modification first suggested by Adrian,[10,11] is capable of

[3] R. Kaptein, *Chem. Phys. Lett.* **2**, 261 (1968).
[4] A. R. Lepley, *J. Amer. Chem. Soc.* **90**, 2710 (1968); *Chem. Commun. P.* 64 (1969).
[5] E. R. Ward, R. G. Lawler, and R. A. Cooper, *J. Amer. Chem. Soc.* **91**, 746 (1969).
[6] H. Fischer and J. Bargon, *Accounts Chem. Res.* **2**, 110 (1969).
[7] G. L. Closs and L. E. Closs, *J. Amer. Chem. Soc.* **91**, 4549 (1969).
[8] G. L. Closs, *J. Amer. Chem. Soc.* **91**, 4552 (1969).
[9] R. Kaptein and L. J. Oosterhoff, *Chem. Phys. Lett.* **4**, 214 (1969).
[10] F. J. Adrian, *J. Chem. Phys.* **53**, 3374 (1970).
[11] F. J. Adrian, *J. Chem. Phys.* **54**, 7912 (1971).

explaining all spectra published to date and will be discussed in detail in this article.

An obvious extension of this mechanism predicts that CIDNP should also be observable in reactions proceeding via diradicals, and several cases have been reported confirming this prediction.[12–15]

Although not covered in this review electron spin polarization induced by chemical reaction has been reported, and as Kaptein and Oosterhoff suggested in 1969 can also be explained by the same basic mechanism.[16]

In the short time since the discovery of the phenomenon a multitude of chemical reactions have been reported to give CIDNP signals.[17] Since no special equipment is required and qualitative interpretation is easily made with the aid of the radical pair theory, a potentially powerful tool has been added to the armamentarium of the chemist concerned with reaction mechanisms.

It is the aim of this article to review the theory and to examine the scope and limitations of the method by discussing several selected examples with the emphasis on simple, uncomplicated organic reactions. This review is not intended to cover all the literature, and if a disproportionately large fraction of the examples is chosen from the author's own laboratory this should be attributed to the fact that these are the experiments with which the author is most familiar and not to a lack of equally or perhaps more illuminating experiments from other laboratories.

II. Experimental

Basically, any high resolution NMR spectrometer can be used to study CIDNP effects. The most convenient method involves the recording of the spectra while the reaction is in progress inside the NMR spectrometer probe. This poses no special problems if the reaction rate can be adjusted by choosing the temperature to result in half-lives of several minutes. If the rate is too fast it may be necessary to regenerate the solution with the aid of a flow system since the polarization at the end of the reaction

[12] R. Kaptein, M. Frater Schroder, and L. J. Oosterhoof, *Chem. Phys. Lett.* **12**, 16 (1971).

[13] G. L. Closs, *J. Amer. Chem. Soc.* **93**, 1546 (1971); *Ind. Chim. Belge* **36**, 1064 (1971).

[14] G. L. Closs and C. E. Doubleday, *J. Amer. Chem. Soc.* **94**, 9248 (1972).

[15] G. L. Closs and C. E. Doubleday, *J. Amer. Chem. Soc.* **95**, 2735 (1973).

[16] R. Kaptein and L. J. Oosterhoff, *Chem. Phys. Lett.* **4**, 195 (1969).

[17] For other reviews covering reactions not discussed in this article, see H. Fischer, *Top. Curr. Chem.* **24**, 1 (1971); H. R. Ward, *Accounts Chem. Res.* **5**, 18 (1972); R. G. Lawler, *Accounts Chem. Res.* **5**, 25 (1972); G. L. Closs, *Proc. 23rd Int. Congr. Pure Appl. Chem. Spec. Lect.* **4**, 19 (1971); "Chemically Induced Magnetic Polarization" (A. R. Lepley and G. L. Closs, eds.). Wiley (Interscience), New York, 1973.

dies out with the spin lattice relaxation time. Flow systems allowing spinning samples have been described.[18,19]

Many photochemical reactions yield radicals and often CIDNP spectra can be observed in the final products.[20] In these cases it is necessary to modify the probe of the spectrometer to allow an intense light beam to enter the sample. Most of the photochemical systems in operation have been provided with a line-of-sight entry port through the axis of the transmitter coil in commercial cross-coil spectrometers such as Varian HA-60, HA-100, XL-100, and Bruker HX-90. At least one research group has reported good results obtained on a Varian A-60A spectrometer in which the light beam was admitted through the long axis of the sample tube with the aid of a light pipe.[21] A special system is in operation where a capillary lamp is immersed in the 16 mm spinning sample tube. To avoid interference the lamp is operated in a time sharing mode with the rf receiver.[22]

Although most experiments reported in the literature to date have been recorded in the cw mode, Fourier transform spectroscopy has an obvious advantage in recording a time-dependent phenomenon such as CIDNP. In addition, all the advantages of signal-to-noise improvements obtained for regular spectra are also valid for CIDNP. This is particularly important in the study of nuclei of low natural abundance or with small magnetogyric ratios.

To study the field dependence of CIDNP spectra down to very low fields it may be necessary to transfer the sample from the low magnetic field in which the reaction has been allowed to run to a high field in which the nuclear spin system can be analyzed by NMR. Since the nuclei in most organic compounds in fluid solutions have relaxation times on the order of many seconds, this can often be accomplished by simple manual transfer from one magnet to the other. More reproducible results are obtained with the aid of flow systems.

III. Theory of CIDNP

A. Kinetic Formulation of the CIDNP Problem

It is useful to define CIDNP as the phenomenon associated with the formation of molecules with a nuclear spin state population which is far from thermal equilibrium. Such an abnormal spin state distribution is easily

[18] M. Lehnig and H. Fischer, *Z. Naturforsch. A* **25**, 1963 (1970).

[19] R. G. Lawler, *Int. Conf. CIDNP, Tallin, Estonia, 1972.*

[20] The first photochemically driven CIDNP spectrum was observed by M. Cocivera, *J. Amer. Chem. Soc.* **90**, 3261 (1968).

[21] M. Tomkievicz and M. P. Klein, *Rev. Sci. Instr.* **43**, 1206 (1972).

[22] E. D. Lippmaa, personal private communication (1972).

detected by observing unusually strong absorptive or emissive NMR lines while the product formation is in progress or immediately after it has been stopped. To describe such a dynamic system one can either use a formalism based on the equation of motion of the density matrix or, alternatively, he may resort to the conventional solution of the coupled kinetic equations for the population numbers of all the nuclear spin levels in the products of the reaction. Since a suitable choice of basis functions renders the two approaches entirely equivalent, we will make use of the latter formalism except when solving relaxation problems.

Because CIDNP spectra are NMR spectra which are normal in every respect except for their line intensities, the theory can be limited to the derivation of a quantitative expression for the intensities.

Using the familiar concept of population numbers of nuclear eigenstates (N_i), the intensity of an NMR line connecting state m and n is governed by

$$\mathrm{Int}_{mn} \propto \Omega_{mn}(N_m - N_n), \tag{1}$$

where Ω_{mn} is the conventional probability for rf-induced transitions. The proportionality factor depends on the magnitude of the rf field (H_1) and the relaxation behaviour of the system under study. Since these factors are common to all NMR experiments and are well understood they need not concern us here.

For a dynamic system, the time dependence of the population number of the nth nuclear spin state of a given product, N_n, is conveniently described by a differential equation,

$$(d/dt)\,N_n = vP_n - \sum_{i \neq n} (W_{n\to i}\,N_n - W_{i\to n}\,N_i), \tag{2}$$

where v denotes the chemical rate with which this product is formed by the reaction and the P's are the normalized probabilities of forming the nth nuclear spin states. The W's denote the nuclear relaxation probabilities per unit time between nuclear states. The subscripts indicate the states connected and the direction of the relaxation transitions. For a system with l nuclei of spin I there are $\kappa = (2I+1)^l$ such coupled differential equations which in matrix notation reduce to

$$(d/dt)\,\mathbf{N} = v\mathbf{P} + \mathbf{RN}. \tag{3}$$

The column vectors $\mathbf{N}$ and $\mathbf{P}$ are of dimension κ, and $\mathbf{R}$ is a $\kappa \times \kappa$ matrix containing the relaxation transition probabilities. Its diagonal and off-diagonal elements are given by

$$R_{nn} = -\sum_{m \neq n} W_{n\to m}, \qquad R_{nm} = W_{m\to n}, \tag{4}$$

where $R_{mn} \neq R_{nm}$ because of the requirement that the system relaxes to finite temperature.

As Eq. (3) shows the problem of calculating the CIDNP spectrum factors into the two separate tasks of finding the CIDNP probability vector **P** and the construction of the relaxation matrix **R**. Once this is accomplished the set of differential equations can be integrated and the population numbers thus obtained for any given time can be used in Eq. (1) to give the time dependence and relative intensities of the desired NMR transitions. Since the theory of nuclear relaxation has been well established[23] no special theoretical problems are associated with the construction of **R** although major difficulties arise in practical problems because the correlation functions needed as input parameters for the calculations of the elements of **R** are rarely available for complicated organic molecules. A useful theory of CIDNP then has to provide a quantitative expression for the derivation of the vector **P**.

B. The Radical Pair Theory of CIDNP

1. *The General Solution*[24,25]

A way of rephrasing the definition of CIDNP given above is to say that the individual components of the vector **P** in Eq. (3) are not governed by the Boltzman law, thus leading to the inequality

$$P_n/P_m \neq e^{-(E_n - E_m)/kT}. \tag{5}$$

The radical pair theory gives a plausible explanation for the dependence of the rate of product formation on the nuclear spin states in certain chemical reactions. It is possible to describe the majority of reactions showing the effect with one general scheme summarized in Scheme 1. The remaining reactions involve diradicals and will be described in a later section. In this scheme M is a molecule which serves as the immediate precursor of a radical pair produced by homolytic cleavage of a bond in M. In the most general sense M may either be a molecule in a ground state which undergoes thermal dissociation or it may be a dissociating excited state. In some cases M can also be an activated complex derived from two molecules in an active collision. Whatever its nature, the only property that matters is its electron spin multiplicity denoted as μ. The radical pair I

[23] R. K. Wangness and F. Bloch, *Phys. Rev.* **89**, 728 (1953); A. G. Redfield *IBM J. Res. Develop.* **1**, 19 (1957); A. G. Redfield, *Advan. Magn. Resonance* **1**, 1 (1965).

[24] The formalism of this section is analogous to that used by R. Kaptein, *J. Amer. Chem. Soc.* **94**, 6251 (1972).

[25] For a related, density matrix based formalism, see J. I. Morris, R. C. Morrison, D. W. Smith, and J. F. Garst, *J. Amer. Chem. Soc.* **94**, 2406 (1972).

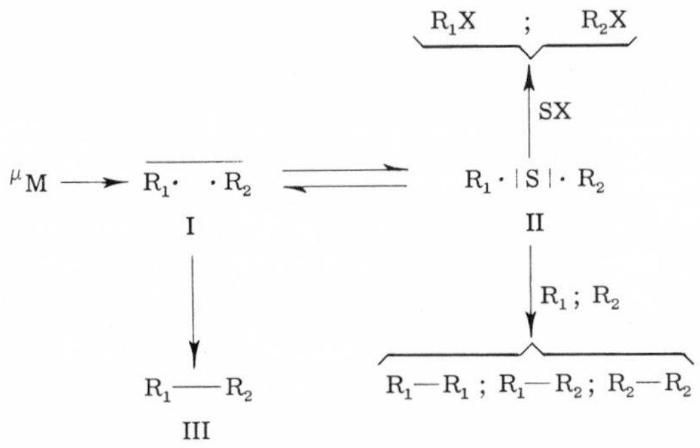

Scheme 1

resulting from the homolytic cleavage of M should be envisioned as two weakly interacting molecules, each in a doublet state. The fate of this so-called geminate radical pair is governed by the probabilities of several processes. First, the two components of the pair may combine or disproportionate to give diamagnetic coupling or disproportionation product III. Competing with this is separation of the components by diffusion leading to a solvent separated pair II. There is a finite probability that the two components of II diffuse together again and give III. Finally the individual components of the pair may each react with the solvent or other substrates to give radical transfer products (R–X), or they can encounter radicals derived from other pairs and form combination and disproportionation products. It is necessary to distinguish between combination (and disproportionation) products formed from the components of a geminate pair on the one hand, and the same product derived from the random encounter of two radicals originating from different precursor molecules on the other. We will refer to the former case as the geminate or correlated coupling product while the latter will be called the random phase or free radical coupling product.

To elaborate further on the different fates of the radical pair it should be pointed out that the lifetime of the initial pair I before diffusive separation begins must be on the time scale of Brownian motion, that is about 10^{-11} to 10^{-12} sec. In contrast, according to diffusion theory, the probability that the two components of a solvent separated pair (II) diffuse together again after times as long as 10^{-9} sec remains sizable enough to be of importance to CIDNP.[26] Finally, in a typical free radical reaction the lifetime of the individual free radicals may be as long as 10^{-4} sec. Also it is worth emphasizing that geminate combination and disproportionation

[26] R. M. Noyes, *J. Chem. Phys.* **22**, 1349 (1954); *J. Amer. Chem. Soc.* **77**, 2042 (1955); *J. Amer. Chem. Soc.* **78**, 5486 (1956); *Progr. React. Kinet.* **1**, 129 (1961).

products can be formed from either I directly (I → III) (primary recombination) or via the route I → II → I → III (secondary recombination). As will be elaborated below the first pathway contributes little if anything to CIDNP, but the probability of geminate product formation via the second route is dependent on the nuclear spin states and forms the basis of the radical pair theory of CIDNP. Before attempting a quantitative description of the theory it may be useful to give a qualitative outline of the treatment.

The processes in Scheme 1 are dependent on the electronic spin functions in several ways. In the precursor molecule M the two strongly coupled electrons of the bond to be broken may be either in a singlet state with the appropriate spin function $S = 2^{-1/2}(\alpha\beta - \beta\alpha)$, or in a triplet state defined by the three spin functions $T_+ = \alpha\alpha$, $T_0 = 2^{-1/2}(\alpha\beta + \beta\alpha)$, and $T_- = \beta\beta$. These functions are no longer eigenfunctions of the Hamiltonian operators describing the stationary states of either radical pair I or II, which must be considered mixtures of these states due to magnetic interactions to be discussed below. However, the dissociation of M occurs on a time scale too short to allow adiabatic transfer from the stationary state of M to the stationary state of I or II. The time required to transverse the region of strong scalar electron exchange coupling between the electrons of the covalent bond in M ($|J| > 10^{15}$ rad/sec, where J is the scalar exchange coupling contrast) to the region of weak or vanishing coupling in I and II respectively ($|J| < 10^{10}$ rad/sec) will be of the order of a molecular vibration followed by separation by Brownian motion ($< 10^{-11}$ sec). Since the interactions that mix singlet and triplet functions in the radical pairs are of the order of 10^8–10^9 rad/sec, application of the theorem of adiabatic processes shows that the radical pairs are formed in nonstationary spin states.[27]

Applying the same argument under time reversal leads to the conclusion that product formation by coupling or disproportionation of a geminate pair must also be a nonadiabatic process. This conclusion is particularly apparent for the special case where III and M are identical.

The consequences of this nonadiabatic behavior as far as the electron spin functions is concerned may be summarized as follows: At the time of its birth, radical pair I, and if diffusion is fast enough also pair II, will be found in a nonstationary state with spin coordinates identical to those of the precursor molecule. If the nature and magnitude of the perturbation which mixes singlet and triplet is known, the wavefunction will evolve with time in a predictable manner. If, for example, the precursor molecule dissociates from a triplet state ($\mu = T$) the electron spin function of the

[27] A. Messiah, "Quantum Mechanics," pp. 742–750. Wiley, New York, 1966.

radical pair is describable at the time of its formation by a pure triplet function. As time passes the pair will acquire singlet character. Since in most cases only the singlet state of the combination product is a bound state, the time-dependent probability of product formation should be proportional to the singlet character of the pair.

One of the perturbations mixing triplet and singlet is the hyperfine interaction which provides a mechanism for the nuclear spin dependence of geminate product formation. Considering the magnitude of hyperfine coupling in typical organic radicals (10^8–10^9 rad/sec), it becomes apparent that the time required for a significant change in the singlet character of the pair via this mechanism is at least 10^{-10} sec or a time almost two orders of magnitude longer than the lifetime of I. It therefore becomes necessary to invoke as the origin of nuclear polarization the relatively small fraction of pairs which have separated to give II and whose random walk leads to a reencounter giving I. The probability of this reencounter will therefore enter a quantitative estimate of the magnitude of polarized product formed.

Once a mechanism has been found to account for the nuclear polarization of the geminate coupling product it is obvious that the remaining products such as radical transfer products can also be polarized if the free radicals are converted to diamagnetic products within the nuclear relaxation time of the radicals. The goal of the theory is therefore to obtain an expression for the polarization in the geminate coupling (or disproportionation) product.

To put the preceding discussion on a quantitative basis one may proceed in several ways. We will present here the simplest formalism which leads to expressions strongly supported by experiments, and which is essentially taken from contributions by Kaptein and Oosterhoff,[24,28,29] Adrian,[10,11] and by ourself.[30,31] Several variations on this method have been published but none differs in any fundamental way from the one to be discussed here.[32-34]

In this formalism the details of the dissociation process of M into the radical pair I are of no consequence except that with respect to the electron spin functions the nonadiabatic or sudden approximation is applied.

[28] R. Kaptein, Ph.D. Thesis, Univ. of Leiden, Leiden, 1971.

[29] R. Kaptein and J. A. de Hollander, *J. Amer. Chem. Soc.* **94**, 6229 (1972).

[30] G. L. Closs and A. D. Trifunac, *J. Amer. Chem. Soc.* **92**, 2183 (1970).

[31] K. Muller and G. L. Closs, *J. Amer. Chem. Soc.* in press (1974).

[32] H. Fischer, *Z. Naturforsch. A* **25**, 1957 (1970).

[33] J. I. Morris, R. C. Morrison, D. W. Smith, and J. F. Garst, *J. Amer. Chem. Soc.* **94**, 2406 (1972).

[34] S. H. Glarum, *in* "Chemically Induced Magnetization" (A. R. Lepley and G. L. Closs, eds.), pp. 1–38. Wiley (Interscience), New York, 1973.

One then proceeds to evaluate the time evolution of the electron spin functions with a time-independent spin Hamiltonian, carrying the time dependence in the coefficients of the wavefunction. Diffusion and return of the components of the pair are evaluated separately with a radical pair distribution function, $f(t)$. From this and the assumption that only singlet pairs can combine, the probability of formation of the geminate coupling product in its different nuclear spin states is obtained.

The spin function of the radical pair states are generated from a basis set which is the direct product of the electron spin functions, $\phi_v(v: S; T_+; T_0; T_-)$ and the nuclear spin function, $\chi_{n'}$, where the prime on the subscript distinguishes the nuclear function in the radical pair from those in the diamagnetic product. Provided the basis set

$$\mathbf{\Psi} = \sum_v \sum_{n'} C_{vn'} \phi_v \chi_{n'} \tag{6}$$

is complete, any choice of a set for $\chi_{n'}$ can be taken because the complimentary principle of quantum mechanics requires all physical observables to be independent of the choice of a complete basis set.[35] It is convenient to choose the basic product functions of the nuclear spins, although frequently they are not eigenfunctions of the nuclear spin Hamiltonian, $\mathbf{H}^P$, of the geminate product. In this case it becomes necessary to project the singlet components of the radical pair states, given by $C_{sn'}(t)$, onto the eigenstates of $\mathbf{H}^P$. To accomplish this we define an eigenvector matrix $\mathbf{S}^P$ by the unitary transformation

$$\mathbf{S}^{P-1}\mathbf{H}^P\mathbf{S}^P = \boldsymbol{\omega}^P \tag{7}$$

which diagonalizes $\mathbf{H}^P$ to give the eigenvalue matrix $\boldsymbol{\omega}^P$ of the nuclear spin states. The direct product of $\mathbf{S}^P$ with the matrix $\mathbf{E}_S$ defined to select the singlet electronic states

$$\mathbf{E}_s = \begin{bmatrix} 1 & & & 0 \\ & 0 & & \\ & & 0 & \\ 0 & & & 0 \end{bmatrix} \tag{8}$$

gives the operator $\mathbf{Q}^P = \mathbf{E}_s \otimes \mathbf{S}^P$, whose product with its transpose forms the operator necessary to accomplish the desired projection.

$$|C_{sn}(t)|^2 = \sum_{n'} |Q^P_{nn'}|^2 C^*_{sn'} C_{sn'}. \tag{9}$$

[35] A. Messiah, "Quantum Mechanics." Wiley, New York, 1966.

Equation (9), although still describing the radical pair, anticipates the nuclear eigenstates in the geminate product and when multiplied with $f(t)$ gives the probability of the formation of the nth nuclear eigenstate of the geminate product. Integration over time and multiplication by the factor λ, expressing the probability of a radical pair I in a pure singlet state to collapse to give product III, gives the total probability vector, **P**, to be used in Eq. (3) for the calculation of the population numbers,

$$P_n = \lambda \int_0^\infty |C_{sn}(t)|^2 f(t)\, dt. \tag{10}$$

The spin Hamiltonian to be used in the description of the pair needs to contain only isotropic terms since in the rapidly tumbling components of the pair any dipolar contributions are assumed to average to zero. Although experimental conditions can be envisioned where this assumption is no longer valid, all experiments described to date are satisfactorily accounted for without any anisotropic interactions. Among the remaining scalar interactions listed

$$\mathscr{H} = \mathscr{H}_{H\cdot S} + \mathscr{H}_{H\cdot I} + \mathscr{H}_{H\cdot L} + \mathscr{H}_{L\cdot S} + \mathscr{H}_{S\cdot S} + \mathscr{H}_{S\cdot I} + \mathscr{H}_{I\cdot I}, \tag{11}$$

several may be deleted from specific consideration. The indirect nuclear coupling term $\mathscr{H}_{I\cdot I}$ is too small to be of any consequence and the nuclear Zeeman interaction $\mathscr{H}_{H\cdot I}$, although of the same order of magnitude as the hyperfine interaction $\mathscr{H}_{S\cdot I}$, can be deleted because detailed consideration shows its inclusion to be of no consequence in the calculation of P_n. The remaining terms are potential contributors to the mechanism and have to be considered. The explicit form of the Hamiltonian used is essentially the sum of the Hamiltonians of two doublet states allowing for a possible weak electron exchange interaction, $\mathscr{H}_{S\cdot S}$, between them.[36] With energy expressed in rad/sec this leads to

$$\mathscr{H} = \beta H \cdot (g_1 S_1 + g_2 S_2) - J(\tfrac{1}{2} + 2S_1 \cdot S_2) + \sum_i a_i S_1 I_i + \sum_k a_k S_2 \cdot I_k, \tag{12}$$

where the first term allows for different Zeeman energies at components 1 and 2 of the radical pair, thus incorporating the field-dependent part of spin orbit coupling, $\mathscr{H}_{S\cdot L}$, and the field–orbital interaction, $\mathscr{H}_{H\cdot L}$, on each component in the corresponding g-factor. S_1 and S_2 are the electron spin operators for the two components. The next term represents the scalar exchange interaction of magnitude $2J$ while the last two terms account for the isotropic part of the hyperfine interactions. Attention is drawn to the assumption that no cross terms are necessary, in other words

[36] D. C. Reitz and S. I. Weissman, *J. Chem. Phys.* **33**, 700 (1960).

that the electron on component 1 is believed to be coupled only to the nuclear spins on that component $(I_i, I_j, \ldots)$ but not to those nuclei located on component 2 $(I_k, I_l, \ldots)$. Within the framework of weak or even vanishing electronic coupling this assumption seems to be justified. As will be backed up later with experimental data, it appears that the best fit is obtained for no exchange interaction during that part of the lifetime of the radical pair that matters for the time evolution of the electron spin functions. Construction of the Hamiltonian matrix is straightforward and yields non-vanishing elements only between zero-order states of identical total spin angular momentum. This allows the factoring of the problem into $2lI+3$ independent matrixes for l nuclei with spin I. If we define for each nucleus a quantum number M_i and stipulate that nuclei $i, j, \ldots$ are located on component 1 and nuclei $k, l, \ldots$ are on component 2 of the radical pair the diagonal and off-diagonal elements of the Hamiltonian can be written as

$$\langle S, M_i M_j \ldots M_k M_l \ldots |\mathscr{H}| S, M_i M_j \ldots M_k M_l \ldots \rangle = J, \tag{13}$$

$$\langle T_0, M_i M_j \ldots M_k M_l \ldots |\mathscr{H}| T_0, M_i M_j \ldots M_k M_l \ldots \rangle = -J, \tag{14}$$

$$\langle T_{\pm}, M_i M_j \ldots M_k M_l \ldots |\mathscr{H}| T_{\pm}, M_i M_j \ldots M_k M_l \ldots \rangle$$
$$= -J \pm \tfrac{1}{2}[(g_1 + g_2)\beta H_0 \pm \textstyle\sum a_i M_i \pm \sum a_k M_k], \tag{15}$$

$$\langle T_0, M_i M_j \ldots, M_k M_1 \ldots |\mathscr{H}| S, M_i M_j \ldots M_k M_1 \ldots \rangle$$
$$= \tfrac{1}{2}[(g_1 - g_2)\beta H_0 + \textstyle\sum a_i M_i - \sum a_k M_k], \tag{16}$$

$$\langle T_{\pm}, M_i \mp 1, M_j \ldots M_k M_l |\mathscr{H}| S, M_i M_j \ldots, M_k M_l \rangle$$
$$= \mp(1/8^{1/2})[I_i(I_i+1) - M_i(M_i \mp 1)]^{1/2}, \tag{17}$$

$$\langle T_{\pm}, M_i \mp 1, M_j \ldots M_k M_1 |\mathscr{H}| T_0, M_i M_j \ldots M_k M_1 \rangle$$
$$= \pm(1/8^{1/2})[I_i(I_i+1) - M_i(M_i \mp 1)]^{1/2}. \tag{18}$$

Of course, there are off-diagonal elements corresponding to Eq. (17) and (18) for all other nuclei carrying the hyperfine interaction in addition to nucleus i. Inspection of the off-diagonal elements connecting T_0 with S [Eq. (16)] shows that these states are mixed from the difference in the Zeeman energies as well as the differences in the sums of the hyperfine interactions on the two components of the radical pair. Furthermore, it is noteworthy that these elements connect states whose nuclear spin functions are identical. In contrast, elements connecting S with $T_{\pm}$ [Eq. (17)] involve the hyperfine interactions only, and are associated with a change in the nuclear spin function. The physical consequences of these characteristics of mixing will be discussed in connection with the expected spectral patterns.

The time evolution of the spin functions is obtained from the time-dependent Schrodinger equation

$$i\frac{\partial}{\partial t}\Psi = \mathscr{H}\Psi, \tag{19}$$

which, on substitution of the wavefunction and the Hamiltonian, results in a set of 4κ coupled differential equations represented by

$$i\frac{\partial}{\partial t}\mathbf{C}(t) = \mathbf{HC}(t), \tag{20}$$

where $\mathbf{C}(t)$ is a column vector containing the time-dependent coefficients and $\mathbf{H}$ is the Hamiltonian matrix. With a specified initial condition, $\mathbf{C}(0)$, representing the radical pair wavefunction at the time of its birth, the general solution of (20) is given by

$$\mathbf{C}(t) = \mathbf{Z}e^{-i\boldsymbol{\omega}t}\tilde{\mathbf{Z}}\mathbf{C}(0), \tag{21}$$

where the columns of $\mathbf{Z}$ are the eigenvectors corresponding to the eigenvalues, $\boldsymbol{\omega}$, of the Hamiltonian and $\tilde{\mathbf{Z}}$ is the transpose of $\mathbf{Z}$.

The exact form of $\mathbf{C}(0)$ depends on the type of reaction to be described, but it is possible to single out two of the most common situations. In these it is assumed that all nuclear states are populated with equal probability, neglecting the small differences within the Boltzmann distribution. The electronic component of the wavefunction is treated either as a pure singlet or as a pure triplet function depending on the spin multiplicity of the precursor molecule. If we define $\mathbf{I}$ as the unit matrix of order κ,

$$\mathbf{C}(0) = \kappa^{-1/2}\mathbf{I} \otimes \mathbf{E}_v, \tag{22}$$

where $\mathbf{E}_v$ is given by (8) if the precursor dissociates from a singlet state and by

$$\mathbf{E}_T = (1/3^{1/2})\begin{bmatrix} 0 & & & \\ & 1 & & \\ & & 1 & \\ & & & 1 \end{bmatrix} \tag{23}$$

for triplet precursors. The definition of E_T neglects to allow for the Boltzmann distribution in the triplet sublevels, an assumption which may be correct for very short lifetimes of the triplet precursor. However, if the triplet levels are known to be equilibrated before dissociation sets in a better definition of E_T is

$$\mathbf{E}_T = (1/3^{1/2}) \begin{bmatrix} 0 & & & \\ & (1-g\beta H_0/kT)^{1/2} & & \\ & & 1 & \\ & & & (1+g\beta H_0/kT)^{1/2} \end{bmatrix}. \tag{24}$$

For practical purposes the use of (24) instead of (23) will have only a negligible effect on the final results.

Multiplication of $\mathbf{C}(t)$ by its complex conjugate and substitution into 9 gives the singlet character of the nuclear eigenstates as

$$|\mathbf{C}_s(t)|^2 = \tilde{\mathbf{Q}}^P \mathbf{Z} e^{-i\omega t} \tilde{\mathbf{Z}} |\mathbf{C}(0)|^2 \mathbf{Z} e^{i\omega t} \tilde{\mathbf{Z}} \mathbf{Q}^P. \tag{25}$$

To evaluate the individual components of the column vector represented by (25) we introduce two matrixes, U^P and ${}^{\mu}Y$, defined by (26) and (27) respectively. In this way summations pertaining to the eigenvectors of the geminate product are separated from those reflecting the initial condition of the radical pair states. With these definitions (25) is transformed to

$$U^P_{nl} = \sum_m Q^P_{mn} Z_{ml}, \tag{26}$$

$${}^{\mu}Y_{lj} = \sum_r Z_{rl} |C_{rr}(0)|^2 Z_{rj}, \tag{27}$$

$$|{}^{\mu}C_{sn}(t)|^2 = \sum_{lj} U_{nl} U_{nj} {}^{\mu}Y_{lj} e^{i(\omega_j - \omega_l)t}. \tag{28}$$

Before further progress can be made a suitable model must be found for the diffusive behavior of the components of the radical pair. As outlined above the total probability of geminate product formation is governed not only by the singlet character of the pair but also by the probability of an encounter of the components [Eq. (10)]. In the early communications on the radical pair theory an exponential distribution function was adopted with marginal success.[8,9,30] As was first pointed out by Adrian[10,11] and later further elaborated by Kaptein,[28] a better function is provided by the diffusion theory developed by Noyes.[26] The model treats the components of a dissociating molecule as hard spheres on a random walk with no potential between them. Although this model is generally believed to be too simplistic for many applications, the CIDNP spectra calculated with its aid are in surprisingly good agreement with experiment.

According to this theory the probability of a reencounter, $f(t)$, of two molecules which have separated from an encounter is given by

$$f(t) = mt^{-3/2}\, e^{-\pi m^2/\beta'^2 t}, \tag{29}$$

where β' is the total probability of at least one reencounter, $\beta' = \int_0^\infty f(t)\,dt$, and m is a constant containing the root mean square diffusive displacement, σ, the encounter diameter, ρ, and the frequency, v, of the displacements. Explicitly β' and m are given by

$$\beta' = 1 - (1/2 + 3\rho/2\sigma)^{-1}, \tag{30}$$

$$m = 1.036(1-\beta')^2(\rho/\sigma)^2 v^{-1/2}. \tag{31}$$

At intervals sufficiently large to be of importance in the time development of the spin functions in the radical pair the exponential part of (29) goes to unity so that a good approximation for $f(t)$ is

$$f(t) = mt^{-3/2}. \tag{32}$$

As a sideline, it may be noted that the same basic reason that permits this simplification for $f(t)$ is probably responsible for the success of the Noyes diffusion theory in the calculation of CIDNP spectra. Because of the small magnitude of the perturbation mixing singlet and triplet, it does not matter how poor a description of the diffusive behavior is given by $f(t)$ at short times, as long as the model gives the correct behavior for long time intervals.[37]

With the adoption of (32) the general solution obtained from substitution of (28) and (32) into (10) and integration is

$$^{\mu}P_n = \lambda \sum_{lj} U^P_{nl}\, U^{P,\mu}_{nj} Y_{lj}(\beta' - m[2\pi\,(\omega_j - \omega_l)]^{1/2}). \tag{33}$$

Before a more detailed discussion of the possible spectral types anticipated from (33) is presented, it will be useful to introduce a practical simplification of the model valid under high field reaction conditions.

2. *The High Field (T_0–S) Approximation*

A considerable simplification of the calculations is possible if the assumption is made that mixing between the T_0 and S states of the radical pair is much more pronounced than mixing among any other electronic states. This condition is fulfilled if the Zeeman splitting is large compared to the

[37] For a discussion of the shortcoming of the model, see J. M. Deutch, *J. Chem. Phys.* **56**, 1004 (1972).

hyperfine and exchange interactions.[8,9] The assumption seems to be valid for radical pairs at sufficiently high fields ($H_0 > 10^3$ G). Although the real justification for this simplification is found in the experimental data, a detailed consideration of the model leads to the same conclusion. As has been pointed out above, the evolution of the wavefunction proceeds sufficiently slowly when measured on the time scale of diffusion that only separated pairs (II) contribute noticeably to the effect. It is only reasonable to expect the exchange coupling to drop to very low or vanishing values once the components of the radical pair have separated by several diffusive steps. As will be shown below, for most, if not all high field cases reported, the best agreement of the calculations with experiment is obtained for a vanishing exchange coupling.

Adoption of the high field approximation makes unimportant all off-diagonal elements connecting states with different z-components of the electron spin angular momentum. The only elements remaining are those of the form (16) connecting states with identical nuclear spin functions. As a result, the Hamiltonian factors into κ 2×2 matrixes and an explicit solution can be written for the time evolution of the wavefunction. Incorporating the initial conditions of the radical pair states, the solutions are

$$|{}^{\mathrm{S}}\mathbf{C}_s(t)|^2 = \mathbf{Q}^P[1-(\mathbf{H}_{ST_0}^2/\boldsymbol{\omega}^2)\sin^2\boldsymbol{\omega}t]\tilde{\mathbf{Q}}^P \tag{34}$$

and

$$|{}^{\mathrm{T}}\mathbf{C}_s(t)|^2 = \tfrac{1}{3}\mathbf{Q}^P[(\mathbf{H}_{ST_0}^2/\boldsymbol{\omega}^2)\sin^2\boldsymbol{\omega}t]\tilde{\mathbf{Q}}^P \tag{35}$$

for singlet and triplet precursors, respectively. In (34) and (35) the nuclear eigenstates of the geminate product have been projected out of the radical pair states with the operator $\tilde{\mathbf{Q}}^P$, in a strictly analogous fashion to that described for the general solution. The appropriate off-diagonal elements, $\mathbf{H}_{ST_0}$ are given by (16) and the eigenvalues are

$$\boldsymbol{\omega} = \pm(J^2+\mathbf{H}_{ST_0}^2)^{1/2}. \tag{36}$$

Using the radical pair distribution function of Noyes, it is now possible to write the probability vector $\mathbf{P}$ in its explicit form for a given initial condition of the radical pair. For a singlet precursor the substitution of (34), (32), and (31) into (10) and integration yields

$${}^{\mathrm{S}}P_n = \lambda\sum_{n'}|Q_{nn'}^P|^2(\beta'-B_{n'}) \tag{37}$$

and

$$B_{n'} = m\pi^{1/2}|H_{ST_{0},n'}|^2/\omega_{n'}^{3/2}, \tag{38}$$

where the summation is over all radical pair states. Similarly, the use of (35) gives the explicit expression for the probabilities of populating the nuclear spin states of the geminate product when the radical pair precursor is a triplet state.

$$^{\mathrm{T}}P_n = \tfrac{1}{3}\lambda \sum_{n'} |Q_{nn'}^P|^2 B_{n'}. \tag{39}$$

The equations are closely related to those proposed by Adrian who assumed a vanishing exchange coupling during the lifetime of the solvent separated radical pair. With this assumption the dependence of the polarization becomes proportional to the square root of the off-diagonal elements in the spin Hamiltonian because $\omega_{n'}$ goes to $H_{ST_0,n}$ under those conditions. This yields the simple relationships

$$^{\mathrm{S}}P_n^0 \propto (a - |H_{ST_0,n}|^{1/2}), \tag{40}$$

$$^{\mathrm{T}}P_n^0 \propto |H_{ST_0,n}|^{1/2}, \tag{41}$$

where the superscript 0 indicates the vanishing exchange coupling.

However, Eqs. (37) and (39) are deficient in the description of the diffusion behavior of radical pairs. The radical pair distribution function of Noyes (29) expresses only the probability of the first reencounter after an initial separation. The neglect of additional reencounters can be justified for pairs with large λ when they are generated from singlet precursors. Under those conditions the fraction of pairs contributing to the geminate product, having undergone two or more successive reencounters, will be small because the probability of combination at the first reencounter is rather high. However, this argument does not hold for pairs derived from triplet precursors. Since the spin functions evolve relatively slowly, the probability of combination at successive reencounters increases and must be taken into account. Kaptein[24] has shown that a sufficiently accurate approximation for this behavior introduces a factor of $(1-\beta')^{-1}$ into (39).

$$^{\mathrm{T}}P_n = [\lambda/3(1-\beta')] \sum_{n'} |Q_{nn'}^P|^2 B_{n'}. \tag{42}$$

At this point it should be mentioned that polarization in coupling and disproportionation products can also result from encounters of two radicals with uncorrelated spins. In reality this occurs whenever free radicals derived from different precursor molecules come to an encounter and form a product. An ensemble of radical pairs with uncorrelated electron spins may be envisioned as an incoherent superposition of the four basis states, ϕ_v. Because of the condition that the coupling product be formed only from singlet pairs, the steady state of the reacting system will have greater than statistical triplet character. This leads to the intuitive prediction that polarization in the coupling product should be qualitatively similar to that observed in reactions with triplet precursors.[38] A more quantitative description is based on the following consideration.

When free radicals with uncorrelated spins meet in an encounter, the probability of forming any of the four basis states is the same. The fraction separating from this first encounter is $1-\lambda/4$. It should be noted that the combination product formed from this first encounter does not carry any polarization and that the initial condition to be used in the equations for the time evolution is the one after the first encounter. This leads to the definition of a third form of $\mathbf{E}_\nu$ [Eq. (22)] given by

$$\mathbf{E}_\mathrm{F} = \tfrac{1}{2}\begin{bmatrix} (1-\lambda)^{1/2} & & & \\ & 1 & & \\ & & 1 & \\ & & & 1 \end{bmatrix}, \tag{43}$$

where the subscript F denotes the application to polarization derived from uncorrelated pairs. Substitution of (43) into (22) and (25) will give the correct time evolution of the spin function.

Within the high field approximation in which T_+ and T_- are thought to contribute very little to the polarization, the explicit expression for the development of singlet character in the radical pair is given by

$$|{}^\mathrm{F}\mathbf{C}_s(t)|^2 = \tfrac{1}{2}[(1-\lambda)|{}^\mathrm{S}\mathbf{C}_s(t)|^2 + |{}^\mathrm{T}\mathbf{C}_s(t)|^2]. \tag{44}$$

Inspection of (44) reveals that the time evolution is a weighted superposition of the evolution of radical pairs generated from the singlet and triplet precursors. Substitution of (37) and (42) into (44) and multiplication by the Noyes distribution function followed by integration gives the components of the vector expressing the probabilities of populating the nuclear states in the coupling product formed from uncorrelated pairs

$${}^\mathrm{F}P_n = \tfrac{1}{4}\sum_{n'} |Q_{nn'}^P|^2 [1 + D(\beta'(1-\lambda) + \lambda B_{n'}], \tag{45}$$

where

$$D = 1/(1-\beta'[1-\tfrac{1}{2}\lambda(1-\lambda)]) \tag{46}$$

The derivation of (45) includes multiple reencounter and has incorporated the unpolarized fraction $\lambda/4$ of the coupling product formed in the first encounter.[24]

With the derivation of expressions for the population vector ${}^\mu\mathbf{P}$ for the three possible initial conditions completed [Eqs. (37), (42), and (45)] it is

[38] G. L. Closs and A. D. Trifunac, *J. Amer. Chem. Soc.* **92**, 2186 (1970).

now possible to calculate the NMR transition intensities of the coupling and disproportionation products by substituting ${}^{\mu}\mathbf{P}$ into (3) and by integrating the simultaneous differential equations resulting from the relaxation matrix **R**. Once the population numbers for the κ nuclear spin states have been obtained the NMR intensities are given by Eq. (1).

As has been pointed out above, free radicals escaping combination or disproportionation can be converted into other diamagnetic products via radical transfer reactions with scavengers RX as shown in Scheme I. These products can also be polarized and the derivation of a quantitative expression is particularly simple within the framework of the high field approximation.[39,40] Since at high field the only important contributions are obtained from T_0–S mixing which does not affect the z-component of the nuclear spins, the sum of the vector ${}^{\mu}\mathbf{P}$ and of the corresponding vector ${}^{\mathrm{r}}\mathbf{P}$, denoting the probabilities of the individual nuclear spin states being distributed in the free radical transfer products, must be

$$ {}^{\mu}\mathbf{P} + {}^{\mathrm{r}}\mathbf{P} = \mathbf{I}/\kappa, \tag{47}$$

where **I** is the unit vector of dimension κ. Equation (47), however, is based on the assumption that during the lifetime of the free radicals nuclear relaxation is unimportant. While nuclear relaxation can be safely neglected in the treatment of radical pairs because of their short lifetime, this is no longer permissible in the free radicals. Here lifetimes of the order of 10^{-4} sec are not uncommon, the same order of magnitude as expected for nuclear relaxation in organic radicals. To include relaxation in the escaping free radicals we set up the coupled differential equations for the population numbers of the free radicals, ${}^{\mathrm{r}}N$, as

$$ (d/dt)^{\mathrm{r}}\mathbf{N} = v(\mathbf{I}/\kappa - {}^{\mu}\mathbf{P}) - [k_{\mathrm{sc}}[\mathrm{RX}] + {}^{\mathrm{r}}\mathbf{R}]^{\mathrm{r}}\mathbf{N}, \tag{48}$$

where v is the rate of radical formation, k_{sc} is the rate constant of the radical scavenging reaction, and ${}^{\mathrm{r}}\mathbf{R}$ is the nuclear relaxation matrix for the radical. The equations can be easily integrated but for most practical purposes it suffices to use the solution obtained from the steady state assumption which is the correct solution for a constant rate.

$$ {}^{\mathrm{r}}\mathbf{N}^{\mathrm{ss}} = v(\mathbf{I}/\kappa - {}^{\mu}\mathbf{P})/(\mathbf{I}k_{\mathrm{sc}}[\mathrm{RX}] - {}^{\mathrm{r}}\mathbf{R}). \tag{49}$$

Using this equation the population numbers of the diamagnetic escape product are then given by

$$ (d/dt)\,{}^{\mathrm{e}}\mathbf{N} = k_{\mathrm{sc}}[\mathrm{RX}]\,v(I/\kappa - {}^{\mu}\mathbf{P})/(\mathbf{I}k_{\mathrm{sc}}[\mathrm{RX}] - {}^{\mathrm{r}}\mathbf{R}) - {}^{\mathrm{e}}\mathbf{R}\,{}^{\mathrm{e}}\mathbf{N}, \tag{50}$$

[39] G. L. Closs and A. D. Trifunac, *J. Amer. Chem. Soc.* **92**, 7227 (1970).
[40] R. Kaptein, *J. Amer. Chem. Soc.* **94**, 6257 (1972).

where ${}^{e}\mathbf{R}$ is the matrix describing relaxation in the diamagnetic escape product.

If the elements of ${}^{r}\mathbf{R}$ are very much smaller than $k_{sc}[\mathrm{RX}]$ then (50) tends to

$$(d/dt)\,{}^{e}\mathbf{N} = v(\mathbf{I}/\kappa - {}^{\mu}\mathbf{P}) - {}^{e}\mathbf{R}\,{}^{e}\mathbf{N}. \tag{51}$$

A comparison of (51) with (3) shows that the sums of the population numbers in the escape product and the geminate product are a constant for all nuclear spin states. This is another way of stating that at high fields in the absence of relaxation in the free radicals the polarization when averaged over the whole ensemble will remain identical to the equilibrium value. This fact is the direct result of the sole contribution of T_0–S mixing and the statement is no longer true at very low fields where unequal and finite contributions from T_+ and T_- states can induce a net polarization even in the absence of relaxation. As will be shown in Section V, net polarization is also observed in reactions proceeding through diradicals.

IV. The Characteristics of CIDNP Spectra

A. Qualitative Predictions at High Field

The radical pair theory as outlined above provides qualitative predictions on the intensities of NMR transitions without the necessity of resorting to specific calculations. These predictions are particularly easy to make under conditions where the high field approximation is valid, which is the case for the overwhelming majority of the experiments reported to date. The most fundamental prediction concerns the sign of the NMR transition, that is, whether a line occurs as an enhanced absorption (A) or in emission (E). Before discussing specific examples it is useful to distinguish between two fundamentally different spectral patterns which can be observed in CIDNP spectra.

When a multiplet of NMR lines originating from transitions of a single nucleus or a group of identical nuclei contain both A and E lines, the spectrum is said to exhibit the multiplet effect. In a pure multiplet effect the sum of the integrated line intensities originating from the ith nucleus vanishes after corrections have been made for the transition probability and the equilibrium polarization. This condition is expressed by

$$\sum [{}^{i}\mathrm{Int}_{mn}/\Omega - {}^{i}\mathrm{Int}^{0}_{mn}/\Omega] = 0, \tag{52}$$

where the summation is over all transitions of nucleus i, and where ${}^{i}\mathrm{Int}^{0}_{mn}$ are the intensities observed for a system in thermoequilibrium. For the observation of the multiplet effect it is of course necessary that at least some of the degeneracies of the transitions be lifted by indirect nuclear spin-spin coupling. Condition (52) is contained in the summation

$$\sum_{\kappa} \gamma_i H_0 (N_k - N_k^0) M_i = 0, \tag{53}$$

where M_i is the quantum number of the ith nucleus in state k and the sum extends over all κ nuclear spin states. Inspection of (53) shows that systems exhibiting the multiplet effect can be characterized as having a total nuclear Zeeman energy identical to the equilibrium value.

If (52) and (53) do not describe the spectrum and the population distribution, respectively, the spectrum is classified as showing a net effect, where net emission or enhanced absorption is associated with the lines originating from a given nucleus. There exists a special case of the net effect where the total magnetization of the product is still unchanged from thermoequilibrium but in which the individual spin-spin splitting multiplet show net absorption or emission. This case is expressed by

$$\sum_{i} \sum_{k} \gamma_i H_0 (N_k - N_k^0) M_i = 0, \tag{54}$$

where the summation extends over all states and all nuclei. It should be pointed out that condition (54) is a necessary but not sufficient condition for the multiplet effect.

These spectral patterns are more easily understood with the aid of Fig. 1 where a two-spin system ($I = 1/2$) is given as an example. A population distribution as shown in Figs. 1a and 1b corresponds to a pure multiplet effect while the distribution in Fig. 1c gives rise to a net effect although Eq. (54) is still applicable.

It is customary to classify the multiplet effect by its "phase." If the spectrum is displayed in the field sweep mode and is "read" from low to high field the phase is defined as *AE* if enhanced absorption precedes emission and as *EA* if the order is reversed. As is shown in Fig. 1 the phase of the multiplet effect is not only determined by the population pattern of the nuclear energy levels but is also a function of the sign of the nuclear spin coupling constant. This distinguishes CIDNP spectra with multiplet effects from those with strong net effects which, like ordinary NMR spectra, are not sensitive to sign reversal of the spin coupling constants. On some occasions, when all signs of the other parameters which influence the phase of the multiplet effect are known, CIDNP might be useful for the determination of signs of nuclear coupling constants.

The two opposite population patterns leading to reversal of the multiplet phase are either an overpopulation of the states characterized by the smallest absolute value of the z-component of magnetization or a depletion of the population of those states. Since the forcing term in Eq. (3) is

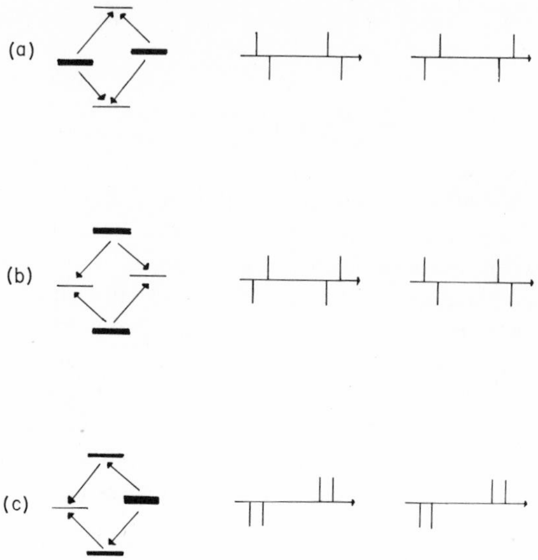

FIG. 1. Schematic representation of energy levels and corresponding CIDNP spectra for a two-spin-1/2 system. Spectra are drawn for positive (middle) and negative (right) spin coupling constants.

governed by the magnitude of the individual components of the probability vector **P**, a qualitative discussion can be restricted to an evaluation of the magnitude of B_n [Eq. (38)], a choice of the appropriate precursor condition, and the selection of the correct equations describing either the geminate combination product or the free radical escape product.

Within the high field approximation B_n contains the magnitude of the off-diagonal elements connecting the S state and the T_0 state with identical nuclear spin states, raised to some power the exact form of which depends on the assumption made about the magnitude of the exchange coupling constant J. As shown above [Eq. (16)],

$$H_{ST_0,n'} = \tfrac{1}{2}\left[\beta H_0(g_1 - g_2) + \sum_i a_i M_i - \sum_k a_k M_k\right] \tag{55}$$

where the first summation is over all the hyperfine interactions on component 1 and the second over all interactions on component 2 of the radical pair. If we consider first cases in which the g-factors of the two radical pair components are identical, it is readily seen that B_n for the nth nuclear spin state depends on the absolute signs of the hyperfine interactions, a_i and a_k, and on the distribution of the individual nuclei contributing to the spectrum among the two components of the radical pair. For example, if a two-proton spectrum is observed in the geminate coupling product, it is necessary to know whether the two nuclei were on the same or on opposite components of the radical pair because in the former case the hyperfine interactions must be added and in the latter they are subtracted to obtain the magnitude of B_n for each of the four possible spin states. Since only the magnitude of B_n is of importance it is easily seen that

$$|B_{m_z}| = |B_{-m_z}|, \tag{56}$$

where the index m_z characterizes the nuclear spin state by its total z-component of the magnetization. With the components of **P** being a direct function of B_n the condition (56) is equivalent to the condition (53) which is pertinent to pure multiplet effect spectra. We can therefore state as a rule: *If the two components in the radical pair have identical g-factors the observed polarization pattern is a pure multiplet effect.* Inspection and comparison of Eqs. (40), (42), and (45) shows that the phase of the multiplet effect will be opposite for the geminate product derived from radical pairs with singlet and triplet precursors and will be the same for the product from triplet precursors and those derived from random pair combination (F-reactions). Similarly, the multiplet effect in radical transfer products (escape products) will have a phase opposite to that observed for the corresponding geminate combination product.

These various relationships governing the phase of the multiplet effect are conveniently summarized in a simple algorithm[41]

$$\mu\, a_i\, a_j\, J_{ij}\, \sigma_{ij}\, \varepsilon \} \, EA \tag{57}$$

in which the various parameters have either positive or negative signs according to the convention

μ: + for triplet and F-reactions
– for singlet reactions

$a_i; a_j$ + absolute sign of hyperfine coupling positive
– absolute sign of hyperfine coupling negative

J_{ij} + absolute sign of nuclear spin coupling positive
– absolute sign of nuclear spin coupling negative

σ_{ij} + when nuclei i and j are on same component of radical pair
– when nuclei i and j are on opposite components of radical pair

ε: + for coupling and disproportionation products
– for escape products.

[41] R. Kaptein, *Chem. Commun.* p. 732 (1971).

With the additional definition of $-EA = AE$ the rule is applied by assigning the appropriate sign to each of the parameters in (57) and by determining the sign of the product of the six signs. The phase of the multiplet effect is EA for a positive product and AE for a negative product.

If the components of the radical pair differ in their electron Zeeman energies ($g_1 - g_2 \neq 0$), then Eq. (56), and, by analogy to the considerations elaborated above, Eq. (52) are no longer true, and the spectrum will exhibit net effects. The prediction of the sign of the net effect can be summarized in an equally efficient algorithm[41]

$$\mu\, a_i\, \Delta g\, \varepsilon \} A_i, \tag{58}$$

which predicts whether the transitions originating from the ith nucleus (or group of equivalent nuclei) in the product will show a net enhanced absorption (A) or net emission (E). Three of the parameters in (58) are the same as in (57) and are defined accordingly. The fourth, Δg, is given by the sign of the difference, $g_1 - g_2$, where g_1 is always the g-factor of that component of the radical pair on which the ith nucleus is located. With the additional definition, $-A = E$, the sign is predicted in the same fashion as outlined for the multiplet effect.

The qualitative predictions derived from the radical pair theory summarized in (57) and (58) were recognized quite early[8,9,30] and a number of experiments were designed to test them. Perhaps the most striking and chemically most useful prediction is the dependence of the phase of the multiplet effect and of the sign of the net effect on the spin multiplicity of the precursor molecule. A reaction scheme designed to test this and several other predictions of the theory is centered around the benzyl-benzhydryl radical pair (VI) which can be generated by several different routes encompassing all three precursor types. The reaction sequence is summarized in Scheme 2. Radical pair VI may be generated via hydrogen abstraction

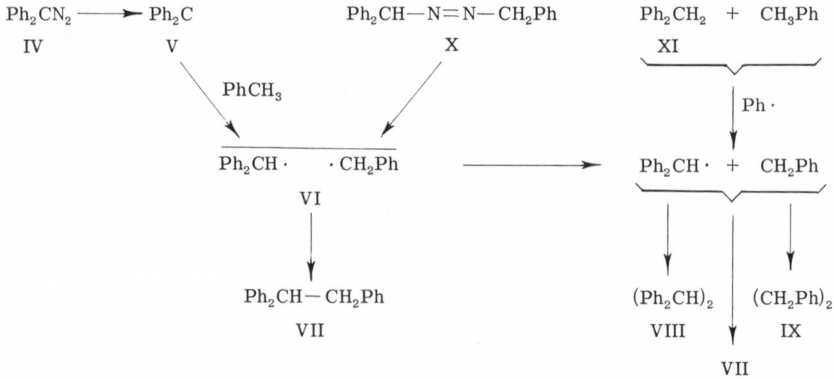

Scheme 2

from toluene by diphenylmethylene (V), a molecule known to have a triplet ground state, and generated from diphenyldiazomethane (IV) by photolysis or pyrolysis. The geminate coupling product 1,1,2-triphenylethane (VII) is formed together with the random phase coupling products 1,2-diphenylethane (IX) and 1,1,2,2-tetraphenylethane (VIII). A substantial fraction of VII is also formed via the F-reaction pathway because the amount of geminate combination in reactions with triplet precursors is always small because the evolution of the singlet character in the pair is slow compared to diffusion. Neglecting the aromatic protons, the geminate coupling product (VII) forms a three spin, A_2B type spectrum which can be analyzed in first order with the A_2X approximation. Since the benzyl and benzhydryl radicals are expected to have almost identical g-factors, close to the free electron value, one predicts a multiplet effect for the coupling product. The parameters required in (57) are either known or can be extrapolated with considerable confidence from analogous cases. The precursor is a triplet molecule (μ: +), the product of the signs of the two hyperfine interactions is almost certainly positive, as is the sign of the nuclear coupling constant in VII. The location parameter is negative because the protons which give rise to the spin-spin coupling multiplet are on different components of the pair. Finally, one is looking at the combination product, giving ε a positive sign. The signs of the parameters, listed in the same sequence as in (57), are

$$++++-+EA = -EA = AE$$

predicting an AE multiplet effect. The spectrum is shown in Fig. 2a and bears out the prediction.[7]

The second entry into the system is via the thermal decomposition of the azo compound X. Formally, the same radical pair will be generated

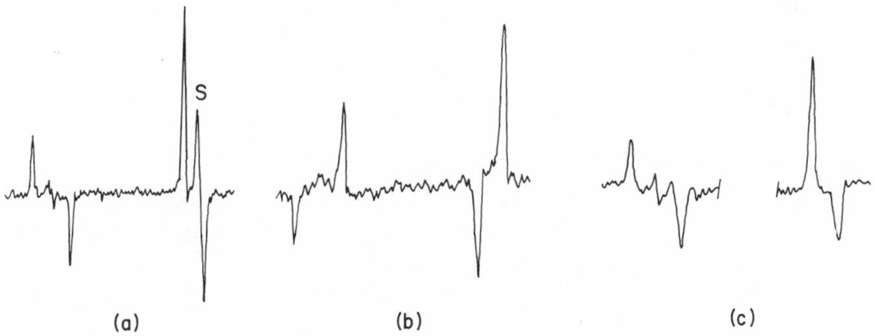

FIG. 2. Spectra at 60 MHz of the benzylic protons of 1,1,2-triphenylethane (VII) generated from (a) a triplet precursor (V), (b) a singlet precursor (X), and (c) by random encounter of benzyl and benzhydryl radicals. In this and all subsequent spectra the field increases from left to right. S denotes ^{13}C satellite of solvent.

this way except that the precursor is now a singlet state. With all the other parameters remaining the same, the geminate coupling product should show an *EA* multiplet effect, a polarization pattern of opposite sign to that displayed in the reaction of diphenylmethylene. Comparison of Figs. 2a and 2b shows that this prediction is again confirmed by experiment.[42]

Finally it is possible to generate the pair VI by diffusion of uncorrelated radicals. Hydrogen abstraction by phenyl radicals, generated from dibenzoylperoxide, provides a way to generate the radicals from toluene and diphenylmethane (XI) and assures uncorrelated spins. Figure 2c shows the spectrum of the combination product VII with an *AE* multiplet effect as is predicted by rule (57).

Other examples have been reported where changes in precursor multiplicity cause a reversal of the observed polarization pattern. Notable in this context are experiments involving the photosensitized decomposition of benzoylperoxide (XII). The thermal decomposition of this compound is of interest as the first reaction in which the CIDNP effect was observed by Bargon *et al.*[1] Although the decomposition of benzoylperoxide is a fairly complicated process,[43] the salient features of the reaction are summarized in the simplified Scheme 3.

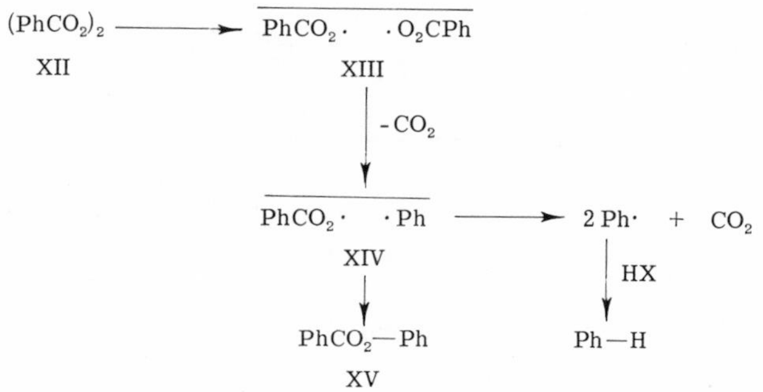

Scheme 3

Homolytic cleavage of the peroxide bond in XII generates the short-lived radical pair XIII which is the precursor of pair XIV. Geminate combination of XIV gives phenyl benzoate in competition with the hydrogen abstraction of the escaping phenyl radicals from the solvent. The *g*-factor difference in pair XIV gives rise to a net effect in phenyl benzoate, XV, and benzene.

[42] G. L. Closs and A. D. Trifunac, *J. Amer. Chem. Soc.* **91**, 4554 (1969); *J. Amer. Chem. Soc.* **92**, 2186 (1970).

[43] A detailed review of the decomposition of aryl peroxides is given by R. C. P. Cubbon, *Progr. React. Kinet.* **5**, 29 (1970).

The prominent emission line of the escape product, benzene, is predicted by rule (58). A singlet precursor, a positive hyperfine coupling of the ortho protons in the phenyl radical and a negative Δg, together with a negative ε give $-A = E$. Although the benzoyloxy radical is too short-lived to be observed directly by ESR, it appears reasonable that its g-factor should be larger than that of phenyl ($g = 2.0022$). Photosensitized decomposition produces a spectrum with opposite polarization if the sensitizers are ketones, thus implying a decomposition from the triplet state of the peroxide.[44]

The influence of g-factor differences in radical pairs was studied in a series of experiments centered around the hydroxybenzyl benzhydryl radical pair (XVI) which is easily generated via hydrogen abstraction by photoexcited benzaldehydes from diphenylmethanes.[45] Substitution in the para position by halogens provides a way to change g-factors in a predictable manner since it is well known that introduction of heavy nuclei at carbon atoms with positive spin density raises the g-factor above the free electron value in benzyl-type radicals.[46] The reaction and the substituent combinations studied are shown in Scheme 4, (a)–(e).

$$^{3}\mathrm{ArCHO} + \mathrm{CH_2AR'_2} \longrightarrow \overline{\mathrm{Ar\dot{C}HOH} \quad \mathrm{\dot{C}HAR'_2}} \longrightarrow \text{Free radical products}$$

XVI

$$\downarrow$$

$$\mathrm{ArCH(OH)CHAr'_2}$$

XVII

(a) Ar = p-Br-C_6H_4- ; Ar′ = p-H-C_6H_4-

(b) Ar = p-Cl-C_6H_4- ; Ar′ = p-H-C_6H_4-

(c) Ar = p-H-C_6H_4- ; Ar′ = p-H-C_6H_4-

(d) Ar = p-H-C_6H_4- ; Ar′ = p-Cl-C_6H_4-

(e) Ar = p-H-C_6H_4- ; Ar′ = p-Br-C_6H_4-

Scheme 4

The benzylic protons of the geminate coupling product form a four-line subspectrum of the AB type and are shown in Fig. 3a–e. The qualitative interpretation of these spectra is straightforward with the aid of rules (57) and (58). The signs of the parameters are easily assigned by reasoning from analogy and from inspection of Scheme 4. It is known that hydrogen abstraction by excited aromatic carbonyl compounds involves the triplet

[44] S. R. Fahrenholtz and A. Trozzolo, *J. Amer. Chem. Soc.* **93**, 251 (1971); M. Lehnig and H. Fischer, *Z. Naturforsch. A* **24**, 1963 (1970).

[45] G. L. Closs, C. E. Doubleday, and D. R. Paulson, *J. Amer. Chem. Soc.* **92**, 2185 (1970).

[46] J. Sinclair and D. Kivelson, *J. Amer. Chem. Soc.* **90**, 5074 (1968).

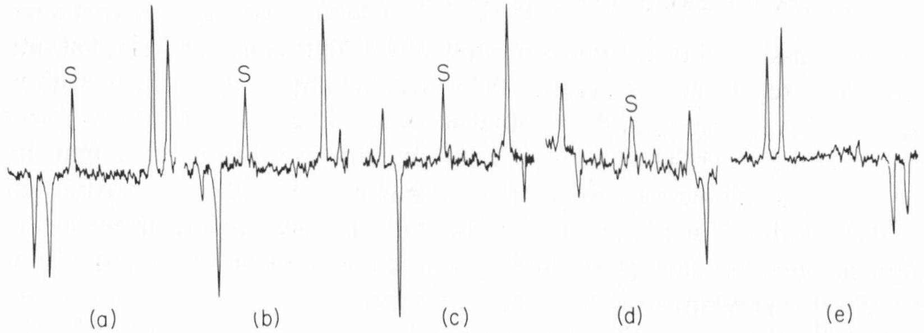

FIG. 3. Spectra of the benzylic protons of 1,2,2-triarylethanol (XVII) at 60 MHz. The spectra (a)–(e) correspond to substituent combinations (a)–(e) in Scheme 4. S denotes ^{13}C satellite of solvent.

state and that vicinal nuclear coupling constants are positive. Similarly, the hyperfine coupling constants of the benzylic protons are known to be negative.

Inspection of Fig. 3 shows that reactions (a) and (e) give almost pure net effect spectra with the expected signs. The low field proton in reaction (a) is predicted to emit because $\Delta g > 0$ in the appropriate pair. In reaction (e) the g-factor difference is reversed and so is the polarization. To predict the polarization of the high field proton it must be remembered that the definition of rule (58) includes the statement that g_1 always refers to the component of the radical pair which carries the nucleus whose net effect is to be predicted. Therefore in reaction (a) $\Delta g < 0$ if the high field proton is to be examined!

Reactions (b), (c), and (d) produce spectra which are superpositions of net and multiplet effects. By applying rule (57) with the appropriate parameters the multiplet effect is predicted to be *AE*, in agreement with the observation. It is noteworthy that reaction (c) with no halogen substituents still shows some net effect, the sign of which indicates a slightly larger g-factor for the hydroxybenzyl than for the benzhydryl radical. The larger spin-orbit coupling parameter of the oxygen atom compared to carbon atoms accounts for the small difference.

A similar series of g-factor changes has been studied in the benzhydryl–benzyl radical pair (VI) generated with the reactions discussed above (Scheme 2). By introducing p-chloro and p-bromo substituents at both components of the pair, a series of nine reactions was generated. Evaluation of the CIDNP spectra gave g-factors in good agreement with those obtained by ESR.[47]

[47] G. L. Closs, *Proc. 23rd Int. Congr. Pure Appl. Chem. Spec. Lect.* **4**, 55 (1971).

B. Quantitative Predictions at High Field

The discussion of quantitative aspects of CIDNP is usefully divided into predictions regarding the relative intensities of the transitions in a given spectrum on the one hand and the absolute intensities or enhancement factors on the other. Much experimental information on relative intensities is currently available and good progress has been made in accounting for the data within the framework of the radical pair theory. The state of affairs is much less satisfactory with regard to the absolute intensities. Here, considerable experimental difficulties have prevented in the past the evaluation of reliable enhancement factors, and the radical pair theory provides much less certain predictions. Nevertheless data have begun to accumulate slowly and with the advent of Fourier transform spectroscopy in particular several of the experimental problems can be overcome.

1. *Relative Intensities of Transitions in CIDNP Spectra*

As inspection of Eq. (1)–(3) shows, relative as well as absolute intensities of CIDNP spectra can be expected to change as a function of time. In practice, however, it is found that more often than not the relative intensities of a multiline spectrum will change very little over the major fraction of the reaction and that it is permissible to introduce the concept of a "quasi-steady state" to describe the reacting system. The justification for this becomes apparent if Eqs. (1) and (2) are combined and written as

$$\frac{d}{dt}\frac{\text{Int}_{mn}}{\Omega_{mn}} \propto \frac{d}{dt}(N_m - N_n) = v(P_m - P_n) + \sum_{i \neq m} (W_{i\to m} N_i - W_{m\to i} N_m) - \sum_{i \neq n} (W_{i\to n} N_i - W_{n\to i} N_n). \quad (59)$$

Introducing the equilibrium population numbers, N_i^0, which are functions of time because the product is continuously generated, we define

$$\Delta N_i = N_i - N_i^0; \qquad W_{i\to m} + W_{m\to i} = 2W_{im}. \quad (60)$$

Using (60), (59) can then be rewritten to give

$$(d/dt)(\text{Int}_{mn} - \text{Int}_{mn}^0)/\Omega_{mn} \propto (d/dt)(\Delta N_m - \Delta N_n) = v(P_m - P_n) - 2W_{mn}(\Delta N_m - \Delta N_n) + \sum_{i \neq n,m} [W_{im}(\Delta N_i - \Delta N_m) - W_{in}(\Delta N_i - \Delta N_n)]. \quad (61)$$

If the components of the forcing term, $v(P_m - P_n)$, deviate strongly from the Boltzman distribution, Int^0 will remain small compared to the total intensity over a substantial fraction of the reaction. Under those conditions a "quasi-steady state" can be observed.

Equation (61) serves the purpose of elaborating the fact that the relative intensities in CIDNP spectra are not only governed by the elements of **P**, but can be influenced in a major way by the relaxation matrix **R** as well. For a simple two-level system the steady state solution to (61) is given by

$$\Delta N_m - \Delta N_n = v(P_m - P_n)\,T_{1,mn}, \tag{62}$$

where T_1 is the spin lattice relaxation time of the system. In a multilevel system T_1 loses its meaning because the relaxation behavior can no longer be described by a single exponential. Similarly, the individual lines will not only be determined by the forcing term $v(P_m - P_n)$ in (61), but the terms of the summation which constitute part of the relaxation matrix can be expected to contribute to different extents to the individual lines of the spectrum. An experimental example of the importance of solving the whole relaxation problem for predicting the relative intensities in CIDNP spectra will be discussed further below.

At this point it will be profitable to examine the behavior of the forcing term in (61) for several characteristic spin systems. To facilitate discussion we will make use of a graphic display method in which the differences, $P_m - P_n = P_{mn}$, are plotted against a suitably chosen variable expressing the characteristics of the system.[48] Within the high field approximation the elements of **P** are essentially determined by the precursor multiplicity, the magnitude of the exchange coupling, and by the off-diagonal elements of the spin Hamiltonian. The magnitude of λ and β' have no influence on the relative magnitude of the P_{nm}'s because they influence all nuclear spin states in the identical way. Since H_{ST_0} contains the electronic Zeeman energy difference and hyperfine couplings, a_i, it is advantageous to plot P_{nm} as a function of a variable G which expresses the Zeeman energy difference in units of hyperfine coupling

$$G = 2\beta H_0(g_1 - g_2)/a_i. \tag{63}$$

Plots of P_{mn} versus G have the advantage of combining the forcing term for any pair of Δg and a_i values in a single graph. Figure 4 shows a plot of P_{mn} versus G for a simple two-level system with and without the inclusion of a finite exchange coupling. The specific curves shown are calculated from Eq. (42) applicable for the behavior of the geminate coupling product derived from a triplet precursor. To obtain the corresponding curve for a singlet precursor a 180° rotation around the G axis is required. The ordinate of the graph is in arbitrary units since no specific

[48] K. Muller, *Chem. Commun.* p. 45 (1972).

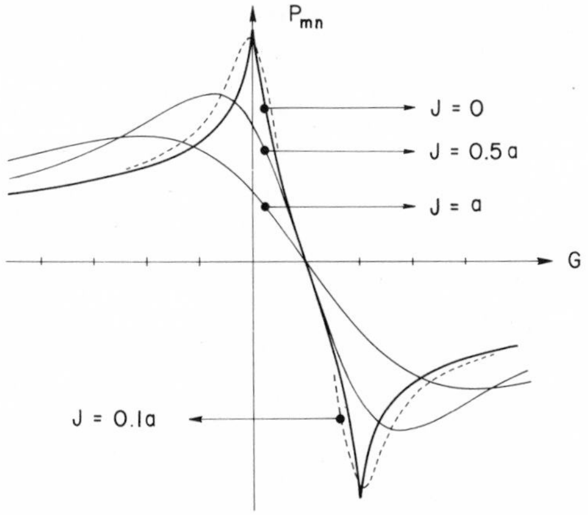

FIG. 4. Plot of the forcing term of Eq. (61) for a one-proton system as a function of G.

values for β' and λ are included. Because of the applicability of Eq. (62) to a two-level system the curves reflect the signal intensities as a function of G. The sharp extrema at $G = \pm 1$ in the curve obtained for no exchange interaction becomes blunted and move away from the center with increasing J.

It is more instructive to examine such diagrams for multilevel systems, particularly those containing frequently encountered first-order spin-spin splitting multiplets of the $A_m X_n$ type. In those cases it is possible to construct a diagram for each of the forcing terms corresponding to lines in the transition multiplets originating from a group of identical nuclei. For example Figs. 5 and 6 show the diagrams for the doublet and triplet of an A_2X system generated as the geminate coupling product from a triplet precursor where the A_2 group has a hyperfine coupling twice as large as the X nucleus. The curves have been calculated using Eq. (42) with vanishing exchange coupling. An example of such a system is the formation of 1,1,2-triphenylethane (VII) from diphenylmethylene and toluene in the reaction discussed above. As the figures show, all curves in the diagrams have identical characteristics and their amplitudes are weighted by the binomial coefficients characterizing the relative line intensities of the corresponding multiplet at thermoequilibrium. Since the system is characterized by two hyperfine coupling constants it is possible to define two different variables, G_A and G_X, where the subscript indicates which coupling constant is used in (63). It is advantageous to plot the terms for the A transitions versus G_X and the X transitions versus G_A.

It is possible to make the connection between these diagrams and the qualitative rules (57) and (58) by defining a "phase" and a numbering

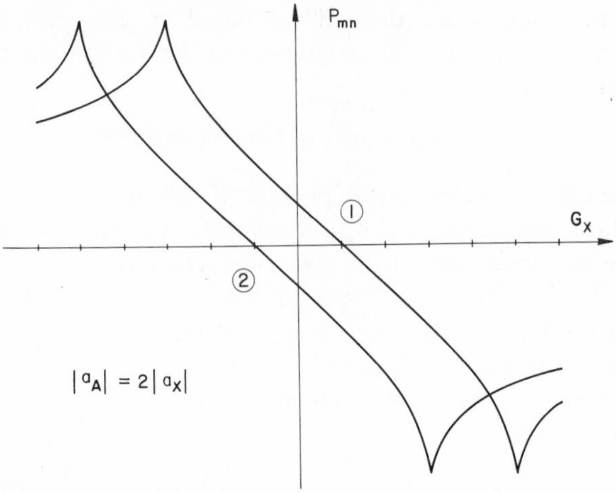

FIG. 5. Plot of the forcing terms, P_{mn}, against G_X for the A resonances of an A_2X geminate coupling product generated from a triplet precursor. The A and X protons are located on different components of the radical pair and $a_A = 2a_X < 0$.

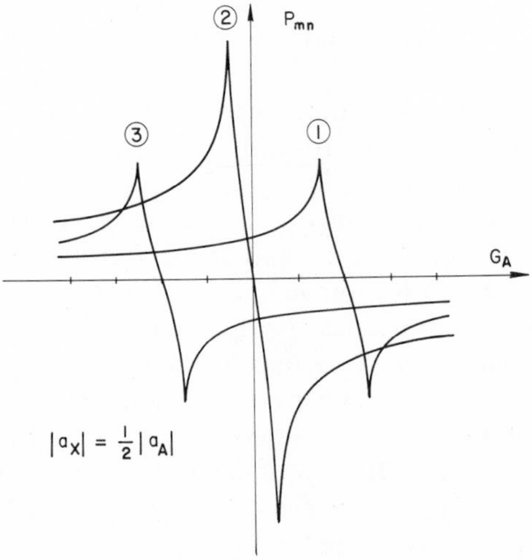

FIG. 6. Plot of the forcing terms, P_{mn}, of the X resonances of the A_2X system described in the legend of Fig. 5.

sequence connecting the curves with the multiplet lines in the spectrum. The phase of the curves is defined by the sign of the slope at their zero point and is given by

$$\mu \cdot a_X \cdot \varepsilon = \text{phase}, \tag{64}$$

where the parameters are defined as stated in the discussion of the qualitative rules. Similarly, the assignment of the lines to the curves in the diagrams can be expressed by

$$a_A J_{AX} \varepsilon_{AX} = \text{sign of labeling sequence,} \tag{65}$$

where the sequence is defined as positive if the numbers increase in the same direction and negative if in the opposite direction compared to the labels of the multiplet lines in the spectrum. These diagrams together with (64) and (65) can serve as a "back of the envelop method" to estimate the polarization pattern expected for a given reaction, and may be used as an extension of the qualitative rules.[48]

Since relative intensities of spin-spin splitting multiplets are easy to measure, they have been used extensively to estimate the magnitude of the parameters of the spin Hamiltonian. With hyperfine coupling constants and g-factors often available from ESR measurements, the parameter of greatest uncertainty is the exchange coupling constant J. A particularly suitable system to test for the magnitude of J is the A_3X type spectrum generated from geminate combination of radical pairs of the type

$$R\cdot \; R'\dot{C}H{-}CH_3.$$

Figure 7 shows diagrams for the X part of the spectrum with and without exchange interaction. Just as in the case of the simple two-level system, the introduction of an exchange coupling moves the extrema toward larger magnitudes of G. In addition, the point corresponding to a pure multiplet ($G = 0$) shows a smaller difference in the forcing terms representing the outer and inner lines. In many cases the actual line intensities correspond more closely to the diagram which has been constructed with the inclusion of a sizable J. Some authors have failed to make the distinction between the relative magnitude of the forcing terms and the line intensities and found it necessary to include large values for J to simulate the spectra.[28,29,49,50] This conclusion is erroneous and it can be shown that if the spectral simulations include the explicit treatment of relaxation as well as the proper summations in cases where a large number of degenerate transitions exist, the best fit is obtained for $J = 0$. Considering the slow evolution of the wavefunction this is eminently reasonable because the time span in which the components of the pair are close enough for a sizable exchange interaction to exist must be small compared to the time required to change the singlet character of the pair to a significant extent.

[49] H. Fischer and M. Lehnig, *J. Phys. Chem.* **75**, 3410 (1971).

[50] R. Kaptein, *J. Amer. Chem. Soc.* **94**, 6262 (1972); H. Shindo, K. Maruyama, T. Otsuki, and T. Maruyama, *Bull. Chem. Soc. Jap.* **44**, 2789 (1971).

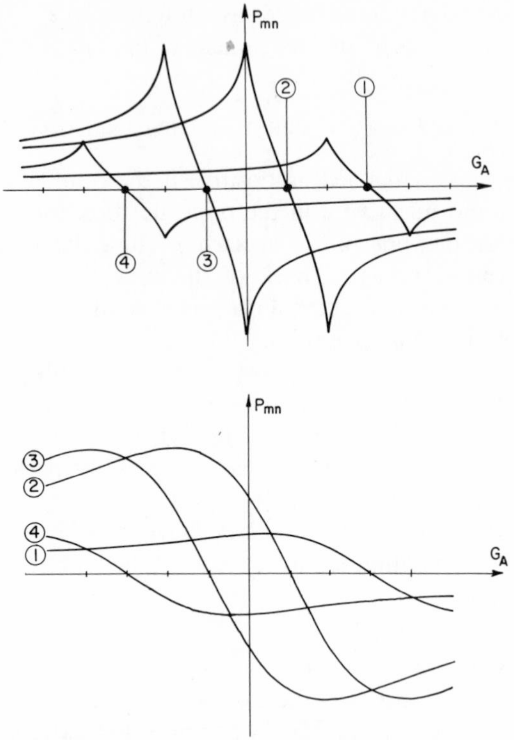

FIG. 7. Plots of the forcing terms, P_{mn}, against G_A for the X resonances of an A_3X geminate coupling product generated from a triplet precursor. The A and X protons are located on the same component of the radical pair and $a_A = a_X < 0$. (a) J = 0; (b) $J = 0.9a_X$.

How relaxation and a large number of degenerate transitions give the appearance of an apparent J can be demonstrated with the example of the X part of an A_3X spectrum obtained from reactions containing the α-phenylethyl radical. A typical reaction involving this radical as one of the components of the radical pair is the photolytic cleavage of α-phenylethylphenyl ketone (XVIII) summarized in Scheme 5. The ketone is known to dissociate from the lowest triplet state giving the benzoyl-α-phenylethyl radical pair (XIX). A measurable fraction of this pair undergoes geminate combination, regenerating the reactant, which shows an A_3X subspectrum for the CH_3–CH group. The spectrum is shown in Fig. 8 where the transitions originating from the methyl protons are obscured by the solvent.[51] Geminate disproportionation gives rise to benzaldehyde and styrene whose transitions are seen at low field. The three-line emission feature centered

[51] K. Muller and G. L. Closs, *J. Amer. Chem. Soc.* **94**, 1002 (1972).

$$^3PhCO-CH(CH_3)Ph \longrightarrow \overline{PhC\dot{O} \quad \cdot CH(CH_3)Ph} \longrightarrow \text{Free radical products}$$

XVIII   XIX

hν

PhCO—CH(CH₃)Ph   Ph—CH=CH₂ + PhCHO

$$\overline{PhCO \quad \cdot CH(CH_3)Ph} + CCl_4 \longrightarrow \overline{\cdot CCl_3 \quad \cdot CH(CH_3)Ph} + PhCOCl$$

XIX   XX

Free radical products   Ph—C(CCl₃)CH₃

XXI

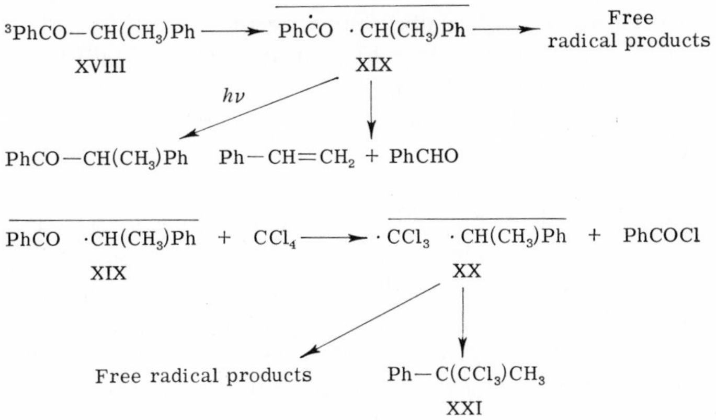

Scheme 5

at 4.6 ppm (TMS) can be assigned to the quartet of XVIII in which the lowest field line is missing.

In carbon tetrachloride (lower trace of Fig. 8) an additional reaction takes place. Benzoyl radicals react readily with the solvent in what one might call a radical pair substitution reaction. The newly formed trichloromethyl-α-phenylethyl pair (XX) is still correlated and forms the geminate coupling product 1,1,1-trichloro-2-phenylpropane (XXI). Its methine proton quartet

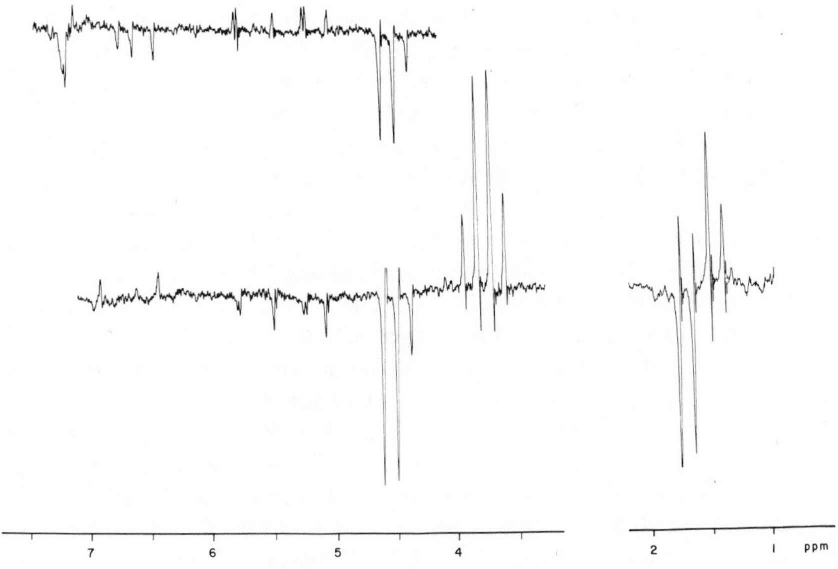

FIG. 8. Spectrum of irradiated α-phenylethyl phenyl ketone at 60 MHz. Upper trace in cyclohexane, lower trace in carbon tetrachloride as solvent.

is seen in enhanced absorption at 3.7 ppm together with the methyl doublet emission signal at 1.7 ppm. The enhanced absorption doublet at 1.4 ppm originated from the ketone XVIII. The opposite signs of the CH_3–CH signals in XVIII and XXI demonstrate the reversal of signs of Δg in the radical pairs XIX and XX.

The diagram in Fig. 7 has been calculated for the X quartet of XVIII with the parameters describing this reaction, and if the forcing terms were identical to the relative line intensities one would conclude $G = 2$ for this system. This together with the known hyperfine coupling of 18 G would lead to a value of $g_1 - g_2 = 3 \times 10^{-3}$, which does not agree with ESR measurements which give a value of 1.8×10^{-3}. The reason for the apparent discrepancy is the neglect of degenerate transitions and relaxation. We will discuss the influence of these effects in sequence.

a. *The Effect of Degenerate Transitions.* Very often a system under study contains a number of paramagnetic nuclei contributing to the hyperfine interaction in the radical pair which do not affect the number of observable lines in the multiplet of the diamagnetic product because they do not interact with the nuclei causing the spin multiplet via nuclear coupling. These nuclei may not even be located on the same molecule in the final product, but they do have a significant effect on the relative line intensities of the spectra. An example is afforded by the aromatic protons in the α-phenylethyl radical which carry substantial hyperfine coupling in the radical pair but do not change the number of resolvable nmr lines in the X quartet of the coupling product XVIII. It is not permissible to calculate the spectrum by only including the hyperfine interaction of the nuclei which give rise to the observable multiplet, but it is essential to include all nuclei carrying the hyperfine interaction in the radical pair. Although this appears to be obvious, it must be pointed out that this has rarely been done. Instead, very often a sizable exchange coupling has been included to account for the observed spectra.[28,29,49] Comparison of Fig. 7 with Fig. 9 shows that a finite exchange coupling produces similar effects, as does the inclusion of additional nuclei. The curves of Fig. 9 are calculated with the same parameters as Fig. 7 except that the hyperfine interaction of the ortho and para protons of the α-phenylethyl radical have been included in the summation of the off-diagonal elements. The result of adding three more nuclei of spin 1/2 is that at $G = 0$ the difference between the intensities of the outer and inner lines of the quartet becomes less pronounced and that the extrema of the curves move further away from the center of the diagram. These are qualitatively the same effects as are produced by the inclusion of a finite exchange coupling.[52]

[52] K. Muller and G. L. Closs, *J. Amer. Chem. Soc.* in press (1974).

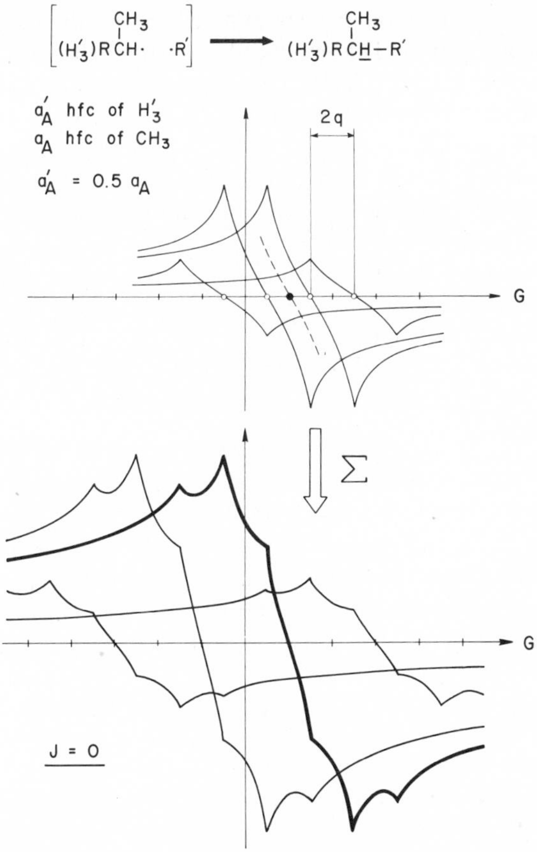

FIG. 9. Plot of the forcing terms, P_{mn}, against G_A for the X resonances of an $A_3X(Y_3)$ system. The Y protons are not coupled to the A and X protons and carry a hyperfine interaction in the radical pair of magnitude $a_Y = \frac{1}{2}a_X$. The remaining parameters of the system are identical to those given in the legend of Fig. 7a.

This simularity in the appearance of the polarization pattern seemingly caused by unrelated origins becomes understandable upon a more detailed examination of the time evolution of the radical pair wavefunction. As Eqs. (37) and (42) show, the polarization pattern is essentially governed by the elements of the vector **B**. In the limit of $J = 0$, **B** is proportional to $\mathbf{H}_{ST_0}^{1/2}$, while a very large J changes this dependence toward $\mathbf{H}_{ST_0}^{2}$.

The addition of other nuclei, $M_k, N_l, \ldots, Q_P$, to a system containing an A_mX_n subspectrum whose lines are to be observed has essentially the same effect. Since we are not explicitly interested in the evolution of all nuclear substates in the radical pair but only in those which are experimentally accessible by measuring the A_mX_n transitions, we may carry out the summation over a subensemble which combines all nuclear states which have the same spin function for the A and X nuclei. This leads to

$$\langle|C_{sn^*}(t)|^2\rangle = \sum_i \sum_j \cdots \sum_h |C_{sn^*(i,j,\ldots,h)}(t)|^2, \tag{66}$$

where the n^* designates a single basic product function of the A_mX_n system and the summations are over all possible states of the groups $M_k, N_l, \ldots, Q_P$. In essence this corresponds to a summation of a large number of sinusoidal functions [Eqs. (34) and (35)] with different frequencies. At the birth of the radical pair all functions are in phase, but due to interference will get out of phase on a time scale short compared to any individual oscillation. This behavior is further illuminated by Fig. 10 where the time evolution of a single nuclear spin state of a "pure" A_3X system

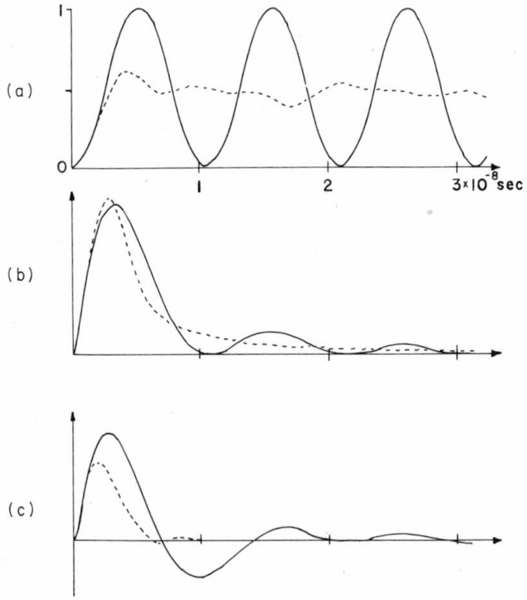

FIG. 10. Plots of the functions, (a) $|C_{s,n}(t)|^2$, (b) $|C_{s,n}(t)|^2 f(t)$, and (c) $|C_{s,n}(t)|^2 - |C_{s,m}(t)|^2 f(t)$ against time. The solid curves represent a radical pair with an A_3X nuclear spin system. The state n whose time evolution is shown in a superposition of the three degenerate levels in which the A_3 group is represented by $m_z = 1/2$ and the X proton by $m_z = 1/2$. State m differs from n by having the X proton in a level characterized by $m_z = -1/2$. The parameters of the Hamiltonian are: $a_A = 13\,G$, $a_X = -18\,G$, $g = 0$, $J = 0$. The dashed lines represent a $A_3X(Y_3)(Z_2)$ system with the same parameters for A and X and $a_Y = -5\,G$ and $a_Z = 25\,G$. The Z nuclei have a spin of 1. (An example of such a pair would be C_6H_5–N=Ṇ CH_3ĊH–C_6H_5·.) In the diagram the time evolution is averaged over all states of the Y and Z nuclei and the subsystem shown has the same spin states of the A_3X protons as described above for the solid curves.

is compared with that calculated for a system containing additional nuclei. Figure 10a shows the singlet character of the subensemble characterized by identical basic product functions of the A_3X nuclei. While the pure system undergoes the expected sinosoidal oscillations, the system with additional nuclei shows no recurrence and after an initial rise deviates little from the stationary state. Figure 10b shows the product of the singlet character with the damping function $f(t)$, representing the probability of forming a geminate coupling product as a function of time. This function falls slowly enough that the "pure" system lives through several recurrence times characterized by the minima in the curve. Because of the interference effects the subensemble in the system with additional nuclei never recurs and the probability of forming a geminate product after longer times is solely determined by $f(t)$. The result of this behavior is that the forcing terms for the A_3X transitions are mainly determined by the behavior of the functions at short time intervals, where the sine of $\mathbf{H}_{ST_0}$ [Eqs. (34) and (35)] may be approximated by the argument. In the limit, integration of this modified function will yield a dependence on $\mathbf{H}_{ST_0}^2$. It is interesting to note that in the early publications the damping function $f(t)$ was approximated by an exponential with a characteristic lifetime of the radical pair, τ.[8,9,28] This model gives a poor description of the diffusive behavior of radical pairs but gives reasonable agreement with experiment if lifetimes of the order of 10^{-10} sec are chosen. In this case the recurrences are prevented by the fast falling damping function and the dependence of the forcing terms is strictly proportional to $\mathbf{H}_{ST_0}^2$.

It is interesting to note the similarity which exists between the effect caused by the addition of further paramagnetic nuclei in CIDNP and radiationless processes occuring in electronically excited states of molecules. In both cases the ensemble is prepared in a nonstationary state and "crossing over" to another state is caused by interactions between the zero-order states. In the case of radiationless transitions the breakdown of the Born–Oppenheimer approximation causes resonance transitions between the initial states and usually a dense manifold of a second electronic state. The higher the density of states in this second state the longer the recurrence time will be because of the interference effects mentioned above, and the process may become for all practical purposes irreversible. In strictly analogous fashion the addition of more and more nuclei to the radical pair in effect increases the density of states to which the subensemble of identical A_mX_n functions is coupled, and the crossover rate can be described in the limit by a "golden rule" expression containing the summation over the square of all interactions between the zero-order states.[53]

[53] K. F. Freed, *Top. Curr. Chem.* **31**, 105 (1972).

b. *The Effect of Relaxation on the Relative Intensities.* As has been pointed out on several occasions above, a complete solution to the line intensity problem of CIDNP spectra must include relaxation in the diamagnetic product. It should be recalled that as long as one observes only geminate coupling products, nuclear relaxation in the radicals can be neglected because it is too slow to be of any consequence in coupling reactions which proceed in time spans of less than 10^{-8} sec.

Although no new theory is required to treat the relaxation in the products, in practice it appears to be the most difficult part of the CIDNP problem to solve satisfactorily. This arises from the fact that nuclear relaxation is caused by several mechanisms, the relative importance of which is not easily established and may vary from sample to sample. Although it is in principle possible to establish the relaxation rates for each process this requires the knowledge of motional correlation functions which for complex molecules may become very complicated.[54] To make any progress in estimating the effect of nuclear relaxation it is therefore necessary to reduce the problem to a drastically simplified model. This has been done for the relaxation behavior of ketone XVIII obtained in the reaction sequence outlined in Scheme 5, and is elaborated here as an example.

To calculate the relative line intensities of any CIDNP spectrum it is necessary to obtain the elements of **R** in Eq. (3). Basically the matrix **R** may be thought of as the sum of several matrixes, $\mathbf{R}_i$, each representing a particular relaxation mechanism. It is reasonable to assume that in a molecule of the size of XVIII spin rotation interaction will contribute only to a minor extent to the proton relaxation. This leaves the various dipole-dipole interactions as the major contributing mechanisms. For example the nuclei of the CH_3CH group in XVIII will be subjected to at least three different relaxation mechanism, all based on dipolar interactions. The matrix may be written as

$$\mathbf{R} = \mathbf{R}_{\text{intra}} + \mathbf{R}_{\text{inter, n}} + \mathbf{R}_{\text{inter, e}}, \tag{67}$$

where the first term represents the intramolecular dipole-dipole interaction among the protons of the A_3X group. The second term describes relaxation due to intermolecular interaction with the solvent and other solute nuclei, while the last term represents relaxation due to intermolecular dipolar coupling with the unpaired electrons of the free radicals in solution. Although in theory it is possible to calculate the elements of all of these matrices by choosing the appropriate model, the generally poor results obtained do not warrant the effort. Instead, one may use the very much simpler approach of introducing weighing factors whose magnitudes are to

[54] For example, see R. Freeman, S. Wittenkoek, and R. R. Ernst, *J. Chem. Phys.* **52**, 1529 (1970).

be adjusted by experiment. It is then possible to combine the elements of $\mathbf{R}_{\text{inter,n}}$ with $\mathbf{R}_{\text{inter,e}}$ into one matrix, $\mathbf{R}_{\text{inter}}$, covering all intermolecular dipolar mechanisms without specifying the fraction of each specific contribution. This is permissible because both mechanisms contribute only to relaxation transitions characterized by $\Delta m_{\text{I}} = \pm 1$. In contrast $\mathbf{R}_{\text{intra}}$ contains elements corresponding to transition probabilities for processes where $\Delta m_{\text{I}} = 0, \pm 1, \pm 2$. The model to be adopted for the intramolecular relaxation problem will depend strongly on the structure of the molecule. For example, if the multiplet originating from the CH_3–CH group in ketone XVIII is to be calculated, a further simplication can be achieved by neglecting intramolecular interactions between the methyl and the methine protons. The error introduced this way will be small because the dipole-dipole interaction falls off rapidly with distance and the major contribution of intramolecular dipole interaction will be limited to the protons of the methyl group. These admittedly somewhat crude simplifications break the relaxation problem of the A_3X part of XVIII down to

$$\mathbf{R} = \mathbf{R}_{\text{inter}}^{\text{X}} + q\mathbf{R}_{\text{inter}}^{\text{A}} + p\mathbf{R}_{\text{intra}}^{\text{A}}. \tag{68}$$

Here q is the weighing factor allowing empirical adjustment of the relative weights of the coupling to the lattice of protons A and X, and p adjusts the admixture of intramolecular relaxation within the A_3 group.

The relative magnitude of the elements of the matrices in (68) are obtained from calculations based on Redfield's density matrix method.[55] However, the calculation of the spectral densities requires a model for the motional behavior of the various protons. In the specific example under discussion the simplest model was adopted, assuming random isotropic motion of the whole molecule and neglecting internal rotation of the methyl group. Cross-correlation among the methyl protons was taken into account by the method outlined by Hubbard.[56]

The procedure of fitting the weighing parameters is demonstrated with the aid of Fig. 11.[51,52] The stick diagram in Fig. 11a represents the relative intensities of the X quartet of XVIII calculated by equating the forcing terms of Eq. (3) with the spectral intensities, making use of g-factors and hyperfine interactions determined by ESR. This clearly incorrect procedure corresponds to neglecting the summation in Eq. (61) and is the one used in most published spectral simulations. Comparison of the observed spectrum with the stick diagram shows a very poor agreement in the intensity of line 1. Figure 11b is calculated for the steady state of Eq. (61) with no intramolecular relaxation ($p = 0$) and equal coupling of the A and X groups with the lattice ($q = 1$). As the figure shows there is a substantial

[55] A. G. Redfield, *Advan. Magn. Resonance* **1**, 1 (1965).
[56] P. S. Hubbard, *Phys. Rev.* **109**, 1153 (1958); *Phys. Rev.* **111**, 1746 (1958).

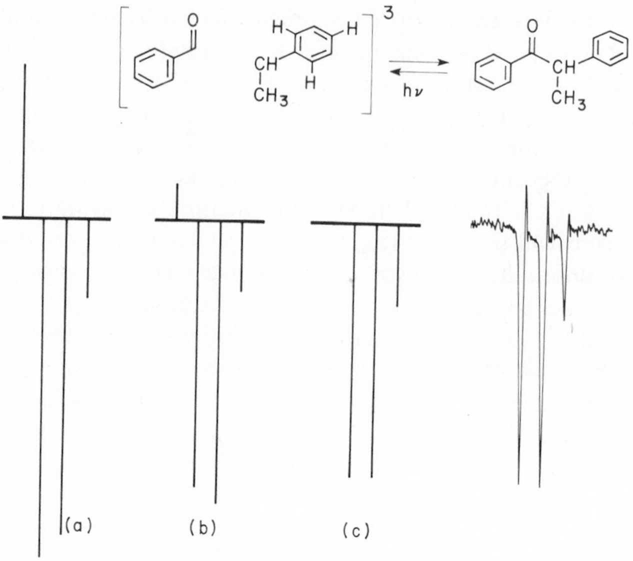

FIG. 11. Stick diagram of the X transitions of an $A_3X(Y_3)$ geminate coupling product generated from a triplet precursor. The parameters are chosen for the pair XIX and are $a_A = 18\,G$, $a_X = -18\,G$, $a_Y = -5\,G$, $g = 1,8 \times 10^{-3}$. (a) Intensities are set equal to the forcing terms in Eq. (61). (b) Calculated with Eq. (61) but neglecting intramolecular relaxation. (c) R_{intra} has been adjusted to the point where the low field line vanishes.

effect on the relative intensities caused by the fact that each line borrows intensity from all other multiplet lines, the mixing effect between adjacent lines being dominant. It is possible to allow for different relaxation probabilities for the A and X nuclei ($q \neq 1$) but the general result is the same.

Figure 11c shows the calculated line intensities after intramolecular relaxation within the methyl group has been introduced. This has the effect of cutting the intensity of line 1 still further. The weighing factor p has been adjusted to the point where line 1 vanishes to make the calculated intensities agree with the 60 MHz spectrum. In general the effect of finite p is to further increase the mixing of line intensities within the multiplet and especially to reinforce the borrowing of intensity from second next nearest neighbor lines.

As expected, the general effect of relaxation is to "even out" intensity differences in multiplets and can be expected to lead to largest deviations in relative intensities from those calculated by only using the forcing term in Eq. (3) in those spectra where both emission and enhanced absorption lines occur within the same multiplet. In spectra with very large net effects the relative intensities are not much affected by relaxation processes. The combined effect of relaxation and degenerate transitions caused by additional nuclei is summarized in Fig. 12 showing the relative line intensities of the X transitions of the A_3X system versus *G*. A comparison of this diagram with that of Fig. 7 in which a large exchange coupling has been used shows that the general appearance of the two diagrams is quite similar. This coincidence is responsible for the assumption by many authors that exchange couplings are necessary to simulate spectra.

The procedure of adjusting the weighing parameters of the various relaxation processes in Eq. (68) by fitting the calculated spectra to the experimental data has the disadvantage of possible ambiguities, particularly in cases where the parameters of the spin Hamiltonian are not available from independent experiments. In those cases it can be helpful to run the experiment at different magnetic fields. Because of the field dependence of the Zeeman energy differences, the relative intensities of the multiplets may show considerable variation. This is equivalent to determining the CIDNP spectrum for different values of *G* in the diagrams. Again, this procedure is demonstrated with the X quarter of ketone XVIII whose proton spectrum was recorded at 25, 60, and 90 MHz. As can be seen in Fig. 13 good agreement with the calculated and experimental line intensities is observed in all three spectra, thus raising hopes that a reasonably adequate approximation for the relaxation problem has been found.

Double resonance experiments may also be helpful in obtaining additional information on the relaxation problem. As has been elaborated above relaxation has the effect of intensity borrowing in multiplet spectra. By

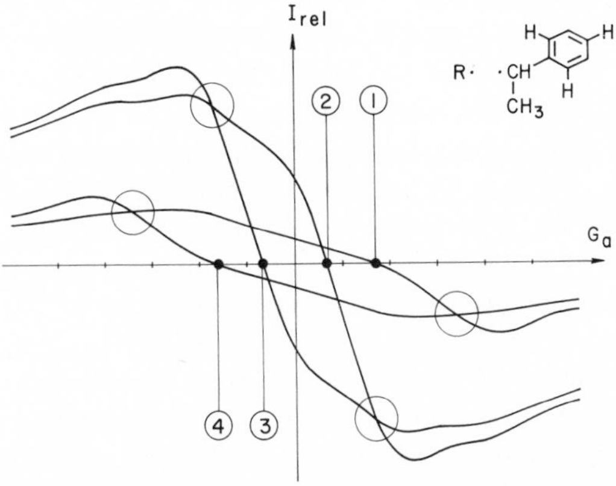

FIG. 12. Plot of the intensities of the $A_3X(Y_3)$ system, which has been described in the legend of Fig. 11, against G_A. Relaxation is included with the same parameters as chosen for the stick diagram shown in Fig. 11c.

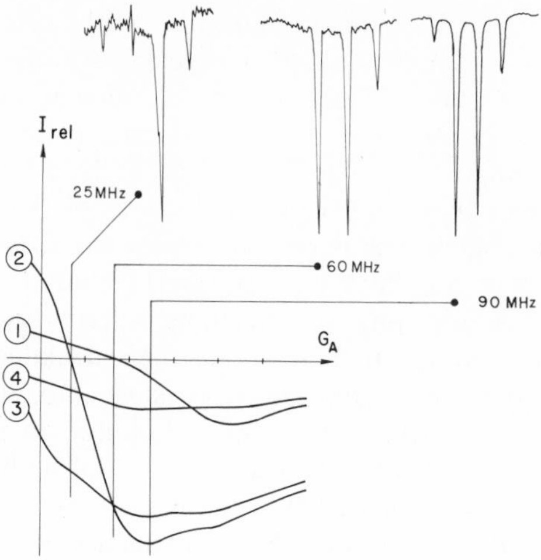

FIG. 13. Transition of the methine protons of irradiated α-phenylethyl phenyl ketone at 25, 60, and 90 MHz, respectively. The diagram below is the same as shown in Fig. 12 and has been included for comparison with the experiment.

annihilating certain select population differences with rf-induced transitions it is often possible to assess the magnitude of this borrowing. This effect is illustrated again by the X quartet of ketone XVIII. As shown in Fig. 11a the forcing term corresponding to line 1 of the quartet is positive and intensity borrowing via relaxation from lines 2 and 3 are thought to reduce it to zero intensity. Application of an H_2 field to line 2 or 3, small enough to avoid significant perturbation of the levels but sufficient to diminish the population difference between the levels, should cause line 1 to appear in absorption. As Fig. 14 shows this qualitative prediction is nicely supported by experiment.[51,52] If the magnitude of the H_2 field is known, additional quantitative checks on the relaxation parameters which have been used in Eq. (68) are possible. A related double resonance experiment involves the application of the H_2 field to either or both of the two A_3 transitions while monitoring the X part of the spectrum. This leads to spectra shown in Fig. 15 which compare well with calculations assuming that a complete averaging of the populations of the levels connected by the transition has been achieved.

Detailed consideration of Eq. (61) leads to the interesting prediction that the relative intensities of multiplet lines in CIDNP spectra should change during the course of the reaction. At the beginning of the reaction the intensities are dominated by the forcing term because relaxation is less important at reaction times short compared to the nuclear relaxation times. This pre-steady state has a counterpart in the post-steady state period after

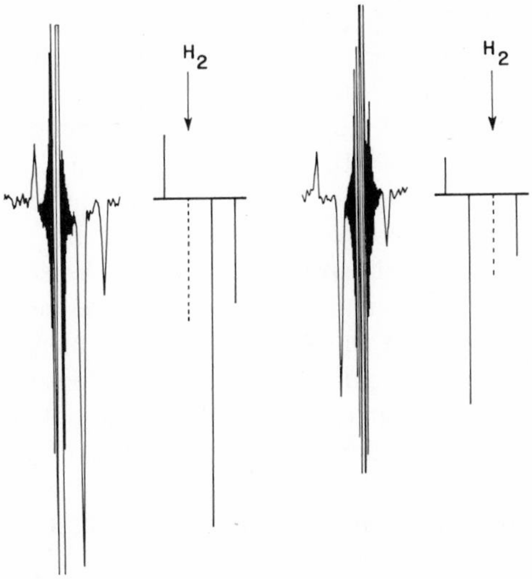

FIG. 14. Double resonance experiment on irradiated α-phenylethyl phenyl ketone at 60 MHz. The H_2 field has been applied at line 2 and line 3 of the methine quartet respectively. The stick diagrams have been calculated assuming equalization of the populations of the levels connected by H_2.

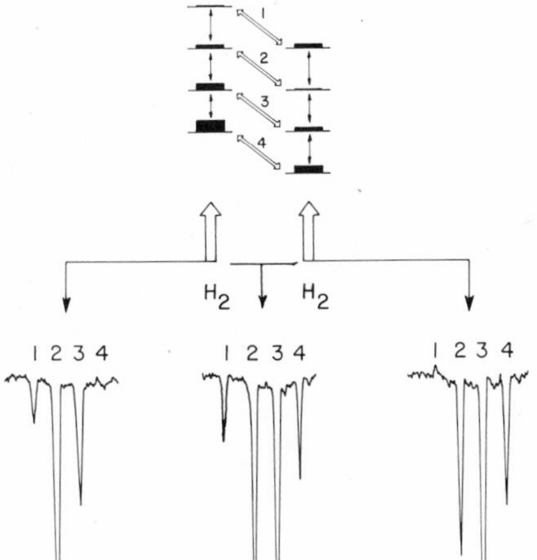

FIG. 15. Double resonance experiment on irradiated α-phenylethyl phenyl ketone at 60 MHz. Left: the H_2 field is applied at the low field line of the methyl doublet; middle: the H_2 field is applied at both lines of the methyl doublet; right: the H_2 field is applied at the high field line of the methyl doublet.

the reaction has been stopped, where the intensities must be governed solely by the relaxation terms in Eq. (61). The pre- and post-steady state spectra are relatively easily obtained in photochemical reactions where the reaction can be started and stopped simply by turning the irradiation on and off, respectively. Figure 16 shows the intensity of line 1 in the X quartet of XVIII as a function of time, and above it the calculated behavior. It should be pointed out that with the advent of readily available Fourier transform spectrometers the pre-steady state spectra are readily obtainable by simply selecting a pulse repetition rate short compared to the relaxation times. Any residual transverse magnetization may be destroyed by a short pulse across the shim coils of the spectrometer. Such experiments can be carried out in thermo-initiated reactions as well and have the advantage of diminishing the influence of relaxation on the relative line intensities. The observed spectra will more closely approximate those calculated by using only the forcing term of Eq. (61). Post-steady state experiments by Fourier transform can take the form of a pulse sequence in which the spin system is prepared by a chemical reaction. In photochemical reactions this is easily accomplished by applying a light pulse of several seconds followed, after a suitable delay, by a 90° pulse. Figure 17 shows such a "light-τ-90" sequence experiment at 90 MHz with the system outlined in Scheme 5. Such sequences may be used to study relaxation in CIDNP.

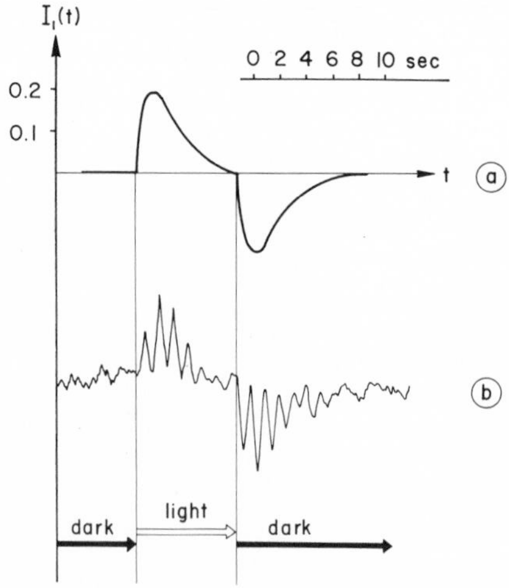

FIG. 16. Experimental and calculated time evolution of the first line of the methine quartet of irradiated α-phenylethyl phenyl ketone at 60 MHz. The pre- and post-steady state spectra were obtained by repetitive scanning after light has been admitted and after light had been turned off, respectively.

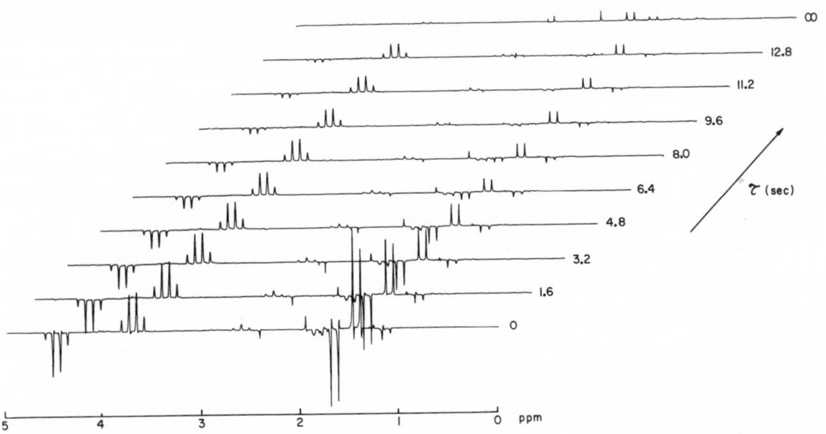

FIG. 17. 90 MHz Fourier transform spectra of irradiated α-phenylethyl phenyl ketone (XVIII) in CCl_4. The time axis indicates the delay, τ, between the end of the uv irradiation period and the 90° rf pulse. The irradiation period has been kept at 10 sec in all spectra.

It has been pointed out above that the relaxation problem in CIDNP is a particularly important and difficult problem which has to be overcome to gain a quantitative insight into the CIDNP problem proper. It might be said that relaxation is a liability in CIDNP. With the same justification it might also be said that CIDNP is an asset to relaxation. The fact that it is easily possible to create systems with multiplet effects makes CIDNP a unique method to prepare systems which cannot be described by a spin temperature and which cannot be prepared by conventional rf pulse methods. Following the decay of the individual lines in a CIDNP multiplet can be expected to give much more accurate information on the relaxation rates and cross-correlation in multilevel systems than a nonexponential decay from a system with negative spin temperature can ever provide. No experiments with this goal have been reported to date!

2. *Absolute Intensities in CIDNP Spectra*

The absolute intensities of the transitions in CIDNP spectra are best expressed as the enhancement factor, V_{mn}. There has been some confusion as to how to define this quantity in a dynamic system so that it is both easy to determine experimentally and is a meaningful quantity. What one is interested in is the magnitude of the magnetization of the reaction product at the moment when it is formed. Since this quantity is not directly measurable it has to be deduced from the signal intensities and additional parameters describing the kinetic behavior of the reaction as well as the relaxation of the nuclei in the product. The theoretical enhancement factor of a system containing κ nuclear spin states is defined by

$$V_{mn} = [(P_m - P_n)/\textstyle\sum_i P_i]\,\kappa(kT/\gamma H_0) - 1. \tag{69}$$

This factor can be related to the experimentally accessible line intensities by Eq. (61). An accurate determination of this factor requires the solution of the whole relaxation problem in any system containing more than two nuclear states. However, if cross relaxation by intramolecular dipole-dipole interactions is not too severe and if the spectrum shows a strong net effect it is possible to neglect the last terms of Eq. (61) and treat the system in the same way as a two-level system to obtain approximate enhancement factors, V'_{mn}.

Transition intensities are related to population differences by

$$[(\Delta N_m - \Delta N_n)/(N_m^0 - N_n^0)]_t = [(\mathrm{Int}_{mn} - \mathrm{Int}_{mn}^0)/\mathrm{Int}_{mn}^0]_t. \tag{70}$$

Under "quasi-steady state" conditions the numerator of (70) is given by (62) and the equilibrium population difference at time t is

$$(N_m^0 - N_n^0)_t = (\textstyle\sum P_i/\kappa)(\gamma H_0/kT)\displaystyle\int_0^t v(t)\,dt. \tag{71}$$

Substitution of (62) and (71) into (70) yields

$$\begin{aligned}[(\mathrm{Int}_{mn} - \mathrm{Int}_{mn}^0)/\mathrm{Int}_{mn}^0]_t &= [(P_m - P_n)/\textstyle\sum P_i]\,\kappa(kT/\gamma H_0)T_{1_{mn}}v(t)\Big/\displaystyle\int_0^t v(t)\,dt \\ &= (V'_{mn} + 1)T_{1_{mn}}v(t)\Big/\int_0^t v(t)\,dt.\end{aligned} \tag{72}$$

To evaluate V'_{mn} from (72) it is necessary to know Int_{mn}^0 at the time the total intensity, Int_{mn}, is measured. Unless the reaction can be instantaneously quenched this quantity is not experimentally obtainable. It is more practical to relate the intensity at time t to the intensity of the same transition when the reaction is completed and all states have come to equilibrium. If the product is formed with a fractional yield of Y from the precursor M,

$$\int_0^\infty v(t)\,dt = \mathrm{Y}[\mathrm{M}]^0, \tag{73}$$

giving (74) as the expression for the enhancement factor.

$$V'_{mn} = \frac{\mathrm{Int}_{mn}(t)\,\mathrm{Y}[\mathrm{M}]^0 - \mathrm{Int}_{mn}(\infty)\int_0^t v\,dt}{\mathrm{Int}_{mn}(\infty)\,v(t)\,T_{1_{mn}}} - 1. \tag{74}$$

It is often possible to neglect the equilibrium polarization when the measurement is made early in the reaction and large enhancement factors are observed. It is then more convenient to relate the observed intensities to a standard, ST, containing the product in known concentration, giving as a good approximation

$$V'_{mn} = \mathrm{Int}_{mn}(t)[\mathrm{ST}]/\mathrm{Int}^{\mathrm{ST}}_{mn}\, v(t) T_{1_{mn}}. \tag{75}$$

Several other approximations have been proposed including a maximum enhancement factor, $V_{\max}$, which however suffers from the fact that it cannot be universally defined.[57]

When comparing experimental enhancement factors evaluated using Eq. (74) [or (75)] with theoretical factors calculated from (69) and the P_i's calculated from the radical pair theory, it must be remembered that the experimental values refer to the total product formed. For example, a specific product may be formed via geminate coupling and random phase combination (F-reaction). While the theory distinguishes between the two pathways the experimental measurement does not unless a control experiment is run in which the *F*-reaction is prevented with the aid of free radical scavengers.

The reactions of Scheme 3 are good examples of how a multiple reaction path yielding the same product affects the enhancement factor. The triphenylethane VII can be formed via geminate combination and from free radical combination. In the reaction of diphenylmethylene, where the precursor is a triplet state, the enhancement factor calculated from Eqs. (69) and (42) with the experimentally measured hyperfine interactions[58] is +31,000 for the first line of the low field multiplet. However, using the same equation with $\lambda = 0.4$ and $\beta' = 0.5$, only 0.3% of the geminate coupling products are calculated to be formed. The experimentally determined yield is however 50%. From this it appears that by far the major fraction of VII is generated via the free radical coupling. Since F-reactions generate identical polarization patterns as geminate coupling from triplet precursors, both effects reinforce themselves. The experimentally measured enhancement factor is 760 ± 100 which leads to an estimated 570 for the enhancement factor of the F-reaction.

The thermal decomposition of the diazo compound X gives VII with an enhancement factor of -90 ± 20 which is again the weighted average of the geminate coupling and F-reaction enhancement factors. In this case, however, it was possible to eliminate the F-reaction with a free radical scavenger and measure the geminate coupling product separately. With a

[57] J. Bargon and H. Fischer, *Z. Naturforsch. A* **22**, 1556 (1967); R. Kaptein, *Chem. Phys. Lett.* **2**, 261 (1968).
[58] G. L. Closs and A. D. Trifunac, *J. Amer. Chem. Soc.* **92**, 7227 (1970).

singlet precursor the enhancement factors of the F-reaction and geminate coupling products are of opposite signs, yielding a larger enhancement in the reaction with scavengers (-350 ± 70) than without (-90). The calculated factor for VII derived from a singlet precursor in a geminate coupling is 510 if a value of 0.5 is assumed for β'.

Substitution of Eq. (42) into (69) shows that in reactions proceeding from a triplet precursor the enhancement factor is independent of the diffusion parameters m and β'. However, the line intensities can be expected to grow with increasing β' and m because the yield of the geminate coupling product will be enhanced. In contrast, reactions proceeding from singlet precursors as well as F-reactions will show a dependence of the enhancement factor on the diffusive behavior of the radical pair. Although these relationships were recognized quite early in the development of the theory, no experiments with the goal of studying diffusion by CIDNP have as yet been reported.

To date very little work has been reported which was specifically aimed at a comparison of experimental with calculated enhancement factors. The few values in the literature are often compared with values calculated with the earlier version of the radical pair model in which diffusion was treated incorrectly. Also, a correct calculation of the enhancement factor must involve the treatment of all nuclei carrying hyperfine interactions and not only those which undergo the observable transition in the diamagnetic product. In general, the presence of degenerate transitions will reduce the theoretical enhancement factor because of the interference effects in the oscillations in the radical pair states.

The enhancement factors of free radical transfer, or escape products, are a function of the lifetime of the free radicals and the relaxation rates of the nuclei in the radical. Using Eq. (50) with its steady state solution and approximating the relaxation matrices ${}^{\mathrm{r}}\mathbf{R}$ and ${}^{\mathrm{e}}\mathbf{R}$ with "effective" nuclear relaxation times ${}^{\mathrm{r}}T_1$ and ${}^{\mathrm{e}}T_1$ for the radical and escape product, respectively, the intensities of the transitions are

$$\left(\frac{{}^{\mathrm{e}}\mathrm{Int}_{mn} - {}^{\mathrm{e}}\mathrm{Int}^0_{mn}}{{}^{\mathrm{e}}\mathrm{Int}^0_{mn}}\right)_t = -\frac{(P_m - P_n)\kappa kT\,{}^{\mathrm{e}}T_{1_{mn}}\,v(t)}{\sum(\mathrm{I}/\kappa - P_i)\gamma H_0 \int_0^t v(t)\,dt} \frac{k_{\mathrm{sc}}[\mathrm{RX}]}{(1/{}^{\mathrm{r}}T_{1_{mn}}) + k_{\mathrm{sc}}[\mathrm{RX}]}. \qquad (76)$$

From (76) one can define a theoretical enhancement factor for the escape product

$${}^{\mathrm{e}}V_{mn} = -\frac{(P_m - P_n)\,kT}{\sum_i(\mathrm{I}/\kappa - P_i)\gamma H_0} \cdot \frac{k_{\mathrm{sc}}[\mathrm{RX}]}{(1/{}^{\mathrm{r}}T_{mn}) + k_{\mathrm{sc}}[\mathrm{RX}]}, \qquad (77)$$

incorporating the relaxation behavior in the radical as well as the scavenging rate.

Equation (77) points to the interesting fact that the CIDNP spectra of escape products will vary with the scavenger concentration. Since the nuclear relaxation time in free radicals is a function of the dipolar coupling between electron and nuclear spin, substantial differences can be expected for ${}^{r}T_1$ within the same molecule. This leads to the possibility that intensity ratios of transitions originating from different nuclei within the same molecule can vary with the reaction condition. An example of such behavior is given by the reaction summarized in Scheme 6.[59] Photoexcited *p*-chlorobenzaldehyde can abstract a hydrogen atom from another aldehyde

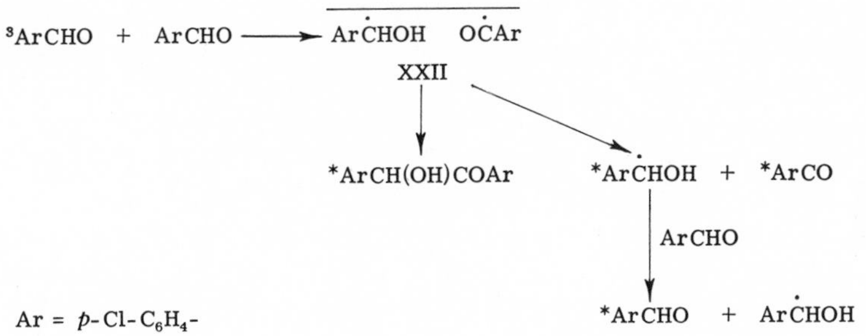
³ArCHO + ArCHO
ArĊHOH OĊAr
XXII
*ArCH(OH)COAr
*ArĊHOH + *ArĊO
ArCHO
*ArCHO + ArĊHOH
Ar = p-Cl-C₆H₄-

Scheme 6

*denotes polarized molecules

molecule to form the *p*-chlorohydroxybenzyl-*p*-chlorobenzoyl radical pair (XXII). Geminate combination gives polarized *p*-chlorobenzoin (XXIII). The escaping hydroxybenzyl radicals exchange a hydrogen atom with an aldehyde resulting in a polarization transfer to the reactant. This mechanism was deduced from the spectrum shown in Fig. 18 and several additional experiments including quenching tests to establish the triplet multiplicity of the precursor. When the spectra were recorded at different *p*-chlorobenzaldehyde concentrations it was noted that the intensities of the transitions of the aromatic protons and of the aldehyde protons responded to a different degree to the change in aldehyde concentration. The ratios of the corresponding transitions are plotted in Fig. 19 as a function of the inverse aldehyde concentration. An evaluation of the curvature and slope of the intensity function relates the two relaxation times in the radical with the rate constant of hydrogen transfer. Assuming a relaxation time of 10^{-4} sec for the benzyl proton in the hydroxybenzyl radical, the best fit was obtained

[59] G. L. Closs and D. R. Paulson, *J. Amer. Chem. Soc.* **92**, 7229 (1970).

with 1.5×10^{-3} sec for the relaxation time of the ortho protons and $8 \times 10^{4}\ M^{-1}\ sec^{-1}$ for the rate constant of hydrogen atom transfer.

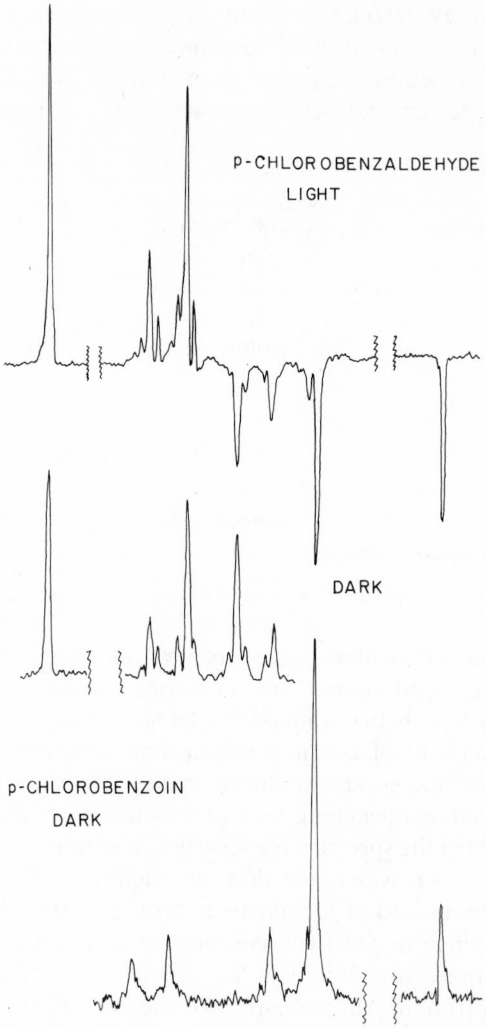

FIG. 18. Upper trace: 60 MHz spectrum of irradiated p-chlorobenzaldehyde in cyclopropane. Left section: aldehyde signal; middle section: aromatic protons; right section: benzylic proton of *p*-chlorobenzoin. The dark spectra of *p*-chlorobenzaldehyde and *p*-chlorobenzoin are shown below.

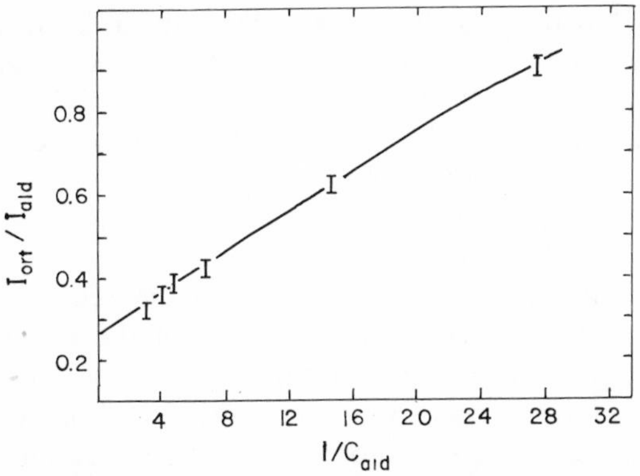

FIG. 19. Ratios of the most intense ortho proton and aldehyde proton transitions as a function of inverse aldehyde concentration. The curve represents the function calculated with $T_1 = 1 \times 10^{-4}\,\text{sec}^{-1}$ for the benzyl proton in the hydroxybenzyl radical and $1.5 \times 10^{-3}\,\text{sec}^{-1}$ for the ortho protons. The rate of hydrogen transfer is $8 \times 10^4\,\text{M}^{-1}\,\text{sec}^{-1}$.

C. CIDNP AT LOW AND ZERO FIELD

Because of the field dependence of the spin Hamiltonian both relative and absolute intensities in CIDNP spectra are a function of the magnetic field. Experiments at low field are afflicted with the difficulty of obtaining well resolved NMR spectra at fields below 5000 G. This complication can be circumvented by running the reaction in a separate field of magnitude H_r and then transferring the sample into the high field H_0 of a conventional spectrometer where the spectrum can be measured. The success of such an experiment depends on being able to quench the reaction effectively before transfer from H_r to H_0 is attempted. Also the transfer time must be shorter or of the same order of magnitude as the relaxation times of the system to be studied.

The interpretation of the resulting spectra is usually based on the assumption that the magnetization of the sample is transferred adiabatically from one field to the other. This condition is met if

$$(d/dt)\,H \ll \gamma H^2, \tag{78}$$

which means the transfer will be adiabatic if the transfer time is much longer than $(\gamma H)^{-1}$. This condition will always be met in practice. If the sample contains nuclei which are coupled among themselves by indirect nuclear spin-spin coupling of magnitude J_{ij} the eigenstates of the system in H_r may differ from those in H_0. In these cases the transfer will be adiabatic if the time spent in the field region where the states change drastically is longer than $(\pi J_{ij})^{-1}$. In systems with very small coupling constants this condition may not always be fulfilled.

Still another difficulty arises from the necessity to correlate the eigenstates at H_r with those at H_0. In complicated multilevel systems this can be a considerable task if H_r is small. A practical solution to this problem involves multiple calculations of the nuclear eigenvalues and eigenvectors at fields interpolating between H_r and H_0. If the field increments are chosen properly it is possible to construct a complete correlation diagram between H_r and H_0.

The general solution for the components of the vector expressing the probabilities of populating the nuclear states of the geminate coupling product was derived in Section III,B,1 and is given by Eq. (33). No special assumptions concerning the magnetic field went into this derivation and therefore Eq. (33) should be valid through all field strengths including zero field. The solution is again a function of the multiplicity of the precursor contained in the matrix ${}^{\mu}\mathbf{Y}$. For a singlet precursor it takes the form

$$ {}^{\mathrm{S}}Y_{lj} = \sum_{r'} Z_{1r',l} Z_{1r',j}, \tag{79} $$

where the index labels electron and nuclear functions and the summation runs over all nuclear states in the radical pair. Substitution into (33) gives for the components of P

$$ {}^{\mathrm{S}}P_n = \lambda\left(\beta' - m \sum_{r'} \sum_{1,j} Z_{1r',l} Z_{1r',j} U^{P}_{nl} U^{P}_{nj} [2\pi(\omega_j - \omega_1)]^{1/2}\right). \tag{80} $$

Similarly for a triplet precursor

$$ {}^{\mathrm{T}}Y_{lj} = \tfrac{1}{3} \sum_{e=2}^{4} \sum_{r'} Z_{er',l} Z_{er',j}, \tag{81} $$

where the first summation is over the three triplet states. Substitution in Eq. (33) gives

$$ {}^{\mathrm{T}}P_n = \lambda m/[3(1-\beta')] \sum_{e=2}^{4} \sum_{r'} \sum_{l,j} Z_{er',l} Z_{er',j} U^{P}_{n,l} U^{P}_{nj} [2\pi(\omega_j - \omega_l)]^{1/2}. \tag{82} $$

That the polarizations obtained from singlet and triplet precursors are of opposite signs is seen by combining (80) and (82) giving

$$ {}^{\mathrm{S}}P_n + 3(1-\beta')\,{}^{\mathrm{T}}P_n = \lambda\beta'. \tag{83} $$

Finally, random combination of free radicals (F-reactions) gives with Eqs. (43) and (46)

$$ {}^{\mathrm{F}}P_n = \tfrac{1}{4}\lambda\left[1 + D\left\{\beta'(1-\lambda) + \lambda m \sum_{r'} \sum_{l,j} Z_{1r',l} Z_{1r',j} U^{P}_{nl} U_{nj} [2\pi(\omega_j - \omega_l)]\right\}^{1/2}\right] \tag{84} $$

Here again the polarization is opposite to that derived for a singlet precursor, analogous to the results obtained in the high field approximation.

The products derived from escaping free radicals via radical transfer are more difficult to treat at low field than within the high field approximation. This is due to the fact that substantial contributions may arise via escape from T_+ and T_-. With the definition of a new operator

$$Q^e = E \otimes S^e, \tag{85}$$

where E is the unit matrix of order 4 and S^e is the direct product of the two nuclear eigenvector matrices for the two escape products from the radical pair, a matrix

$$U^e_{nl} = \sum_m Q^e_{mn} Z_{ml} \tag{86}$$

can be defined characterizing the escape product. The probabilities of forming the nuclear states in the escape product are then given by

$${}^{\mu}P^e_n = \sum_{f=1}^{4} \sum_{l,j} U^e_{fn,l} U^e_{fn,j} {}^{\mu}Y_{l,j}\{\beta' - m[2\pi(\omega_l - \omega_j)]^{1/2}\} - {}^{\mu}P_n. \tag{87}$$

In Eq. (87) ${}^{\mu}P_n$ is given by either Eq. (80) or (82) depending on whether a singlet or triplet precursor is involved, but it is necessary to use S^e instead of S^P in the calculation.

While the characteristics of low and zero field spectra are contained in Eqs. (80), (82), (84), and (87) they give very little insight into the qualitative features of the model. It is therefore advantageous to analyze the main features of low field spectra by qualitative arguments based on first-order perturbation theory.

As has been elaborated in Section IV, A, within the high field region the field dependence of the CIDNP spectra is almost exclusively governed by the Zeeman energy difference ($\Delta g\beta H_0$) of the two components of the radical pair. At lower fields, where the magnitude of the Zeeman splitting may become close to the effective exchange coupling ($2J$) or the hyperfine coupling, the high field approximation breaks down and it is necessary to consider contributions from all three triplet levels. Using first-order perturbation theory, the transition probability per unit time between zero-order states is approximated by

$$k_{ij} \propto H_{ij}^2/(E_i - E_j)^2, \tag{88}$$

where H_{ij} is the off-diagonal element connecting states i and j. Considering a one-proton radical pair generated from a triplet precursor Eq. (88) may be used to estimate the relative weight of contributions arising from transitions between the three triplet levels and the singlet state.

With decreasing field contributions to polarization from differences between probabilities of the transitions, $|T_0\alpha\rangle \rightleftharpoons S\alpha\rangle$ and $|T_0\beta\rangle \rightleftharpoons |S\beta\rangle$ diminish because $|H_{ST_0}, \alpha|^2 \to |H_{ST_0}, \beta|^2$ at low fields. In contrast the transitions $|T_+\beta\rangle \rightleftharpoons |S\alpha\rangle$ and $|T_-\alpha\rangle \rightleftharpoons |S\beta\rangle$ proceed with comparable probability at high field and contribute nothing to the polarization because $(E_{T_+} - E_S)^2 \simeq (E_{T_-} - E_S)^2$ and $|H_{ST_+,\alpha}|^2 = |H_{ST_-,\beta}|^2$. However, if the exchange interaction is not negligible relative to the Zeeman splitting, as might be the case at low fields, net contributions can arise from $T_\pm \rightleftharpoons S$ transitions. Specifically, for $J < 0$ (singlet below triplet), $(E_{T_+} - E_S)^2 > (E_{T_-} - E_S)^2$, causing the $T_- \rightleftharpoons S$ process to predominate and the spectrum to appear in emission. Conversely, if $J > 0$ enhanced absorption lines will be caused by the predominant contribution from $T_+ \rightleftharpoons S$ transitions.

But even if $J = 0$ a net polarization can result from unequal $T_\pm \rightleftharpoons S$ contributions when the field approaches the magnitude of the hyperfine coupling. This arises from the fact that for a positive hyperfine coupling $(E_{T+\beta} - E_{S\alpha})^2 < (E_{T-\alpha} - E_{S\beta})^2$, giving enhanced absorption, while the opposite relationship holds for a negative coupling.

These qualitative predictions are confirmed by the full calculations, the results of which are shown in Fig. 20 for the case of a one-proton radical pair generated from a triplet precursor with and without inclusion of a sizable exchange interaction. It is noteworthy that the change of sign

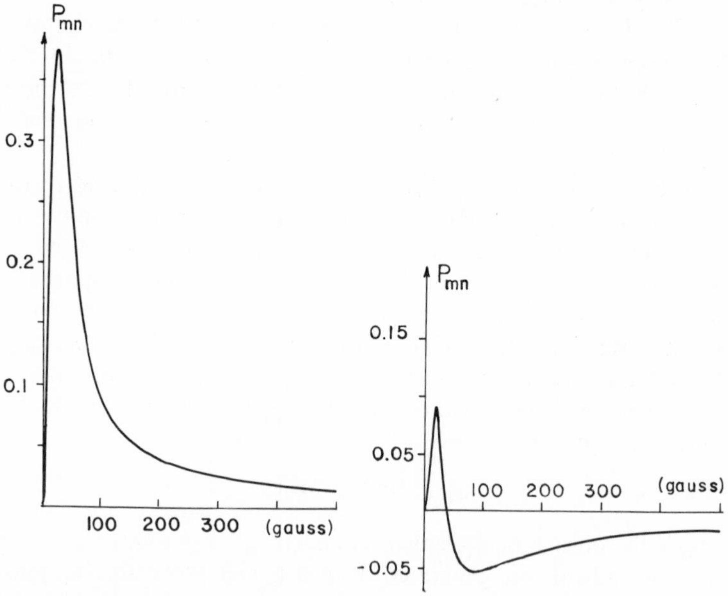

FIG. 20. P_{mn} as a function of H_r calculated for a one-proton radical pair with $a = -25$ G (left), $a = +25$ G (right), and $J = -1.5 \times 10^8$ rad/sec.[29]

for the case of $a > 0$ and $J < 0$ is not immediately obvious from perturbation arguments of the kind used above, except that one arrives at the conclusion that hyperfine and exchange coupling contributions will produce opposite signs in the net effect at low field.[28,29,33]

Low field experiments are expected to be sensitive probes on the assumptions extending into radical pair theory. Of special interest again is the question whether it is necessary to include a finite exchange coupling to simulate experimental results. Low field experiments should be particularly sensitive to this parameter.

As of this writing the question is not entirely settled although it appears that those cases which have been examined in detail do not require the inclusion of J. A particularly interesting example is the one reported by Kaptein and den Hollander who photolyzed diisopropyl ketone in carbon tetrachloride.[29] High field experiments had shown that the key intermediate is the isopropyl-trichloromethyl radical pair (**XXIII**) which is formed in an ill understood process from a singlet precursor (Scheme 7). Disproportionation of the pair gives propene and chloroform as geminate products while

$$(CH_3)_2CH-CO-CH(CH_3)_2 \ + \ CCl_4 \longrightarrow \overline{(CH_3)_2\dot{C}H \quad \dot{C}Cl_3} \ + \ (CH_3)CHCOCl$$

XXIII

Free
radical products

$$CH_3-CH{=}CH_2 \ + \ CHCl_3$$

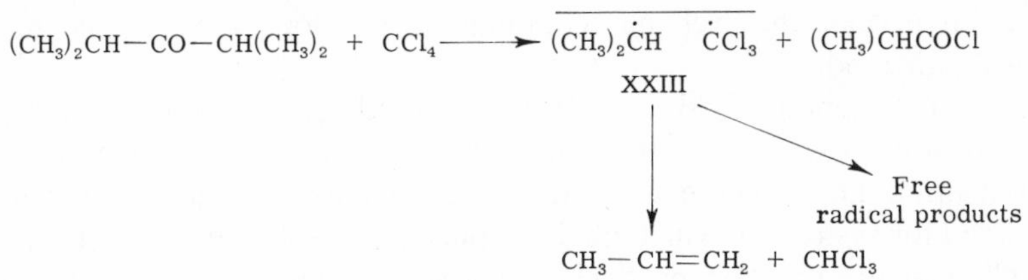

Scheme 7

the escaping isopropyl radical is converted to isopropyl chloride. The field dependence of the chloroform proton resonance shows the double inversion of sign at low field as shown in Fig. 21. The authors were able to

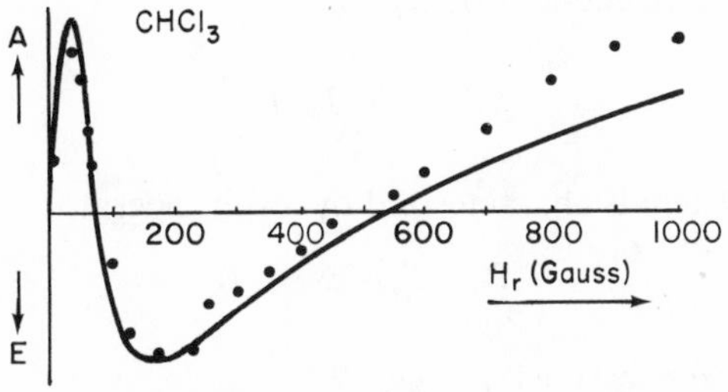

FIG. 21. Polarization of the chloroform proton generated in the reaction outlined in Scheme 7. The curve was calculated by including a J of 1×10^8 rad/sec and the appropriate hyperfine coupling constants. However, see text for the more recent calculations in which J has been set to zero.[29]

simulate this behavior using the appropriate hyperfine coupling of the isopropyl radical and the g-factor difference expected for the pair. It was necessary to adjust J to a value of 1×10^8 rad/sec to obtain the correct crossover points in the calculations. However, the calculation neglected the nuclear spins in the chlorine atoms of the trichloromethyl radical. This does not appear permissible because although nuclear relaxation in chlorine atoms is fast, it is not nearly fast enough when measured on the time scale of the lifetime of the pair. More recently, upon the urging of this author den Hollander has repeated the calculation including the sizable hyperfine interaction of both chlorine isotopes. In this complete treatment it was found that the entire field dependence of the chloroform line as well as the propene spectrum can be simulated very satisfactorily with vanishing J.[60]

Other spectra which were initially simulated with sizable exchange couplings need to be reexamined because in each case the calculations were based on fewer nuclear spins than were actually present in the radical pair or they neglected to take relaxation of multiplet lines into account.[33,61] Just as in the case of high field spectra it appears safe to predict that once this is done the exchange coupling will no longer be needed in the radical pair model.

At zero magnetic field chemically induced polarization of a single nucleus or groups of uncoupled nuclei vanishes because a preferred axis for quantization no longer exists. However, zero field polarization can be observed for systems in which nuclear spins are coupled by indirect nuclear coupling in the reaction products. Such spectra have been observed and were first analyzed by Glarum.[29,34,62] The Hamiltonian of the radical pair at zero field is given by

$$\mathscr{H} = -J(\tfrac{1}{2} + 2S_1 \cdot S_2) + \sum_i a_i S_1 \cdot I_i + \sum_k a_k S_2 \cdot I_k \tag{89}$$

and that of the diamagnetic product by

$$\mathscr{H}^P = \sum_{i<j} J_{N_{ij}} I_i \cdot I_j. \tag{90}$$

The total spin system is characterized by a spin operator

$$F = S_1 + S_2 + K \tag{91}$$

[60] J. A. de Hollander, personal communication (1972).

[61] J. F. Garst, *in* "Chemically Induced Magnetic Polarization" (A. R. Lepley and G. L. Closs, eds.), p. 275. Wiley (Interscience), New York, 1973.

[62] H. R. Ward, R. G. Lawler, H. Y. Loken, and R. A. Cooper, *J. Amer. Chem. Soc.* **91**, 4928 (1969).

with

$$K = \sum_i I_i. \tag{92}$$

The nuclear Hamiltonian commutes with K^2 and K_z and the radical pair Hamiltonian commutes with F^2 and the components of F. It has been shown that the population vector, **P**, at zero field is independent of the eigenvalues of K_z and that the probability of forming a geminate product is identical within each component of the K manifold. This finding is intuitively reasonable because without a field the system must be rotationally invariant.

When the product formed in a zero reaction field is adiabatically transferred to an analyzing field one observes multiplet spectra in which certain lines are missing. If the spin system can be characterized at high field as a first-order spectrum of the A_mX_n type, the multiplets are reduced by one line each and the spectra have been referred to as $N-1$ multiplets.[34] This is shown more clearly by the example of a two-proton radical pair forming a geminate product with a nuclear AX system. At zero field the total spin functions factor into singlet ($F = 0$), triplet ($F = 1$), and quintet ($F = 2$) states. Because of the commutation relationship outlined above, transitions in the radical pair will occur only between levels with the same value of F^2 and F_z. Depending on the signs of the hyperfine coupling constants and on the location of the two protons, transitions will be favored either between states characterized by $F = 1$ or $F = 0$ (the states with $F = 2$ are eigenstates of the system and do not connect to any other state). Transferred to the geminate coupling product, which can only form from the electronic singlet state, overpopulation will be observed for nuclear states with $K = 1$ or $K = 0$. After adiabatic transfer to high field the population distribution will appear as shown in Fig. 22 where excess population has been assumed for $K = 1$ states. The spectrum will be a two-line emission–absorption spectrum in which the inner lines of the AX quartet are missing. Among the features of the $N-1$

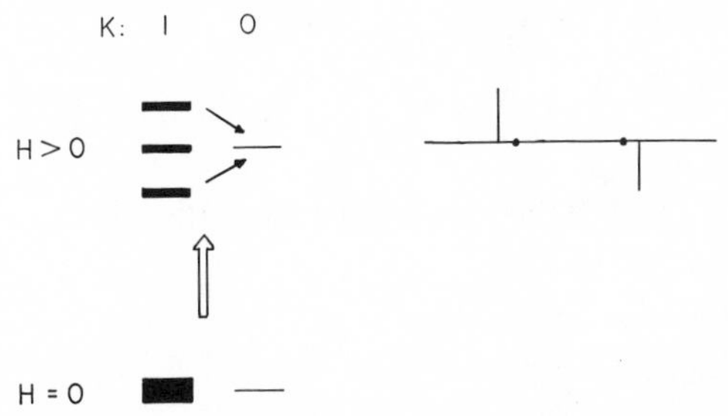

FIG. 22. Population distribution and schematic spectrum of the geminate product of a two-proton radical pair generated at zero field and transferred adiabatically to high field.

multiplet spectra resulting from A_mX_n nuclear spin systems is the absence of the lines between states corresponding in the zero field limit to identical K values which places the missing transitions at the adjacent "inner" positions of the two multiplets. Also, one of the multiplets is in pure absorption and the other is in emission with no net polarization in the total spin system.

It is possible to summarize the prediction as to which spin multiplet will appear in absorption in a simple rule using the same parameters as defined for the multiplet rule [Eq. (57)]

$$\mu A_i A_j J_{ij} \sigma_{ij} \varepsilon\} E/A, \qquad -E/A = A/E, \tag{93}$$

except that E/A stands for emission for the low field multiplet and enhanced absorption for the high field group of lines.[29]

The few reported zero field experiments are in general agreement with the theoretical predictions. A nice example is the formation of ethyl bromide in the decomposition of propionyl peroxide in bromotrichlormethane (Scheme 8).[29] Ethyl radicals escaping the pair are converted by bromine exchange to ethyl bromide which at zero field gives a five-line spectrum. The low field methylene group shows an enhanced absorption triplet while the methyl protons give a doublet in emission. In a similar experiment, 2-iodopropane was generated from isopropyl radicals and iodine transfer from an alkyl iodide. In this case the isopropyl radicals were escaping from random phase encounters and the low field $N-1$ sextet occurred in emission while the methyl group showed enhanced absorption. The low field line of the methyl group actually showed weak emission which may have been caused by incomplete shielding of the earth's magnetic field.

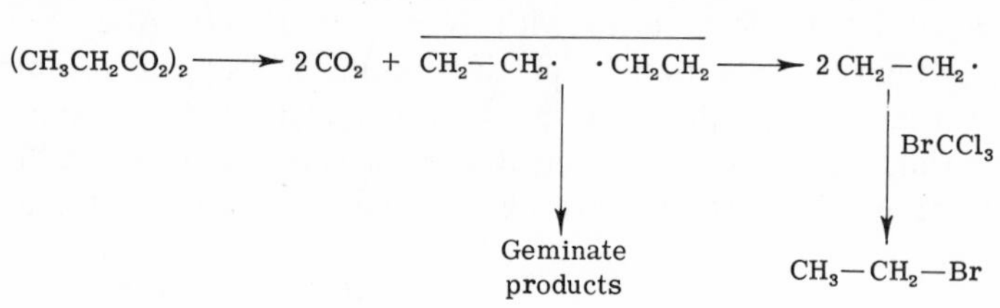

Scheme 8

V. CIDNP Generated from Biradicals

A. Definition and Experiments

Chemically polarized NMR spectra have been observed in reactions proceeding through biradical intermediates.[12-15] The close connection between these experiments and the more common radical pair based

CIDNP becomes obvious if biradicals are defined as molecules with two essentially unpaired electrons which experience only very weak or even vanishing interactions between themselves. It is then possible to visualize biradicals to be special cases of radical pairs in which diffusive separation is limited to a maximum distance determined by the structure of the molecule. A simple example of a biradical is provided by a hydrocarbon chain of variable length with two essentially localized unpaired electrons on each terminal carbon atom. The low energy electronic states of the molecule are conveniently analyzed by visualizing the bond dissociation process of the corresponding cycloalkane yielding the biradical

$$\begin{array}{c} \overset{|}{C}-\overset{|}{C} \\ (\,|\quad|\,) \\ (CH_2)_n \end{array} \rightarrow \cdot\underset{|}{\overset{|}{C}}-(CH_2)_n-\underset{|}{\overset{|}{C}}\cdot \tag{94}$$

Using the admittedly oversimplified picture of a two-center bond, the molecular orbitals of the dissociating bond may be written as a linear combination of the atomic orbitals

$$\Psi_+ = 2^{-1/2}(\phi_1 + \phi_2), \tag{95}$$

$$\Psi_- = 2^{-1/2}(\phi_1 - \phi_2). \tag{96}$$

This gives rise to four electronic configurations

$$\Psi_+^2, \quad 2^{-1/2}(\Psi_+\Psi_- - \Psi_-\Psi_+), \quad 2^{-1/2}(\Psi_+\Psi_- + \Psi_-\Psi_+), \quad \Psi_-^2, \tag{97}$$

of which the first is a reasonable first approximation of the cyclic molecule in the ground state. On stretching the bond in the dissociation process the symmetric configurations Ψ_+^2 and Ψ_-^2 will mix giving the following states

$$\begin{aligned} 1A &= \xi\Psi_+^2 - (1-\xi^2)^{1/2}\Psi_-^2, \\ {}^3B &= 2^{-1/2}(\Psi_+\Psi_- - \Psi_-\Psi_+), \\ {}^1B &= 2^{-1/2}(\Psi_+\Psi_- + \Psi_-\Psi_+), \\ 2A &= \xi\Psi_-^2 + (1-\xi^2)^{1/2}\Psi_+^2, \end{aligned} \tag{98}$$

where ξ goes from 1 in the cyclic form to $2^{-1/2}$ in a large biradical where all electron interactions have vanished. A schematic representation of these states as a function of the distance between the two terminal carbon atoms is shown in Fig. 23. The steep rise of the one-dimensional cross section of the potential energy surface at the extreme separation corresponds to the stretching of still another two-center bond. As indicated in the figure, if the chain is long enough, the ground state and the ^3B state can become near degenerate when overlap and exchange interactions become negligible.

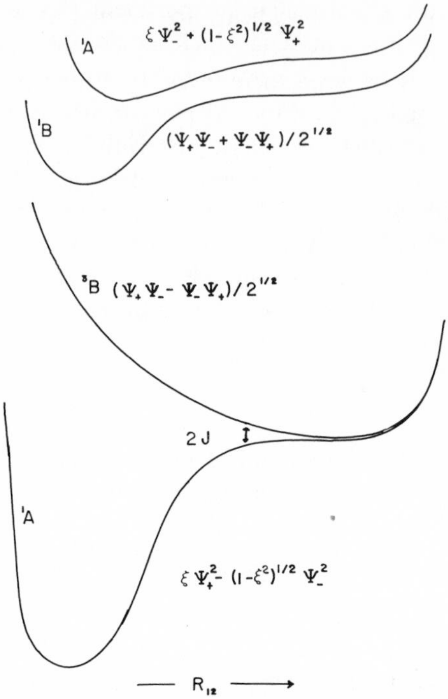

FIG. 23. Schematic one-dimensional representation of the low-lying states of a cycloalkane undergoing cleavage of one ring bond. The energy is plotted against the distance of the separating termini.

Depending on the relationship of exchange and overlap it is possible that at large distances the triplet may even go below the energy of the singlet ground state. In any event, what matters is the existence of a near degeneracy of singlet and triplet states and the associated mixing of the states by hyperfine interaction.

An ideal experiment involves the photoexcitation of the cyclic molecule from the ground state to the ^1B state, followed by intersystem crossing into the ^3B state. Vibrational relaxation within the triplet manifold forces the molecule into its extended biradical configuration. From there intersystem crossing may again occur via spin-orbit coupling and, if the exchange interaction is small enough, hyperfine coupling induced intersystem crossing can compete and provide the basis for CIDNP. Vibrational relaxation within the ground state manifold will reform the cyclic molecule with a nonequilibrated nuclear spin distribution. A variation of this nonradiative decay experiment can be expected if the ^3B state can be near degenerate to the ground state surface of another molecule. In this case the process corresponds to a photochemical reaction.

The difficulty with this idealized experiment lies in the problem of exciting a cycloalkane into a state in which the excitation energy is localized in a single bond. This problem does not exist with cycloalkanones which are known to undergo photolytic cleavage of the bond adjacent to the carbonyl group. Also, chemical evidence indicates that the cleavage occurs from the triplet n–π^* state. The resulting biradical can either reform the cyclic ketone or disproportionate to give the corresponding alkenal (Scheme 9). Although the triplet state of the ketone is known to have a

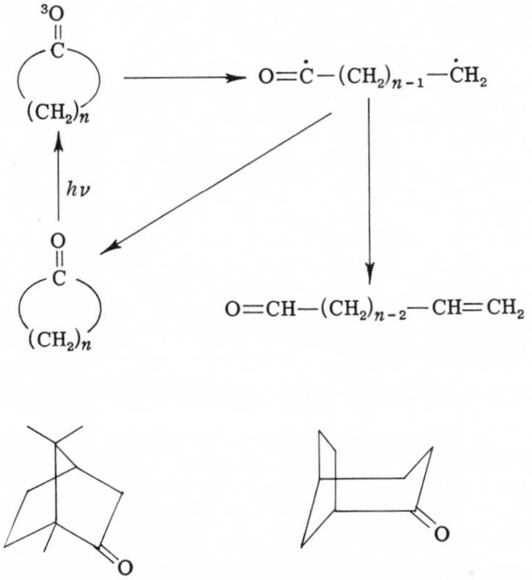

XXIV XXV

Scheme 9

finite lifetime and therefore is not entirely dissociative at small bond stretching, the schematic behavior of the energy levels is not much different from those depicted in Fig. 23.

A series of cycloalkanones of various ring sizes was irradiated inside an NMR spectrometer probe and gave CIDNP spectra with different intensities. A typical spectrum, that obtained from irradiated cycloheptanone, is shown in Fig. 24. Lines originating from both the reactant ketone and from the unsaturated aldehyde occur in emission, lending support to the reaction sequence shown in Scheme 9. It is noteworthy that all lines occur in emission regardless of the signs of the hyperfine interactions of the corresponding protons in the biradical. The emission intensities at high field (14.200 G) increase from cyclohexanones to cycloheptanone and then

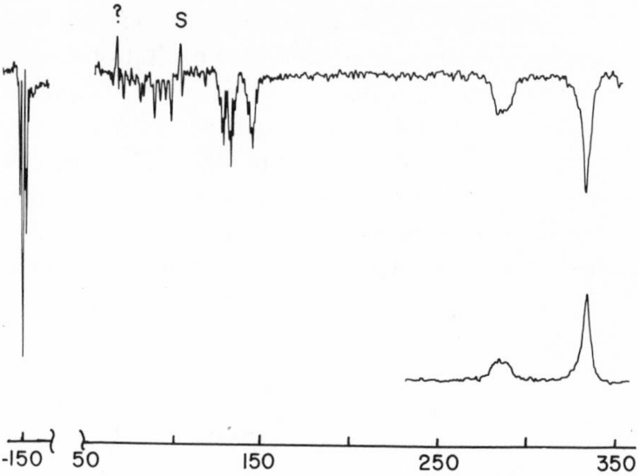

FIG. 24. Spectrum obtained at 60 MHz upon ultraviolet irradiation of cycloheptanone in chloroform. The low field emissions are assigned to the aldehyde and vinyl protons of the heptenal. The high field emissions arise from regenerated cycloheptanone. The chemical shift scale is in Hz from chloroform. The lower trace is the dark spectrum.

diminish with increasing ring size. No signals were observed for cyclopentanone although camphor (XXIV) containing a five-membered ketone ring did show a weak CIDNP signal. Similarly, the bicyclic ketone XXV which contains a cyclohexanone ring gave a much stronger signal than α-methylcyclohexanone.[63]

A similar series of cyclic ketones has been investigated by ^{13}C resonance.[64] Again, emission lines were observed for the carbon atoms carrying hyperfine interaction in the biradical intermediates.

Additional information was obtained from the field dependence of the aldehyde proton resonance in the alkenals.[14] Figure 25 shows the intensities for the series ranging from C_7 through C_{11} ketones as a function of the reaction field, H_r. There are two striking features to the curves: (i) the maximum intensities are shifted to lower fields with increasing ring size, corresponding to increasing chain length in the biradical, and (ii) the width of the observed curves increases in an inverse relationship with ring size. Both features provide a test of the theory which must be able to account for the curves at least in a semiquantitative way. As will be shown below the position of the maximum intensity is mostly determined by the average singlet–triplet splitting in the biradical while the width of the curves is related to the lifetimes of the biradical states.

[63] Cyclohexanone, in contrast to α-methylcyclohexanone, actually gives a mixed absorption emission spectrum. At the time of writing the behavior of this system is not yet understood.
[64] R. Kaptein, personal communication (1972).

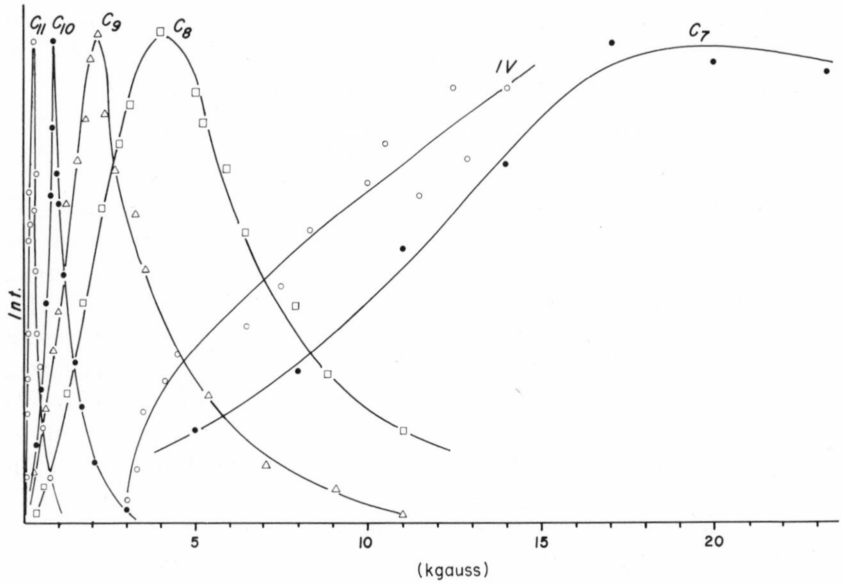

FIG. 25. Intensities as a function of H_r of the aldehyde proton emission signals of alkenals of the indicated chain length obtained by irradiation of the corresponding cycloalkanone.

B. The Biradical Theory of CIDNP

CIDNP based on biradical intermediates can be treated theoretically in a fashion closely related to the radical pair based effect. The major difference is the presence of a sizable exchange interaction in the Hamiltonian and the absence of irreversible diffusive separation of the centers carrying the unpaired electrons. The basic assumption of nonadiabatic formation of the intermediate is carried over to the biradical, as is the postulate that the probability of product formation will be proportional to the singlet character of the biradical. With these assumptions the spin Hamiltonian [Eq. (12)] describing the radical pair is carried over to the biradical and the same basis set [Eq. (6)] used for the pair states will describe the zero-order biradical states. It is important to realize that because of the possibly large exchange coupling in biradicals the high field approximation of the radical pair theory cannot be carried over to biradicals, that is the calculation must always include the whole basis set, even at high fields. As a matter of fact, in first order, polarization in biradical reactions can be traced either to predominant T_-–S or T_+–S mixing while T_0–S mixing does not contribute to any major extent. This arises from the fact that in a biradical no competition by diffusion exists and T_0–S mixing does not involve the z-component of the nuclear spin states. It does not matter what the rate of intersystem crossing of the individual nuclear substates of T_0 is, since all molecules will eventually reach the corresponding S states without a change in the nuclear quantum numbers. This is of course no longer the case with $T_{\pm}$–S intersystem crossing because this process involved a nuclear "spin flip" of $\Delta m_z = \pm 1$.

The absence of diffusion creates yet another, more serious problem in the theoretical treatment. In the radical pair model correlation between the electron spins is lost by diffusive separation and the correlation function, $f(t)$, provides a natural "sink" giving the system dissipative behavior. This can no longer be used in the biradical. It has been suggested to use a simple exponential decay function in which the time constant is characterized as the lifetime of the biradical.[28] Such a treatment corresponds to irreversible disappearance of all radical pair states with the same rate constant and is in direct conflict with the model which specifically requires product formation only via the singlet state of the biradical.

An additional problem arises from the likelihood that triplet–singlet intersystem crossing in biradicals is not only caused by hyperfine induced mixing. Overlap between the orbitals carrying spin densities may never vanish completely and it is likely that spin-orbit coupling will contribute substantially to the intersystem crossing mechanism.

Finally, long chain biradicals can be expected to have considerable conformational flexibility which, within the picture of the one-dimensional energy surface of Fig. 23, translates into a densely packed vibrational manifold within the shallow potential curve. Transitions between these levels during the lifetime of the biradical corresponds to a time-dependent spin Hamiltonian because the exchange interaction, J, is both a function of the separation of the terminal atoms as well as of the relative orientation of the atomic orbitals ϕ_1 and ϕ_2.

It should be apparent at this point that the development of a complete quantum statistical model for the radical pair behavior will be a formidable task. On the other hand the considerable success of the radical pair theory which also contains some drastic assumptions make it worth while designing a severely simplified model which might perhaps simulate the salient features of the experimental results.

One of several conceivable models[65] neglects the conformational motion of the biradical and replaces the time-dependent exchange coupling with an average value. With this assumption the spin Hamiltonian will be the same as used in the radical pair model [Eq. (12)] except that the average J will always be finite and may become quite large for short chain molecules.

[65] G. L. Closs and C. E. Doubleday, to be published (1974).

Spin-orbit coupling and irreversible depopulation of the biradical singlet state can be incorporated into an effective Hamiltonian

$$\mathscr{H}_{\text{eff}} = \mathscr{H}_{\text{spin}} + \mathscr{H}_{\text{so}} + \Gamma, \tag{99}$$

where $\mathscr{H}_{\text{so}}$ describes the spin orbit coupling and Γ is the damping term for the singlet states. The latter takes the form of a diagonal nonhermitian matrix with imaginary elements, $-i2\pi k_{\text{v}}$, for all zero-order singlet states and zero for all triplet states.[66] This method of treating irreversibility corresponds to the assumption of a quasi continuum to which the singlet state is coupled and depopulated with a rate constant k_v. This simulation of vibrational relaxation within the singlet manifold can be expected to be a reasonable approximation because of the high density of states which constitute the singlet biradical.

A detailed consideration of the elements of $\mathscr{H}_{\text{so}}$ leads to the unsatisfactory conclusion that the matrix elements are again strongly dependent on conformational changes and intelligent guesses as to their magnitude are risky at best. It is therefore advantageous to treat intersystem crossing by spin-orbit coupling also as an irreversible process occuring with a rate constant k_{so}, the magnitude of which is to be adjusted by experiment. Within this model $\mathscr{H}_{\text{so}}$ will then be a diagonal matrix with elements $-i2\pi k_{\text{so}}$ for all triplet states and zero for the singlet levels. Essentially this treatment assumes that the triplet levels are coupled via spin-orbit coupling to the same continuum as is the singlet state of the radical pair. If this continuum is now considered to be the highly vibrationally excited ground state of the product, it is possible to write the equation for the probability of populating the nuclear spin states of the ground state as

$$(d/dt)\,\mathbf{P} = k_{\text{v}}|\mathbf{C}_s(t)|^2 + k_{\text{so}}\sum_j |\mathbf{C}_{\text{T}_j}(t)|^2, \qquad j = T_+;\, T_0;\, T_-. \tag{100}$$

The remainder of the treatment is straightforward and equivalent to that of the radical pair theory except that the Hamiltonian operator is no longer hermitian and its diagonalization will give complex eigenvalues of the form

$$\omega_r = \omega_r' - i\gamma_r. \tag{101}$$

Similarly, the eigenvector matrices will have complex elements and will be nonorthogonal reflecting the "shrinking" of the vector in Hilbert space, caused by the decay of the biradical states. With this restriction the time

[66] For a general review on the treatment of irreversible systems, see M. Bixon, J. Jortner, and Y. Dothan, *Mol. Phys.* **17**, 109 (1969).

evolution of the biradical states projected on the nuclear eigenstates of the product are given by

$$|\mathbf{C}(t)|^2 = \tilde{\mathbf{Q}}\mathbf{Z}^* e^{i\omega' t}\mathbf{Z}^{-1*}|\mathbf{C}(0)|^2 \mathbf{Z} e^{-i\omega' t}\mathbf{Z}^{-1}\mathbf{Q}. \tag{102}$$

This equation is equivalent to Eq. (25) which described the radical pair singlet state except that Q is defined by

$$\mathbf{Q} = \tfrac{1}{2}\mathbf{E} \otimes \mathbf{S}^P, \tag{103}$$

where $\mathbf{E}$ is the unit matrix of order four. This way the coefficients of the triplet states are also projected out and can be used in Eq. (100). The evaluation of the elements of (102) is given by

$$|C_n(t)|^2 = \sum_{lj}\sum_{m}\sum_{r} |Q_{mn}|^2 |C_{rr}(0)|^2 Z^*_{ml} Z_{mj} Z^{-1*}_{rl} Z^{-1}_{rj} e^{-i\omega_{lj}t} e^{-\gamma_{lj}t}, \tag{104}$$

where the real part of the eigenvalues has been separated from the imaginary part and the definitions

$$\omega_{lj} = \omega_l - \omega_j \tag{105}$$

$$\gamma_{lj} = \gamma_l + \gamma_j \tag{106}$$

have been introduced for brevity. Inspection of (103) reveals how the oscillatory behavior is damped by the γ's which are close in magnitude to the rate constants $2\pi k_v$ and $2\pi k_{so}$ for well separated zero-order states.

Substitution of (104) and (100) and integration gives the probabilities of populating the nuclear spin states in the product as

$$\begin{aligned} {}^{BR}P_n = \sum_{lj}\sum_{m}\sum_{r} |C_{rr}(0)|^2 \{k_v |Q_{s,mn}|^2 Z^*_{ml} Z^{-1*}_{mj} Z_{rl} Z^{-1}_{rj} \gamma_{lj}/(\gamma^2_{lj} + \omega^2_{lj}) \\ + k_{so} |Q_{T,mn}|^2 Z_{ml} Z^{-1*}_{mj} Z_{rl} Z^{-1}_{rj} \gamma_{lj}/(\gamma^2_{lj} + \omega^2_{lj})\}, \end{aligned} \tag{107}$$

where $\mathbf{Q}_T$ and $\mathbf{Q}_s$ are defined by

$$Q_T = E_T \otimes S^P, \qquad Q_s = E_s \otimes S^P. \tag{108}$$

Substitution of (108) into (3) and (1) gives the population numbers and intensities of the biradical CIDNP spectra, respectively.

It must be recalled that the derivation of (107) is based on the assumption that intersystem crossing from triplet to singlet by the spin-orbit coupling mechanism is irreversible. This is not a bad assumption if the coupling is weak and the irreversible depopulation of the singlet state is fast. However, it restricts the application of (107) to cases where the biradical is generated from a triplet precursor. A proper model for a biradical descendent from a singlet precursor must treat the contribution from all mechanisms to intersystem crossing reversibly since molecules crossing to the triplet levels must return to the singlet biradical state before they can go to product. However, this obvious deficiency of the model is of no consequence because it is easy to show that biradicals generated from singlet precursors should not cause any polarization at all. The argument used here is simply that *all* molecules crossing from the singlet to the triplet manifold have to return to the singlet state to give product. If this cycle is completed on a time scale short compared to relaxation, any polarization caused by the first intersystem crossing must be exactly reversed by the second, leaving no net observable effect. This is in contrast to the behavior of radical pairs, where only a small fraction of the pairs will return to the initial singlet state, the rest being annihilated irreversibly by diffusion.

Another deficiency of the model is the failure to take into account electron spin lattice relaxation. Qualitatively, two extreme cases can exist which will give different relaxation behavior in the biradical. In long chain molecules, where the exchange coupling is extremely weak, the electrons can be considered as essentially uncoupled and the two spin 1/2 systems may be expected to relax independently by the usual mechanisms operative for doublet states. In that case the relaxation transitions will connect all four zero-order states and within the framework of the model the effect will be the same as weak spin-orbit coupling with inter-system crossing rates of 10^5 to $10^6\,\mathrm{sec}^{-1}$. In other words relaxation can be absorbed by adjusting k_{so} accordingly. In shorter chains the motional modulation of the dipolar coupling between the electron spins can be expected to become the major mechanism for electron relaxation. In these molecules, the exchange coupling will also be large and consequently singlet and triplet functions will be almost eigenfunctions of the system and dipolar coupling will contribute to relaxation within the triplet manifold only. If the dipolar coupling as measured by the zero-field splitting parameter, D, exceeds $0.01\,\mathrm{cm}^{-1}$, corresponding to a separation of the biradical termini of less than 5 Å, the dipolar mechanism can be estimated to predominate and may become important in determining the field dependence of biradical CIDNP spectra. The effect of fast relaxation within the triplet manifold will be essentially a predominant drainage of the system through that triplet level which is closest in energy to the singlet state.

Although this additional shortcoming of the model must be kept in mind, it may not be too important in the description of the cycloalkanone derived spectra in which the biradical chains may be long enough to render relaxation by dipolar interactions unimportant.[67] The fact that all lines of the spectra occur in emission is a clear indication that T_{-}–S mixing is predominating in these systems. From perturbation theory it follows that

[67] This may be no longer the case in 1,6 biradicals as generated from the photolytic cleavage of cyclohexanone, and may account for the anomaly of this system. See Ref. 63.

the singlet level lies below the triplet and that the most intense spectra should be obtained if T_- and S levels are degenerate. Since the energy of the T_- state is adjustable by changing the magnetic field, the average singlet–triplet splitting should be revealed by the field dependence of the CIDNP spectra. As Fig. 25 shows the intensity curves exhibit the expected maxima with the proper dependence on the chain length.

The shapes of the intensity–field curves are determined by the relationship of ω_{lj} and γ_{lj}. For a given compound the ω_{lj}'s are functions of the reaction field, H_r. As inspection of (107) reveals the contributions to the P_n's as functions of ω_{lj}, and therefore H_r, follow an approximate Lorentzian function with a width of γ_{lj}. For the eigenstates with predominant singlet character γ_r is close $2\pi k_v$ which will be the major contribution to the uncertainty broadening. Similarly, the lifetime of the triplet states as determined by k_{so} (or spin-lattice relaxation, also absorbed in k_{so}) will have a broadening effect. However, if k_{so} is chosen too large the whole effect will vanish because the hyperfine induced singlet–triplet crossing can no longer compete with the spin-orbit contribution.

These relationships are demonstrated with the aid of Fig. 26 in which field–intensity curves are displayed, calculated from Eq. (107) by choosing the parameters J, k_v, and k_{so} to obtain the best fit with the experimental data. Although only three curves are shown, comparable satisfactory fits have been obtained for the remaining members of the series.

As expected, J determines the position of the maximum while k_{so} has a major effect on the intensities. The width of the curves is quite sensitive to the magnitude of k_v as long as it is chosen to be much larger than k_{so}. The parameters for the calculations of all five curves are summarized in Table I.

TABLE I

FITTING PARAMETERS FOR EFFECTIVE HAMILTONIANS OF BIRADICALS DERIVED FROM CYCLOALKANONES[a]

Chain length of biradical	$2J$ (cm^{-1})	k_v(sec^{-1})	k_{so}(sec^{-1})
7	-1.87 ± 0.28	$4.8 \pm 1.6 \times 10^{10}$	$8.0 \pm 4.8 \times 10^5$
8	-0.374 ± 0.037	$1.2 \pm 0.2 \times 10^{10}$	$6.4 \pm 0.7 \times 10^5$
9	-0.196 ± 0.009	$4.8 \pm 0.8 \times 10^9$	$6.4 \pm 0.7 \times 10^5$
10	-0.085 ± 0.003	$3.2 \pm 0.5 \times 10^8$	$6.4 \pm 0.7 \times 10^5$
11	-0.026 ± 0.005	$1.0 \pm 0.4 \times 10^8$	$6.4 \pm 0.7 \times 10^5$

[a]The hyperfine coupling constants for the two α protons and two β protons were set at -22 and 33 G, respectively, in analogy to the experimentally measured values of the n-propyl radical.[68] The small couplings of the protons adjacent to the carbonyl group were neglected. The g-factor of the carbonyl end of the biradical was set at 2.0007 and that of the alkyl end at 2.0026.

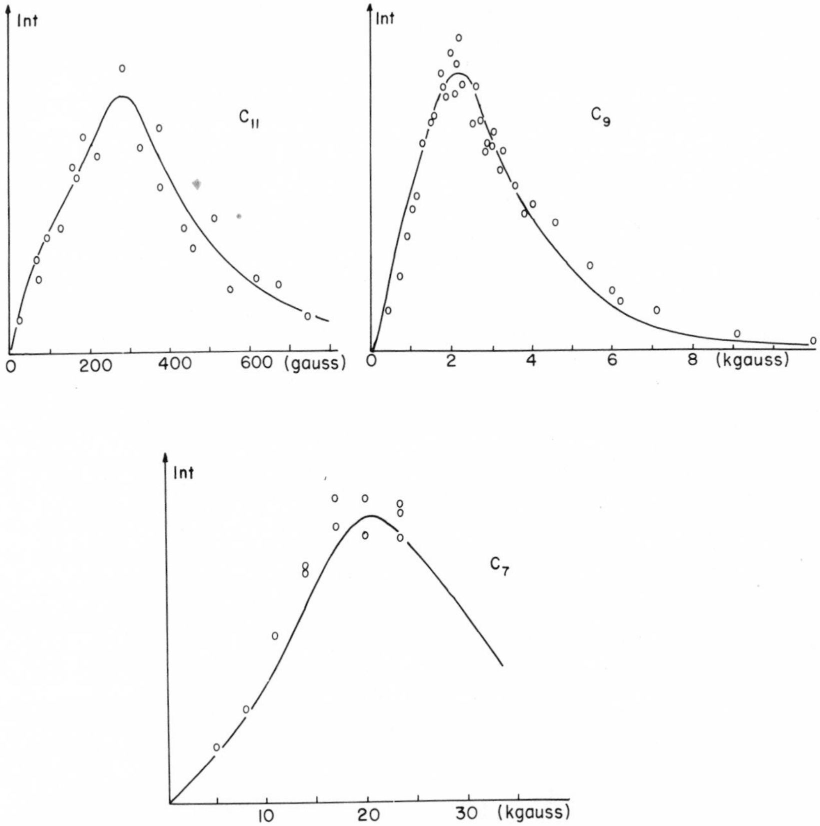

FIG. 26. Calculated and experimental emission intensities as a function of H_r of the aldehyde protons in the alkenals of the indicated chain length. The curves were calculated using parameters given in Table I.

In view of the limitations of the model it is necessary to comment on the question whether these parameters have any physical meaning. The most realistic values are probably the singlet–triplet splittings because it is hard to see how a more sophisticated model would predict an intensity maximum which deviates substantially from the point of degeneracy of the T_- and S levels. However, it must be remembered that the splittings are weighted averages over a large number of conformations. Therefore one expects to find a dependence of these splittings not only on the chain length but also on structural features. That this may be so is indicated by the preliminary results obtained for the bicyclic ketones XXIV and XXV. The biradicals derived from cleavage of these molecules appear to have smaller splittings than the corresponding single chain biradicals with an identical number of carbon atoms. Presumably this reflects the steric effects of the methyl groups in XXVI and the partial rigidity imposed on both XXVI and XXVII by the presence of the five membered ring.

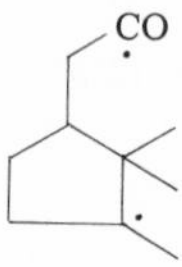

XXVI

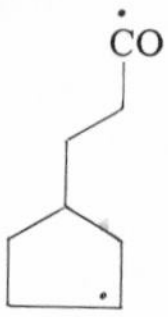

XXVII

The lifetimes of the singlet states ($1/k_v$) and the spin-orbit induced intersystem crossing rates, k_{so}, must be interpreted with caution and should be viewed as exhibiting trends rather than be taken for actual measurements. Nevertheless, the magnitudes of the singlet lifetimes are at least not unreasonable when judged by chemical intuition (the only yardstick available at present, since no measurements have been made by any other method). The magnitudes of k_{so} are of the order of electron spin relaxation rates and are probably lower limits. They should be more accurately determined when enhancement factors have been measured.

VI. Outlook

During the relatively short time of its existence the radical pair theory has been tested by a very large number of experiments. As a result it is probably fair to say that the basic features of the theory are well established and will need little, if any, modification. This does not exclude the possibility of developing a more elegant formalism incorporating less serious assumptions. It should, for example, be possible to incorporate the radical pair correlation function into a true time-dependent Hamiltonian in which the exchange interaction may be treated more explicitly.[68] When more quantitative measurements become available the comparison of predicted enhancement factors with experimental values will reveal whether the rather simplistic treatment of diffusion in the Noyes model is really adequate to describe the systems behavior.

An interesting suggestion has been made by Deutch who proposes an experiment in which the diffusive degrees of freedom are limited to two dimensions.[37] This way CIDNP could be used to gain additional information

[68] One attempt in this direction has been made by G. T. Evans, P. D. Fleming, III, and R. G. Lawler, *J. Chem. Phys.* **58**, 2071 (1973).

on the microscopic theory of diffusion. Although in principle such an experiment could be designed by making use of boundary layers in inhomogeneous systems, the expected experimental difficulties are not trivial.

As has been pointed out in the preceding section, there is ample room for improvement in the quantitative description of biradical derived spectra. Specifically, it would be useful to develop a model in which the irreversible processes are better integrated with the spin-dependent parts. Also, explicit averaging procedures over all conformations of the biradical might be expected to give a more realistic description of the system.

In the meantime, however, CIDNP should continue to be useful to the organic chemist who is interested in reaction mechanisms. The real advantage of the method lies in the fact that information can be obtained on very short-lived molecules. In contrast to the ESR method one is looking at the final reaction product derived from these intermediates and one can be sure that the paramagnetic molecules were indeed on the reaction coordinate. However, that does not ensure that all the product is formed via this pathway, and like most methods CIDNP must be supplemented with other chemical information before firm conclusions about the reaction mechanism can be drawn.

Most of the reported experiments to date involve proton resonance. In principle all nuclei giving high resolution spectra can be examined and useful information has been obtained from ^{19}F,[69] ^{13}C,[70] and ^{15}N.[71] With the increasing abundance of spectrometers with multinuclear capabilities, a greater share of future CIDNP experiments will be on nonprotonic systems.

Acknowledgments

The author is indebted to his students and postdoctoral fellows who contributed almost all the experimental work and a major fraction of the ideas which went into the development of the theory. Specific thanks is due to A. D. Trifunac, C. E. Doubleday, K. Müller, and D. E. Paulson. A note of thanks, however, would be incomplete without mentioning the one man who more than anybody else inspired the author and made him appreciate the beauty of spin mechanics, Clyde A. Hutchison, Jr.

[69] D. Bethell, M. R. Brinkman, and J. Hayes, *Chem. Commun.* p. 475 (1972).

[70] E. Lippmaa, T. Pehk, A. L. Bhuchachenko, and S. V. Rykov, *Chem. Phys. Lett.* **5**, 521 (1970).

[71] E. Lippmaa, *Int. Conf. CIDNP, Tallin, Estonia, 1972.*

Magnetic Shielding and Susceptibility Anisotropies

BERNARD R. APPLEMAN AND
BENJAMIN P. DAILEY

DEPARTMENT OF CHEMISTRY, COLUMBIA UNIVERSITY, NEW YORK, NEW YORK

I. Introduction

The last several years have seen significant advances in both experimental and theoretical techniques for the study of magnetic shielding and magnetic susceptibility. These advances have served not only to expand these areas but have also tended to develop a more complementary relationship and greater interdependence between the sometimes distant theoretical and experimental wings, particularly for shielding. Among the most important elements of this new relationship are the following: (a) a wider adoption of absolute shielding scales; (b) the increased use of *ab initio* calculations for shielding tensors of second period atoms; (c) the development of better experimental techniques for measuring chemical shifts of rare nuclei such as C-13 and N-15; (d) the development of methods which provide more complete information about the chemical shielding tensors.

The magnetic susceptibility anisotropies and tensors have also been reported in greater detail and accuracy than ever before. They are usually considered along with shielding because of the closely related theoretical formulations. Also attempts have been made, using semiempirical theories of the "neighbor anisotropy" effect in the analysis of relative isotropic chemical shifts, to derive sets of bond susceptibility anisotropies. In addition the shielding and susceptibility are often interpreted in terms of similar physical quantities such as current densities.

This article will be primarily concerned with the anisotropies of the magnetic shielding and susceptibility and their relationship to the complete tensor. The anisotropy, being itself a difference, does not require a reference point and can be easily compared among different systems. Section II contains a brief derivation of the most successful *ab initio* method—the perturbed Hartree–Fock theory—of calculating chemical shielding tensors, including the gauge-invariant modification. The theoretical results are presented for ^{13}C, ^{19}F, $^{14,15}N$, and ^{17}O shielding and for susceptibility. The third section discusses the experimental methods of determining shielding anisotropies and presents these quantities for ^{13}C, ^{19}F, ^{31}P, $^{14,15}N$, and miscellaneous nuclei.

The subject of theoretical and experimental *proton* shielding anisotropies has been omitted for the following reasons: (a) the *ab initio* theoretical methods have generally been considerably less successful for proton shielding than for the other nuclei listed above; (The huge number of empirical and semiempirical studies of proton shielding, which are beyond the scope of this article, has been reviewed in detail.[1,2]) (b) the most extensively used technique for determining proton anisotropies, the liquid-crystal method, has been found to be generally unreliable due to large uncertainties and the small range of proton shifts;[3] (c) other techniques have also been less well-suited for accurate measurements of proton anisotropies.

Section IV discusses and compares the experimental techniques for determining magnetic susceptibility anisotropies. The results are tabulated for linear, symmetric top, asymmetric top, and aromatic systems. Section V evaluates the theoretical models with respect to the experimental quantities and discusses the physical similarities and differences between the shielding and susceptibility. The concept of bond anisotropies is also investigated.

[1] J. W. Emsley, J. Feeney, and L. H. Sutcliffe, "High Resolution NMR Spectroscopy," Vol. 2, Chap. 10. Pergamon, Oxford, 1965.

[2] J. A. Pople, W. G. Schneider, and H. J. Bernstein, "High Resolution Nuclear Magnetic Resonance." McGraw-Hill, New York, 1959.

[3] R. A. Bernheim, D. J. Hoy, T. R. Krugh, and B. J. Lavery, *J. Chem. Phys.* **50**, 1350 (1969).

II. Theoretical Models for Shielding and Susceptibility

A. Introduction

The quantum mechanical expressions for shielding and susceptibility can be derived from second-order perturbation theory. The classical energy of interaction between a magnetic moment $\boldsymbol{\mu}$ and a magnetic field $\mathbf{H}$ is given by

$$E = -\boldsymbol{\mu}\cdot\mathbf{H}. \tag{2.1}$$

From this relation and the expressions

$$\boldsymbol{\mu}_{\text{ind}} = \chi\mathbf{H}, \tag{2.2}$$

$$\mathbf{H}_{\text{eff}} = \mathbf{H}_0(\mathbf{1}-\boldsymbol{\sigma}), \tag{2.3}$$

for the induced magnetic moment and effective magnetic field, respectively, it can be shown that the shielding and susceptibility can be expressed as the appropriate second derivatives of the second-order energy:

$$\sigma_{\alpha\beta} = \partial^2 E^2/\partial\mu_\alpha\,\partial H_\beta \qquad (\boldsymbol{\mu}, \mathbf{H} = 0), \tag{2.4}$$

$$\chi_{\alpha\beta} = -\tfrac{1}{2}\partial^2 E^2/\partial H_\alpha\,\partial H_\beta \qquad (\mathbf{H} = 0). \tag{2.5}$$

The total energy of the system is described by the many electron Hamiltonian, Eq. (2.6), which in the presence of a magnetic field is modified by the inclusion of the term $(-1/c)\mathbf{A}_k$ as part of the kinetic energy operator of the kth electron. Atomic units ($\hbar = m_e = e = 1$) are used throughout this section.

$$\mathscr{H} = \tfrac{1}{2}\sum_{k}^{\text{el.}}[-1/c)\nabla_k - (1/c)\mathbf{A}_k(\mathbf{r})]^2 + V(\mathbf{r}), \tag{2.6}$$

$$\mathbf{A}_k(\mathbf{r}) = \tfrac{1}{2}\mathbf{H}\times(\mathbf{r}_{k\text{N}}-\mathbf{r}_0) + (\boldsymbol{\mu}_{\text{N}}/r_{k\text{N}}^3)\times\mathbf{r}_{k\text{N}}. \tag{2.7}$$

$\mathbf{A}_k(\mathbf{r})$ is the vector potential associated with the kth electron; $\boldsymbol{\mu}_{\text{N}}$ and $\mathbf{H}$ are the nuclear moment and the magnetic field, respectively; $\mathbf{r}_{k\text{N}}$ is the distance from the kth electron to the nucleus, while $\mathbf{r}_0$ is the distance from the nucleus to the origin of the vector potential. This origin can be chosen arbitrarily without affecting the exact eigenstates of Eq. (2.6). However, when approximations are introduced in estimating these energies, it is found that in general the choice of origin or gauge is critical to the final results.

Using the standard Rayleigh–Schroedinger procedures, one considers the effects of $\mathbf{H}$ and $\boldsymbol{\mu}$ as small perturbations upon the total energy. The Hamiltonian, wavefunction ψ, and energy are all expanded in a series of multiorder terms in $\boldsymbol{\mu}$ and $\mathbf{H}$. For the magnetic shielding one requires the second-order energy term which is linear in both $\boldsymbol{\mu}$ and $\mathbf{H}$, Eq. (2.8), while for the susceptibility the term of interest is the one which is second order in **H**, Eq. (2.9).

$$E^2(\boldsymbol{\mu}, \mathbf{H}) = \langle \psi^0 | \mathscr{H}^2(\boldsymbol{\mu}, \mathbf{H}) | \psi^0 \rangle + \langle \psi^0 | \mathscr{H}^1(\boldsymbol{\mu}) | \psi^1(\mathbf{H}) \rangle + \langle \psi^0 | \mathscr{H}^1(\mathbf{H}) | \psi^1(\boldsymbol{\mu}) \rangle, \tag{2.8}$$

$$E^2(\mathbf{H}, \mathbf{H}) = \langle \psi^0 | \mathscr{H}^2(\mathbf{H}, \mathbf{H}) | \psi^0 \rangle + 2 \langle \psi^0 | \mathscr{H}^1(\mathbf{H}) | \psi^1(\mathbf{H}) \rangle. \tag{2.9}$$

If the first-order wavefunction $\psi^1(\mu \text{ or } H)$ is expanded in terms of the excited states ($n = 1, \ldots, \infty$) of the unperturbed system, the expressions obtained for the diagonal components of σ and χ, using Eqs. (2.4) and (2.5), are the Ramsey[4] and Van Vleck[5] formulas, respectively, derived for closed-shell states, with similar expressions for the yy and zz components.

$$\sigma_{xx} = \frac{1}{2c^2} \left\langle \psi^0 \left| \sum_k \left(\frac{y_{kN} y_k + z_{kN} z_k}{r_{kN}^3} \right) \right| \psi^0 \right\rangle - \frac{1}{c^2} \operatorname{Re} \sum_{r=1}^{\infty} (E_r - E_0)^{-1} \left\langle \psi^0 \left| \sum \frac{M_{xkN}}{r_{kN}^3} \right| \psi^r \right\rangle \left\langle \psi^r \left| \sum (M_{xk}) \right| \psi^0 \right\rangle, \tag{2.10}$$

$$\chi_{xx} = -\frac{1}{4c^2} \left\langle \psi^0 \left| \sum_k (y_k^2 + z_k^2) \right| \psi^0 \right\rangle + \frac{1}{2c^2} \sum_{r=1}^{\infty} (E_r - E_0)^{-1} \times \left| \left\langle \psi^0 \left| \sum_k M_{xk} \right| \psi^r \right\rangle \right|^2. \tag{2.11}$$

$\mathbf{M}_x$ is the angular momentum about the x axis; the subscript N indicates that the operator is referred to the resonant nucleus as origin. The sums are over the electrons, k, and the excited states, r. Re specifies the real part. The two terms contained in each of the expressions for σ and χ are referred to as the diamagnetic and paramagnetic parts. The former depend only upon the ground state wavefunction and are relatively easy to evaluate. However the paramagnetic terms depend on an infinite sum of excited states including the continuum. In general, to evaluate this term one must rely on some rather uncertain assumptions concerning the availability and accuracy of the excited state wavefunctions and contributions from higher energy states and the continuum.[6]

In spite of these difficulties there has been a large amount of work done utilizing these expressions. In addition a number of simplifying assumptions

[4] N. F. Ramsey, *Phys. Rev.* **78**, 699 (1950).

[5] J. H. Van Vleck, "Electric and Magnetic Susceptibilities." Oxford Univ. Press, London and New York, 1932.

[6] L. C. Snyder and R. G. Parr, *J. Chem. Phys.* **34**, 837 (1961).

has been made of which the most widely used has been the average-energy approximation.

For the purposes of this article the discussion will be concerned primarily with the most recent *ab initio* methods of calculating shielding and susceptibility. Several other reviews [1,2,7] give more detailed discussions of the chronological development of shielding and susceptibility calculations and of some recent developments in the semiempirical and empirical methods of correlation and calculation of these quantities.

Since Eqs. (2.10) and (2.11) have been widely utilized, it is worthwhile to compare the various parts in terms of the type of expectation values required. The most important difference between the terms is the appearance of a $1/r^3$ factor in the two shielding terms. Thus for the paramagnetic shielding one has to evaluate $M_{x_{kN}}/r_{kN}^3$ rather than M_{xk} and the diamagnetic second moments are divided by r_{kN}^3. The effect of the $1/r_{kN}^3$ factor is to increase the importance of the electron density near the nucleus, since the **r** involved is the radial distance from the nucleus.

Another difference between the shielding and susceptibility terms is the location of the origin for which the operators are to be evaluated. However, since the observed shielding and susceptibility are gauge-independent results, it will be convenient for this discussion to choose the resonant nucleus as the gauge origin. With this choice of gauge, it can be seen that division of the shielding term by the factor $(-1/r_{kN}^3)$ for each excited state will give precisely the susceptibility term. There are a number of alternative ways of extracting an effective $1/r^3$ term or using it to correlate $\boldsymbol{\sigma}$'s and χ's, but there are no theoretical justifications for such a procedure and very few successful empirical correlations or predictions. The McConnell–Pople[8,9] theory, which correlates shielding with susceptibility anisotropy, is based upon a further assumption that the electronic currents on neighboring atoms or groups can be replaced by point dipoles. Although this model has been moderately successful in reproducing several observed trends, the point dipole approximation has been shown to give a very poor description for distances smaller than several bond lengths.[10] (See Section V,C.)

An important aspect of many of the semiempirical theories of magnetic shielding has been the emphasis on the reproduction of relative chemical shifts. This approach appeared sensible since most chemical shielding data have been in the form of NMR chemical shifts, which can only be obtained on a relative scale. In addition it seemed reasonable to expect that many of the inaccuracies of the resulting approximation would largely cancel

[7] D. E. O'Reilly, *Progr. NMR Spectrosc.* **2**, 1 (1967).
[8] H. M. McConnell, *J. Chem. Phys.* **27**, 226 (1957).
[9] J. A. Pople, *J. Chem. Phys.* **24**, 1111 (1956).
[10] J. R. Didry and J. Guy, *C. R. Acad. Sci.* **253**, 422 (1961).

when examining small changes among slightly varying systems. In this regard it is noted that the anisotropy is also a quantity which is independent of any absolute reference point for the susceptibility as well as the shielding.

However, from the form of Eqs. (2.10) and (2.11) it is observed that there are no terms within them which are unchanged when the molecular environment is only slightly changed. Starting from these general expressions, it has not been possible to derive explicit expressions for the changes in the shielding or susceptibility. All relative quantities derived from *ab initio* calculations must be obtained by taking differences between the overall quantities computed. Therefore we feel that any theory which purports to describe only the relative parameters between sets of molecules can do so only for a limited number of systems if the basic description of the system is inadequate.

This reasoning should apply to relative changes between molecules, relative changes within a given molecule, or relative changes of the components of a given quantity. It should also be true for semiempirical as well as *ab initio* methods.

A study by Tokuhiro and Fraenkel[11] relating the charge density to the chemical shifts of azines resulted in a good linear correlation which agreed with experiment, but when the method was applied to a more varied series of substituted benzenes,[12] the expected correlation between the shifts and charge density was not obtained. The Ditchfield, Miller, and Pople (DMP)[13] study of ^{13}C chemical shifts relative to methane gave reasonable results for a number of small organic systems but was unable to reproduce the experimental shift between CH_4 and C_2H_6, and also did not give a good agreement with experiment for molecules whose centers of mass differed appreciably.

An interesting interpretation of such results is that reasonably accurate theories would be expected to be best for moderate changes in the observed quantity. For large deviations the inaccuracies of the theory would most likely cause large discrepancies between theory and reality. However there is an inherent difficulty in reproducing very small changes since one is required to measure a difference between two relatively large similar numbers.

The shielding and susceptibility anisotropies have been found to be very sensitive to molecular parameters such as size, substituents, and structure.

In view of this finding and of the previous discussion, we feel that for any theory to give fairly reliable results for the anisotropies it must at least be able to provide good absolute agreement with the average experi-

[11] T. Tokuhiro and G. Fraenkel, *J. Amer. Chem. Soc.* **91**, 5005 (1969).

[12] G. Fraenkel and B. R. Appleman, unpublished observations (1970).

[13] R. Ditchfield, D. P. Miller, and J. A. Pople, *J. Chem. Phys.* **54**, 4186 (1971).

mental values for a large number of systems. The most successful theoretical approach in this aspect is the perturbed Roothaan–Hartree–Fock[14,15] theory for which a very brief outline will be presented. The details are readily available from several sources.[15,16]

B. Perturbed Hartree–Fock Model

1. *Theoretical Development*

The form of the theory outlined here will be that for a closed shell molecule, i.e. one with no net electronic angular momentum. The perturbations considered will be the ones involved in the expressions for shielding and susceptibility, namely **H** (external) and $\boldsymbol{\mu}_N$.

It is assumed that a solution to the unperturbed (zeroth-order) system can be obtained by minimizing the energy defined by

$$E = \langle \psi_{HF} | \mathscr{H} | \psi_{HF} \rangle. \tag{2.12}$$

The total Hamiltonian consists of one- and two-electron operators:

$$\mathscr{H} = \sum_{\mu=1}^{2n} h_\mu + \sum_{\mu<\nu}^{2n} (1/r_{\mu\nu}). \tag{2.13}$$

The Roothaan–Hartree–Fock wavefunction ψ_{HF} is a determinant whose elements φ_i are one-electron LCAO-MO's:

$$\varphi_i = \sum_{\nu=1}^{p} a_{\nu i} \chi_\nu, \qquad i = 1, 2, \ldots, p. \tag{2.14}$$

The linear coefficients of the φ_i's are determined from the solution of the set of coupled one-electron Fock equations.

$$F\varphi_i = \varepsilon_i \varphi_i, \qquad i = 1, 2, \ldots, p, \tag{2.15}$$

where

$$F(\mu) = H(\mu) + G(\mu).$$

The Fock operator $F(\mu)$ is divided into one- and two-electron terms, $H(\mu)$ and $G(\mu)$, respectively, as shown in Eq. (2.16).

$$F(\mu) = -\tfrac{1}{2}\nabla_\mu^2 - \sum_{N}^{\text{nuc.}} (Z_N/r_{\mu N}) + 2\sum_j (2J_j - K_j). \tag{2.16}$$

The coupling between the equations is contained in the Coulomb and exchange operators, J_j and K_j, respectively.

[14] C. C. J. Roothaan, *Rev. Mod. Phys.* **23**, 69 (1951).
[15] W. N. Lipscomb, *Advan. Magn. Res.* **2**, 137 (1966).
[16] R. M. Stevens, R. M. Pitzer, and W. N. Lipscomb, *J. Chem. Phys.* **38**, 550 (1963).

The procedure used is a self-consistent solution of the matrix Roothaan–Hartree–Fock equation.

$$\mathbf{FC} = \mathbf{SCE}. \tag{2.17}$$

The total energy of the system, E_{HF}, is expressed as a sum of orbital energies and two-electron integrals over the n occupied MO's, i.e.

$$E_{\mathrm{HF}} = 2\sum_{i=1}^{n} \varepsilon_i^0 - \sum_{i,j=1}^{n} [2\langle ii|jj\rangle - \langle ij|ij\rangle], \tag{2.18}$$

$$\langle ij\,|\,kl\rangle = \int \varphi_i(1)\,\varphi_j(1)(1/r_{12})\,\varphi_k(2)\,\varphi_l(2)\,d\tau_1\,d\tau_2. \tag{2.19}$$

The quantities $F(\mu)$, φ_i, and ε_i of Eq. (2.15) are expanded to give a series of perturbation equations. The zeroth-order equation is assumed completely determined. The first-order equation is

$$(F^1 - \varepsilon_i^1)\,\varphi_i^0 = -(F^0 - \varepsilon_i^0)\,\varphi_i^1, \qquad i = 1, 2, \ldots, n. \tag{2.20}$$

The perturbed MO's are expanded in terms of the unperturbed virtual (excited) MO's and not the original AO's.

$$\varphi_i^1 = \sum_{p=1}^{m} C_{pi}^1\,\varphi_p^0, \qquad i = 1, 2, \ldots, n. \tag{2.21}$$

The first-order coefficients, C_{pi}^1, can be determined by evaluating the matrix elements of F^1 over the zeroth-order molecular orbital basis and solving the resulting $n \times m$ simultaneous linear equations, which in the presence of the magnetic perturbation operators have the following form:

$$(\varepsilon_p^0 - \varepsilon_i^0)\,\mathbf{C}_{pi}^1 + \mathbf{M}_{pi} + \sum_{j=1}^{n}\sum_{q=1}^{m} [\langle qi\,|\,pj\rangle - \langle ij\,|\,qp\rangle]\,\mathbf{C}_{qj}^1 = 0,$$

$$i = 1, 2, \ldots, n, \qquad p = 1, 2, \ldots, m. \tag{2.22}$$

It can be shown[15] that the second-order energy, which determines $\boldsymbol{\sigma}$ and $\boldsymbol{\chi}$, depends only on the first-order correction to the wavefunction. The final expressions for the components of $\boldsymbol{\chi}$ and $\boldsymbol{\sigma}$ are:

$$\chi_{\alpha\beta} = -(1/2c^2)\sum_i \langle\varphi_i|r^2\delta_{\alpha\beta} - r_\alpha r_\beta|\varphi_i\rangle + (1/c^2)\sum_i\sum_p (C_{pi}^1)_\alpha \langle\varphi_p|M_\beta|\varphi_i\rangle, \tag{2.23}$$

$$\sigma_{\alpha\beta} = (1/c^2)\sum_i \langle\varphi_i|(\mathbf{r}_{\mathrm{N}}\mathbf{r}\delta_{\alpha\beta} - r_{\mathrm{N}\alpha} r_\beta)/r_{\mathrm{N}}^3|\varphi_i\rangle$$

$$-(2/c^2)\sum_i\sum_p (C_{pi}^1)_\alpha \langle\varphi_p|M_{\beta\mathrm{N}}/r_{\mathrm{N}}^3|\varphi_i\rangle. \tag{2.24}$$

The development of the theory followed here is that of Lipscomb, Stevens, and Pitzer.[15,16] An alternate but equivalent derivation of the perturbed Hartree–Fock equations has been given by Pople and co-workers[13,17] who obtain their final working expressions in terms of atomic orbital integrals and zeroth- and first-order density matrices defined by Eqs. (2.25) and (2.26).

$$P^0_{\nu\lambda} = 2 \sum_j^{\text{occ.}} a^0_{\nu j} a^0_{\lambda j}, \tag{2.25}$$

$$P^1_{\nu\lambda} = 2 \sum_j^{\text{occ.}} (a^0_{\nu j} a^1_{\lambda j} - a^1_{\nu j} a^0_{\lambda j}). \tag{2.26}$$

In their scheme the effect of the perturbations upon the molecular orbitals is described in terms of changes in the atomic orbital coefficients. Their expressions for $\boldsymbol{\sigma}$ and $\boldsymbol{\chi}$ corresponding to Eqs. (2.23) and (2.24) are:

$$\chi_{\alpha\beta} = -(1/4c^2) \sum_\nu \sum_\lambda [P^0_{\nu\lambda} \langle \chi_\nu | r^2 \delta_{\alpha\beta} - r_\alpha r_\beta | \chi_\lambda \rangle - 2(P^1_{\nu\lambda})_\alpha \langle \chi_\nu | M_\beta | \chi_\lambda \rangle], \tag{2.27}$$

$$\sigma_{\alpha\beta} = (1/2c^2) \sum_\nu \sum_\lambda [P^0_{\nu\lambda} \langle \chi_\nu | (\mathbf{r}_\mathrm{N} \mathbf{r} \delta_{\alpha\beta} - r_{\mathrm{N}\alpha} r_\beta)/r^3_\mathrm{N} | \chi_\lambda \rangle$$
$$- 2(P^1_{\nu\lambda})_\alpha \langle \chi_\nu | M_{\beta\mathrm{N}}/r^3_\mathrm{N} | \chi_\lambda \rangle]. \tag{2.28}$$

From the above and Eq. (2.21) are derived the following for φ^1_j, by the Pople and Lipscomb methods, respectively.

$$\varphi^1_j = \sum_\nu a^1_{\nu j} \chi_\nu, \tag{2.29}$$

$$\varphi^1_j = \sum_{p=1}^{m} C^1_{pj} \varphi^0_p = \sum_p C^1_{pj} \sum_\nu a^0_{\nu p} \chi_\nu = \sum_\nu \sum_p C^1_{pj} a^0_{\nu p} \chi_\nu. \tag{2.30}$$

From the relationship between the coefficients

$$a^1_{\nu j} = \sum_p C^1_{pj} a^0_{\nu p} \tag{2.31}$$

one can easily show the exact correspondence between the two approaches.

An important difference exists between the two methods in the manners of determining the first-order AO or MO coefficients. In the Lipscomb–Stevens–Pitzer method the coefficients are evaluated from a formal solution of the perturbed Roothaan–Hartree–Fock equations (2.20). In the Pople-group method the Roothaan equations are solved for various finite values of the magnetic field strength, and the first-order density matrix evaluated numerically using finite difference techniques.[17,18]

[17] R. Ditchfield, *in* "International Reviews of Science: Physical Chemistry" (A. Allen, ed.). Med. Tech. Press, New York, 1972.

[18] J. A. Pople, J. W. McIver, Jr., and N. S. Ostlund, *J. Chem. Phys.* **49**, 2960 (1968).

2. *Results for Carbon, Fluorine, Nitrogen, and Oxygen*

As was mentioned previously we have postulated that in order for a theoretical model to give moderately accurate predictions of magnetic anisotropies, it must be capable of providing good absolute agreement with experiment for individual tensor components. The results from the several research groups which have used the perturbed Hartree–Fock method to calculate chemical shieldings and magnetic susceptibilities of molecules are now evaluated.

Tables I–V give the theoretical chemical shielding tensors and anisotropies along with the isotropic experimental shifts when available for a number of molecules containing C, F, N, and O respectively. (For reasons given previously proton shifts are not included.) The experimental shifts have been placed on absolute scales which can be derived from the relative scales by fixing at least one point on each scale. The ^{13}C scale is based on absolute shielding determinations of CO,[19,20] HCN,[21] C_2H_2,[22] and CS_2[23] based on anisotropies[24] or molecular beam data.[25] The ^{19}F scale is due to Hindermann and Cornwell[26] and the ^{15}N scale to Baker *et al.*[27] For those molecules which possess a three-fold axis or higher symmetry the theoretical anisotropies are presented. For certain other molecules for which there are two independent anisotropies, the anisotropies are also given. The experimental anisotropies are presented and discussed in later sections.

The most extensive series of calculations using the perturbed Hartree–Fock has been that of Ditchfield *et al.* (DMP),[13,28] although it was developed and first applied by Stevens, Lipscomb and Pitzer.[16] Arrighini *et al.*[29–31] and Kern and co-workers[32–34] have also done some calculations and for comparison the shielding results obtained by Kolker and Karplus[35] using an extensive variational perturbation method[36] are also presented. Stevens and

[19] I. Ozier, L. M. Crapo, and N. F. Ramsey, *J. Chem. Phys.* **49**, 2314 (1968).

[20] R. M. Stevens and M. Karplus, *J. Chem. Phys.* **49**, 1094 (1968).

[21] F. Millett and B. P. Dailey, *J. Chem. Phys.* **54**, 5434 (1971).

[22] S. Mohanty, Chem. Phys. Lett. **18**, 581 (1973).

[23] A. Pines, W. K. Rhim, and J. S. Waugh, *J. Chem. Phys.* **54**, 5438 (1971).

[24] T. Vladimiroff, *J. Chem. Phys.* **57**, 579 (1972).

[25] W. Huttner and W. H. Flygare, *J. Chem. Phys.* **47**, 4137 (1967).

[26] D. K. Hindermann and C. D. Cornwell, *J. Chem. Phys.* **48**, 4148 (1968).

[27] M. R. Baker, C. H. Anderson, and N. F. Ramsey, *Phys. Rev. A* **133**, 1533 (1964).

[28] R. Ditchfield, D. P. Miller, and J. A. Pople, unpublished observations (1969–71).

[29] G. P. Arrighini, M. Maestro, and R. Moccia, *J. Chem. Phys.* **49**, 882 (1968); G. P. Arrighini and C. Guidotti, *Chem. Phys. Lett.* **6**, 435 (1970).

[30] G. P. Arrighini, M. Maestro, and R. Moccia, *Chem. Phys. Lett.* **7**, 351 (1970).

[31] G. P. Arrighini, C. Guidotti, M. Maestro, R. Moccia, and O. Salvetti, *J. Chem. Phys.* **51**, 480 (1969); **49**, 2224 (1968).

Lipscomb[15,16,20,37–40] carried out calculations for a number of diatomic molecules using very large basis sets for both the zeroth- and first-order wavefunction (e.g. CO[20] zeroth-order wavefunction; 19σ and 8π Slater Type Orbitals (STO's): first-order wavefunction; 2σ, 6π, and 3δ STO's) and the results were usually very close to the available experimental quantities.

The results were also nearly unaffected by changing the vector potential origin, reflecting the accuracy of the wavefunctions. For LiH Stevens and Lipscomb[37] demonstrated explicitly that their basis set was essentially complete.

For polyatomics it is not practical to use such large basis sets, and the results will usually depend to some extent upon the choice of gauge. This gauge dependence is shown in Table I in which the shielding results for several molecules of Kern and co-workers[34] and for CH_3F of Arrighini *et al.*[30] are given at two different vector potential origins. The former group used a set of double-zeta STO's while the latter used a more extended set of STO's including *d* functions on the carbon. DMP[13] evaluated all their shielding tensors using the nuclear center of mass as gauge origin. The basis set used was a slightly extended set of contracted Gaussian-type orbitals (GTO's), and it can probably be assumed that their results are also dependent upon the gauge. The matter of gauge dependence is one important consideration in evaluating the shielding tensors; another of course is the agreement with experiment.

In Tables I and II it is seen that for most of the molecules shown, though not the fluorocarbons, the agreement with the absolute chemical shielding constant is fairly good. The theoretical values are all certainly of the right order of magnitude (which cannot be said for any of the earlier theories) and are generally about 20–50 ppm too large. DMP[13] have shown that the chemical shifts relative to methane agree pretty well with experiment for many small organic compounds. However there are also

[32] T. Tokuhiro, B. R. Appleman, G. Fraenkel, C. W. Kern, and P. K. Pearson, *J. Chem. Phys.* **57**, 20 (1972).

[33] B. R. Appleman, T. Tokuhiro, G. Fraenkel, and C. W. Kern, *J. Chem. Phys.* **58**, 400 (1973).

[34] B. R. Appleman, Ph.D. Thesis, Ohio State Univ., Columbus, 1972; B. R. Appleman, T. Tokuhiro, G. Fraenkel, and C. W. Kern, *J. Chem. Phys.* (in press).

[35] H. J. Kolker and M. Karplus, *J. Chem. Phys.* **41**, 1259 (1964); H. J. Kolker, Ph.D. Thesis, Columbia Univ., New York, 1962.

[36] M. Karplus and H. J. Kolker, *J. Chem. Phys.* **38**, 1263 (1963).

[37] R. M. Stevens and W. N. Lipscomb, *J. Chem. Phys.* **40**, 2238 (1964).

[38] R. M. Stevens and W. N. Lipscomb, *J. Chem. Phys.* **41**, 3710 (1964).

[39] R. M. Stevens and W. N. Lipscomb, *J. Chem. Phys.* **42**, 3666 (1965).

[40] R. M. Stevens and W. N. Lipscomb, *J. Chem. Phys.* **42**, 4302 (1965).

TABLE I

THEORETICAL ^{13}C SHIELDING TENSORS AND ANISOTROPIES[a]: LINEAR AND SYMMETRIC TOP MOLECULES[b]

		Ref.	$\sigma_{\perp}$[c]	$\sigma_{\parallel}$[c]	$\Delta\sigma$[c]	σ_{iso}[d]	σ_{iso}(obs.)[e]
CO		20	−118.4[f]	271.2[f]	389.5[f]	11.5[f]	
		35	298[f]	273[f]	−25[f]	290[f]	12 ± 10[e]
		28	−205.1	271.0	476.1	−46.4	
OCO		28	19.8	283.9	264.1	107.8	70 ± 10[g]
H—C≡N		28	−16.4	278.3	294.7	81.8	76 ± 10[h]
H—N≡C		28	−129.1	272.9	402.0	4.9	—
HC≡CH		28	74.6	279.4	204.8	142.9	
		34	83.5	279.4	195.9	148.8	120 ± 10[i]
		34	106.5[f]	279.4[f]	172.9[f]	164.1[f]	
$\mathrm{H{-}\underset{1}{C}{\equiv}\underset{2}{C}F}$	C1	28	130.1	280.7	150.6	180.3	180 ± 10[j]
	C2	28	132.1	285.4	153.3	183.2	105 ± 10[j]
F—C≡CF		28	231.4	286.6	55.2	249.8	—
F—C≡N		28	63.6	284.2	220.6	137.1	—
CH_4		13	221.0	221.0	0	221.0	
		34	238.6	238.6	0	238.6	196 ± 10[i]
		30	193.9	193.9	0	193.9	
C_2H_6		28	232.5	225.3	−7.2	230.1	
		34	246.6	239.6	−7.0	244.2	188 ± 10[i]
		34	256.0[f]	239.6[f]	−16.4[f]	250.5[f]	
CH_3F		28	162.4	214.4	52.0	179.8	
		33	194.8	232.4	37.6	207.3	
		33	213.5[f]	232.4[f]	18.9[f]	219.8[f]	118 ± 20[k]
		30	125.0[l]	186.1[l]	61.1[l]	145.4	
		30	149.5[f]	186.1[f]	36.6[f]	161.7[f]	
$CH_3O{-}H$[m]		28	(183.8)[m]	219.6	35.8	195.7	
		34	(210.0)[m]	234.4	24.4	218.1	146 ± 10[k]
		34	(221.5)[m]	234.4	12.9	225.8	
CH_3NH_2[m]		28	(208.2)[m]	223.1	14.9	213.2	181 ± 10[k]
$\mathrm{\underset{1}{C}F_3\underset{2}{C}H_3}$	C1	28	232.0	265.7	33.7	243.2	
	C2	28	231.5	268.8	37.3	243.9	
CF_3H		28	195.1	253.9	58.8	214.7	
CF_4		13	249.7	249.7	0	249.7	75 ± 10[n]
$\mathrm{\underset{1}{C}H_3\underset{2}{C}{\equiv}N}$	C1	28	224.4	236.3	11.9	228.4	193 ± 10[k]
	C2	28	9.7	299.4	289.7	106.3	76 ± 10[k]
$CH_3C{\equiv}CH$	C1	28	228.9	233.3	4.4	230.3	192 ± 10[o]
	C2	28	120.5	302.1	181.6	181.0	114 ± 10[o]
	C3	28	71.7	259.8	188.1	134.3	127 ± 10[o]

Footnotes to TABLE I

[a] Units ppm; origin is center of mass except where otherwise indicated.
[b] Numbers below structure indicate C nucleus of interest.
[c] $\Delta\sigma = \sigma_{\parallel} - \sigma_{\perp}$ where the parallel direction is along molecular or C_3 axis.
[d] $\sigma_{iso} = \frac{1}{3}(2\sigma_{\perp} + \sigma_{\parallel})$.
[e] Absolute C—13 scale based on an average of five independent C—13 shielding scales, i.e. CO (theoretical and experimental), HCN, C_2H_2, CS_2, which yield $\sigma(CO) = 12 \pm 10$. Other shifts from relative scale.
[f] Origin at C nucleus.
[g] Ettinger *et al.*[41]
[h] Millett and Dailey.[21]
[i] Ellis *et al.*[42]
[j] Ditchfield and Ellis.[43]
[k] Ditchfield *et al.*[13]
[l] Origin of best gauge, located along C_3 axis near F nucleus. See Arrighini *et al.*[30]
[m] Average of 2 $\sigma_{\perp}$'s, molecule assume to rotate rapidly.
[n] Motell and Maciel.[44]
[o] Strong *et al.*[45]

TABLE II

THEORETICAL ^{13}C SHIELDING TENSORS AND ANISOTROPIES[a]: ASYMMETRIC TOP MOLECULES[b]

Approximate geometry	σ_{xx}[c]	σ_{yy}[c]	σ_{zz}[c]	$\Delta\sigma$	σ_{iso}	σ_{iso}(obs.)[d]
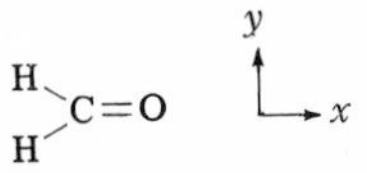	−2.8[b] −33.0[e] −33.0[f]	−136.2[b] −75.8[e] −63.8[f]	181.3[b] 226.2[e] 246.5[f]	250.8[b] 280.6[e] 294.9[f]	14.1[b] 39.1[e] 49.9[f]	−1 ± 10[g]
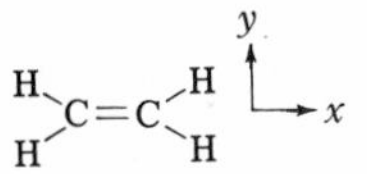	106.4[b] 110.2[h] 110.2[i]	−52.7[b] −13.3[h] 15.3[i]	230.3[b] 257.3[h] 266.5[i]	203.5[b] 208.9[h] 203.8[i]	94.6[b] 118.0[h] 130.7[i]	70 ± 10[j]
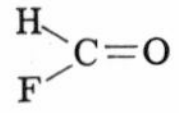	171.8	−125.0	194.6	171.2	80.5	
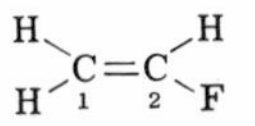 C 1 C 2	146.8 145.5	−47.0 −27.7	212.8 229.4	162.9 170.5	104.2 115.7	105 ± 10[k] 46 ± 10[k]
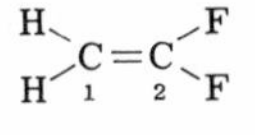 C 1 C 2	167.4 205.5	29.6 −29.3	244.3 253.6	145.8 165.5	147.1 143.3	130 ± 10[k] 32 ± 10[k]
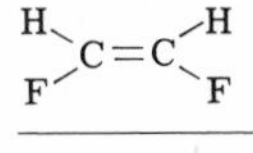	132.3	21.6	230.5	153.6	128.2	57 ± 10[k]

continued

[41] R. Ettinger, P. Blume, A. Patterson, Jr., and P. C. Lauterbur, *J. Chem. Phys.* **33**, 1597 (1960).
[42] P. D. Ellis, G. E. Maciel, and J. W. McIver, Jr., *J. Amer. Chem. Soc.* **94**, 4069 (1972).
[43] R. Ditchfield and P. D. Ellis, *Chem. Phys. Lett.* **17**, 342 (1972).
[44] E. L. Motell and G. E. Maciel, *J. Magn. Resonance* **7**, 330 (1972).
[45] A. B. Strong, D. Ikenberry, and D. M. Grant, *J. Magn. Resonance* **9**, 145 (1973).

TABLE II—continued

Approximate geometry		$\sigma_{xx}{}^{c}$	$\sigma_{yy}{}^{c}$	$\sigma_{zz}{}^{c}$	$\Delta\sigma$	σ_{iso}	$\sigma_{iso}(obs.)^{d}$
F(H)C=C(H)F		205.3	−35.7	268.0	183.2	145.9	49 ± 10[k]
F(H)C₁=C₂F₂	C 1	189.6	46.4	280.5	162.5	172.2	73 ± 10[k]
	C 2	202.4	24.7	307.4	193.9	178.2	35 ± 10[k]
$F_2C=CF_2$		188.9	105.4	348.1	201.0	214.1	49 ± 10[k]
$H_3C_1-C_2(=O)H$ (axes: y up, x right)	C 1	238.0	186.5	215.1	2.9	213.2	162 ± 10[l]
	C 2	−96.3	0.1	231.9	280.0	45.2	−6 ± 10[l]
$H_2N-C(=O)H$ (axes: y up, x right)		−47.0	84.9	225.8	206.9	87.9	28 ± 10[g]
$H-C(=O)-O-H$ (axes: y up, x right)		−96.1	112.2	206.4	198.4	74.2	29 ± 10[g]
$H_2C_1=C_2=O$ (axes: x up, z right)	C 1	129.9	222.7	230.5	54.2	194.4	
	C 2	34.5	−30.4	135.6	133.5	46.6	
$H_2C_1=C_2=CH_2$ (axes: x up, z right)	C 1	32.7	220.1	163.0	36.6	138.6	120 ± 10[j]
	C 2	28.9	28.9	32.7	3.8	30.2	−9 ± 10[j]
$H_3C_1-C_2(H)=C_3H_2$ (axes: x up, z right)	C 1	233.8	211.2	216.3	−6.2	220.4	173 ± 10[j]
	C 2	−31.9	264.3	113.3	−2.9	115.2	58 ± 10[j]
	C 3	−44.7	214.5	128.3	43.4	99.3	77 ± 10[j]
$H_3C_1-C_2H_2-O-H$ (axes: x up, z right)	C 1	244.6	221.6	222.5	−10.6	229.6	175 ± 10[l]
	C 2	247.7	214.8	201.9	−29.4	221.5	135 ± 10[l]
$H_3C-O-CH_3$ (axes: x up, z right)		166.2	167.6	227.3	60.4	187.0	134 ± 10[l]
F_2CH_2 (axes: x up, z right)		198.9	199.9	163.2	−36.2	187.3	
FH_2C-CH_2F (axes: x up, z right)		280.5	248.0	196.9	−67.4	241.8	

Footnotes to TABLE II

[a] Units ppm; origin of vector potential at center of mass unless otherwise indicated. $\Delta\sigma = \sigma_{zz} - \frac{1}{2}(\sigma_{xx}+\sigma_{yy})$.
[b] All calculations except as indicated for H_2CO and C_2H_4 were done by R. Ditchfield, D. P. Miller, and J. A. Pople (unpublished data).
[c] Components given in terms of principal (diagonal) shielding axes, which may differ slightly from assigned axes.
[d] See footnote *e* of Table I.
[e] Tokuhiro *et al.*[32]
[f] Tokuhiro *et al.*,[32] origin of vector potential at C nucleus.
[g] Ditchfield *et al.*[13]
[h] Appleman.[34]
[i] Appleman,[34] origin of vector potential at C nucleus.
[j] Ellis *et al.*[42]
[k] Ditchfield and Ellis.[43]
[l] Emsley *et al.*[1]

TABLE III

THEORETICAL ^{19}F SHIELDING TENSORS AND ANISOTROPIES[a]

	Ref.	σ_{xx}[b]	σ_{yy}[b]	σ_{zz}[b]	$\Delta\sigma$	σ_{iso}	σ_{iso}(obs.)[c]
	35	325.6[d]	325.6[d]	482.9[d]	157.3[d]	378.0[d]	
HF ⟶ z	15	380[d]	380[d]	483[d]	103[d]	414[d]	410 ± 20[e]
	28	272.4	272.4	482.7	210.3	342.5	
LiF	35	334.1[d]	334.1[d]	477.9[d]	143.8[d]	382.0[d]	374 ± 20[f]
	35	180.7[d]	180.7[d]	487.5[d]	306.8[d]	283.0[d]	
F_2	28	−576.3	−576.3	488.4	1064.7	−221.4	−210 ± 25[e]
	38	−543	−543	486	1029	−200	
	38	−658[d]	−658[d]	486[d]	1144[d]	276[d]	
ClF	15	806[d]	806[d]	488[d]	−318[d]	700[d]	637 ± 20[c]
F—C≡N	28	139.4	139.4	489.1	349.7	255.9	346 ± 10[g]
HC≡CF	28	286.4	286.4	489.9	203.5	354.2	284 ± 10[g]
FC≡CF	28	345.0	345.0	490.4	145.4	393.4	—
	28	452.0	452.0	423.1	−28.9	442.4	
H_3C—F (axes x, z)	33	452.0	452.0	442.5	−9.5	448.8	468 ± 15[e]
	33	363.0[d]	363.0[d]	442.5[d]	79.5[d]	389.5[d]	
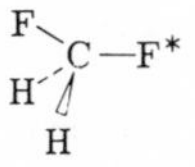	28	473.8	279.3	368.1	—	373.8	341 ± 10[g]
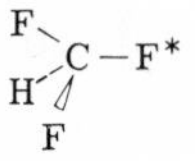	28	319.4	225.2	412.4	—	319.0	274 ± 10[c]
F_3C—F*	28	211.7	211.7	440.1	228.4	287.8	259 ± 10[c]
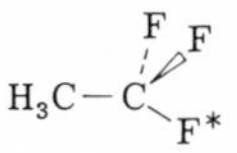	28	418.2	318.8	228.4	—	321.8	255 ± 10[h]

continued

TABLE III—continued

		Ref.	$\sigma_{xx}{}^{b}$	$\sigma_{yy}{}^{b}$	$\sigma_{zz}{}^{b}$	$\Delta\sigma$	σ_{iso}	σ_{iso}(obs.)[c]
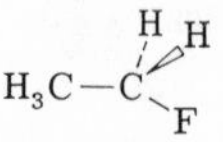		28	416.1	446.7	468.1	—	443.6	401 ± 10[g]
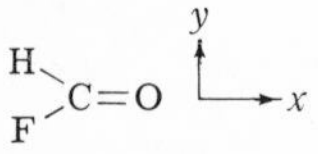		28	−155.4	371.1	307.1	199.3	174.3	148 ± 10[g]
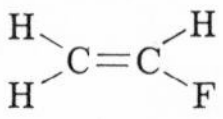		28	232.5	339.4	417.6	131.7	329.8	303 ± 10[h]
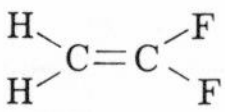		28	242.1	377.5	295.4	−14.4	305.0	270 ± 10[h]
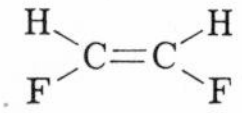		28	261.4	348.6	468.0	163.0	359.3	354 ± 10[h]
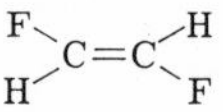		28	350.2	380.4	449.5	84.2	393.4	376 ± 10[h]
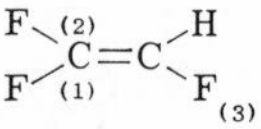	F 1	28	347.6	380.3	300.5	−63.5	342.8	292 ± 10[h]
	F 2		378.3	256.2	335.8	18.6	323.4	316 ± 10[h]
	F 3		372.6	366.0	463.7	94.4	400.7	374 ± 10[h]
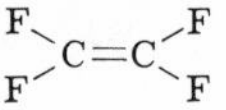		28	342.4	385.3	320.5	−43.4	349.4	325 ± 10[h]
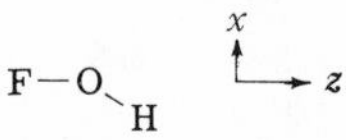		28	−464.7	−63.2	283.5	547.5	−81.5	187 ± 20[i]

[a] Units ppm; origin of vector potential at center of mass unless otherwise indicated. $\Delta\sigma = \sigma_{zz} - \frac{1}{2}(\sigma_{xx} + \sigma_{yy})$.

[b] Components given in terms of principal (diagonal) shielding axes, which may differ slightly from assigned axes.

[c] σ_{iso}(observed) values based on absolute shielding scale of Hindermann and Cornwell,[26] $\sigma(CC1_3F) = 189$ ppm.

[d] Origin of vector potential at F nucleus.

[e] See values given in Table X for σ_{iso}.

[f] Kolker and Karplus.[35]

[g] Dungan and Van Wazer.[46]

[h] Emsley and Phillips.[47]

[i] See Hinderman *et al.*[48] for relative shift with respect to HF.

some large errors in the orders of magnitudes of relative shifts, particularly for fluoromethanes and fluoroethylenes, for which the theory at this level appears inadequate. The observation that the predicted relative shift between C_2H_6 and CH_4 has the wrong sign supports the claim that one cannot expect to reproduce small changes in calculated properties unless the basic description of the system is quite accurate. DMP asserted that their model gives the most reliable results for relative shielding when the systems are of approximately equal size so that the gauge origin remains relatively constant. It has also been found that even for systems of comparable size the agreement may be poor for systems with special features such as highly polar bonds or fluorine substituents.

The qualitative features of the theoretical ^{19}F shielding constants in Table III are similar to those of ^{13}C. The average shielding constants are close to the experimental values but in most cases are too large. The trend for fluorine shielding is much less regular than that for carbon. For the latter most of the theoretical results are in error by a fairly constant amount, a fact which accounts for the good relative shielding predicted by the theory. The F-19 compounds exhibit a wider total range of shielding constants, which is an indication of the greater variety of environments for fluorine and of the larger electron density at the F nucleus. The anisotropies of the fluorine shielding are less well-defined than for carbon because the fluorine atoms are less likely to be located on a symmetry axis. The anisotropies depend upon the manner of selecting the coordinate axes. For the few cases in which there is no ambiguity the theoretical anisotropies should be reliable within order-of-magnitude ranges.

The ^{15}N and ^{17}O theoretical shielding tensors are given in Tables IV and V, respectively. The correlation between theory and experiment is similar to that of ^{19}F. There have been fewer direct experimental determinations of the ^{15}N (or ^{14}N) and ^{17}O shielding tensors. It is interesting that the greatly extended basis set of Laws *et al.*[49] for N_2 gives shielding values which do not agree with the experimental values derived from spin-rotation data. It is also noticed that the N≡C triple bond gives even larger anisotropies for the N shielding than for the C shielding. An interesting feature of the theoretical and experimental ^{17}O shielding constants is the very large range of these quantities and the fact that most of the aldehyde oxygens have large negative shifts. There are only a few

[46] C. H. Dungan and J. R. Van Wazer, "Compilation of Reported F-19 NMR Chemical Shifts." Wiley, New York, 1970.

[47] J. W. Emsley and L. Phillips, *Progr. NMR Spectrosc.* **7**, 1 (1971).

[48] J. C. Hindman, A. Svirmickas, and E. H. Appelman, *J. Chem. Phys.* **57**, 4542 (1972).

[49] E. A. Laws, R. M. Stevens, and W. N. Lipscomb, *J. Chem. Phys.* **54**, 4269 (1971).

TABLE IV

THEORETICAL ^{14}N SHIELDING TENSORS AND ANISOTROPIES[a]

	Ref.	σ_{xx}[b]	σ_{yy}[b]	σ_{zz}[b]	$\Delta\sigma$	σ_{iso}	σ_{iso}(obs.)[c]
N_2 ⟶ z	35	114.4[d]	114.4[d]	339.3[d]	224.9[d]	189.4[d]	
	49	−198.9[d]	−198.9[d]	338.4[d]	537.3[d]	−19.8[d]	−101 ± 20[c]
	28	417.5	417.5	338.3	−79.2	391.1	
H—C≡N	28	−262.7	−262.7	338.7	601.4	−62.2	−37 ± 20[e]
H—N=C	28	−24.3	−24.3	346.7	371.0	99.4	—
F—C≡N	28	−159.3	−159.3	340.6	499.9	7.4	—
H_3C—C≡N ⟶ z	28	−283.0	−283.0	306.1	589.1	−86.6	16 ± 10[c]
H_3N (lone pair) ⟶ z	28	243.9	243.9	244.7	0.8	244.2	175 ± 20[c]
	30	281.4[d]	281.4[d]	254.3[d]	−27.1[d]	272.3[d]	260 ± 20[e]
	34	249.7[d]	249.7[d]	261.4[d]	11.7[d]	253.6[d]	
H_3C—NH_2 ⟶ z	28	(259)[f]		226	−33	247.7	237 ± 10[g]
H_3C—N(CH_3)H (axes x, z)	28	298.0	283.7	214.1	—	265.3	239 ± 10[g]
H_3C—C(H_2)—NH_2	28	226.6	259.5	238.3	—	241.5	
H—C(=O)—NH_2	28	72.5	259.6	246.9	—	193.0	146 ± 10[g]
H_3C—N=O	28	−790.5	223.8	−3530.3	—	−1365.7	
N≡C—O—H	28	−155.0	−206.9	323.4	504.3	−12.8	
HC≡C—NH_2	28	230.3	281.1	266.9	—	259.4	
H_2C=NH (axes y, x)	28	−153.0	−690.5	260.5	682.3	−194.3	
H_2N=NH_2 (axes y, x)	28	168.1	228.0	241.1	43.1	212.4	197 ± 10[c]

[a] Units ppm, origin of vector potential at nuclear center of mass unless otherwise indicated. $\Delta\sigma = \sigma_{zz} - \frac{1}{2}(\sigma_{xx} + \sigma_{yy})$.

[b] Components given in terms of principal (diagonal) shielding axes, which may differ slightly from assigned axes.

[c] Absolute shielding scale from Baker *et al.*[27] $\sigma(N_2) = -101$ ppm.

[d] Origin of vector potential at nitrogen nucleus.

[e] See Gierke and Flygare.[50]

[f] Average of σ_{xx} and σ_{yy}.

[g] See Herbison–Evans and Richards[51] for relative shifts with respect to NH_3.

TABLE V

THEORETICAL ^{17}O SHIELDING TENSORS AND ANISOTROPIES[a]

Approximate geometry	Ref.	σ_{xx}[b]	σ_{yy}[b]	σ_{zz}[b]	$\Delta\sigma$	σ_{iso}
CO (axis z →)	35	357.1[c]	357.1[c]	410.9[c]	53.8[c]	375.0[c]
	20	−109.2[c]	−109.2[c]	411.1[c]	520.1[c]	64.2[c]
	28	−428.6	−428.6	410.5	839.1	−148.9
						[−16±20][d]
OCO (axis z →)	28	−22.8	−22.8	414.7	437.5	123.1
H_2O (axes y ↑, x →)	28	—	—	—	—	267.3
	30	307.0[c]	360.9[c]	325.3[c]	—	331.1[c]
	34	207.7[c]	292.0[c]	247.0[c]	—	248.9[c]
H_3C-O-H (axes x ↑, z →)	28	(320.9)[e]		238.7	−82.2	293.5
	34	(319.0)[e]		274.6	−44.4	304.1
	34	(296.1)[c,e]		267.8[c]	−28.3[c]	286.7[c]
$H_2C=O$ (axes y ↑, x →)	28	−1417.5	−753.4	408.4	1493.9	−587.5
	32	−1511.0	−590.1	436.6	1487.2	−554.8
	32	−1511.0[c]	−531.3[c]	404.2[c]	1425.4	−546.0[c]
						[−375±100][f]
$F(H)C=O$ (axes y ↑, x →)	28	−501.8	−692.8	283.3	880.6	−303.8
$H_3C-C(=O)H$ (axes y ↑, x →)	28	−688.5	−1163.8	369.5	1295.7	−494.3
$H_2C=C=O$ (axes y ↑, x →)	28	−122.0	93.2	−342.6	−328.2	−123.8
$H_2N-C(=O)H$ (axes y ↑, x →)	28	−302.7	335.1	−463.8	−480.0	−143.8
$H_3C-O-CH_3$ (axes x ↑, z →)	28	308.0	316.8	347.4	35.0	324.1
H_3C-CH_2-O-H (axes x ↑, z →)	28	125.9	261.2	440.2	246.6	275.8
$H-C\equiv C-O-H$ (axes x ↑, z →)	28	211.0	224.1	325.8	108.3	253.6

continued

[50] T. D. Gierke and W. H. Flygare, *J. Amer. Chem. Soc.* **94**, 7277 (1972).
[51] D. Herbison-Evans and R. E. Richards, *Mol. Phys.* **8**, 19 (1964).

TABLE V—continued

Approximate geometry	Ref.	$\sigma_{xx}{}^{b}$	$\sigma_{yy}{}^{b}$	$\sigma_{zz}{}^{b}$	$\Delta\sigma$	σ_{iso}
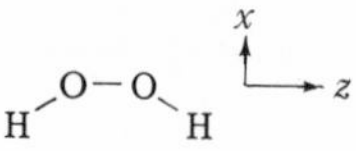	28	180.4	94.5	279.4	—	184.8
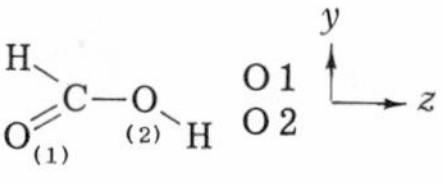 O1	28	−553.9	−428.6	286.1	—	−232.1
O2	28	293.5	−78.7	164.5	—	126.4
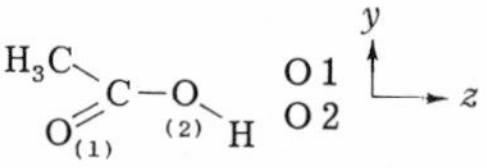 O1	28	−261.6	372.2	39.0	—	49.8
O2	28	90.3	295.1	−206.8	—	59.5

[a] Units ppm, origin of vector potential at nuclear center of mass unless otherwise indicated. $\Delta\sigma = \sigma_{zz} - \frac{1}{2}(\sigma_{xx} + \sigma_{yy})$.

[b] Components given in terms of principal (diagonal) shielding axes, which may differ slightly from assigned axes.

[c] Origin of vector potential at oxygen nucleus.

[d] Experimental value from Gierke and Flygare[50] and Flygare and Weiss.[52]

[e] Average of σ_{xx} and σ_{yy}.

[f] Experimental value from Flygare and Lowe[53] and Neumann and Moskowitz.[54]

molecules which have enough symmetry at the oxygen to give a unique anisotropy. However for nearly all of the molecules, there is a large variation among the individual tensor components of the shielding.

The above results represent a major advance in the field of *ab initio* calculation of chemical shielding constants. However, since the agreement between theory and experiment is somewhat less than quantitative for the average or isotropic value of the chemical shift, it is expected that it will be less reliable in predicting the tensor components of $\boldsymbol{\sigma}$, and hence anisotropies. It is certainly not expected that the relative changes in the principal tensor components be more reliable than the absolute values of the components themselves. The change in environment resulting from a (say) 90° rotation is quite substantial and inaccuracies will in general not be compensated for. The anisotropies given in Tables I–V should provide good order-of-magnitude estimates of the experimental values, probably within about 40 ppm.

[52] W. H. Flygare and V. W. Weiss, *J. Chem. Phys.* **45**, 2785 (1966).

[53] W. H. Flygare and J. T. Lowe, *J. Chem. Phys.* **43**, 3645 (1965).

[54] D. B. Neumann and J. W. Moskowitz, *J. Chem. Phys.* **50**, 2216 (1969).

C. Gauge-Invariant Atomic Orbitals

The perturbed Hartree–Fock theory was able to give very good results using very large basis sets and reasonable results for most small molecules using moderate-sized basis sets. One of the deficiencies of this approach was that the molecular and atomic orbital integrals and hence the shielding depended on the origin of the magnetic vector potential. It is possible to eliminate the gauge dependence of the computed shielding tensor by using atomic orbitals containing explicit vector potential factors[55]:

$$\xi_\nu(\mathbf{H},\mathbf{r}) = \exp[-(i/c)\,\mathbf{A}_\nu(\mathbf{r})\cdot\mathbf{r}]\,\chi_\nu(\mathbf{r}),$$
$$\mathbf{A}_\nu(\mathbf{r}) = \tfrac{1}{2}\mathbf{H}\times\mathbf{r}_\nu = \tfrac{1}{2}\mathbf{H}\times(\mathbf{r}-\mathbf{R}_\nu), \tag{2.32}$$

where $\mathbf{R}_\nu$ is the distance from the arbitrary origin to the atom on which the AO is centered. This approach, first used by London,[56] is known as the gauge-invariant atomic orbitals method (GIAO).

Ditchfield [57,58] has recently developed a theory which utilizes GIAO within the perturbed Hartree–Fock framework. Since the atomic orbitals contain a field dependence, the overlap, core, and two-electron integrals, as well as the expansion coefficients and the density matrix, will depend explicitly on the magnetic field. The total energy of the system in the presence of the perturbations $\boldsymbol{\mu}_{\mathrm{N}}$ and $\mathbf{H}$ (using Pople's formalism) is

$$E(\mathbf{H},\boldsymbol{\mu}_{\mathrm{N}}) = \sum_{\mu,\nu} P_{\mu\nu}H_{\mu\nu}^{\mathrm{core}} + \tfrac{1}{2}\sum_{\mu,\nu} P_{\mu\nu}G_{\mu\nu}. \tag{2.33}$$

Recalling the definitions of χ and $\boldsymbol{\sigma}$ as second derivatives of the energy, one sees that the use of gauge-dependent AO's will yield several additional terms to their expressions. The expression corresponding to Eq. (2.28) for the magnetic shielding is[17]

$$\begin{aligned}\sigma_{\alpha\beta} = (1/2c^2)\sum_\nu\sum_\lambda \{&P_{\nu\lambda}^0\langle\chi_\nu|\mathbf{r}_{\mathrm{N}}\cdot(\mathbf{r}-\mathbf{R}_\nu)\delta_{\alpha\beta} - r_{\mathrm{N}\alpha}(r-R_\nu)_\beta|\chi_\lambda\rangle\\ &+ P_{\nu\lambda}^0\langle(\partial\xi_\nu/\partial H_\alpha)|M_{\beta\mathrm{N}}/r_{\mathrm{N}}^3|\chi_\lambda\rangle + \langle\chi_\nu|M_{\alpha\mathrm{N}}/r_{\mathrm{N}}^3|(\partial\xi_\lambda/\partial H_\beta)\rangle\\ &+ (P_{\nu\lambda}^1)_\alpha\langle\chi_\nu|M_{\beta\mathrm{N}}/r_{\mathrm{N}}^3|\chi_\lambda\rangle\}.\end{aligned} \tag{2.34}$$

Since one requires only a single derivative with respect to $\mathbf{H}$ for the first order density matrix, $(P_{\nu\lambda}^1)_\alpha$, it is sufficient to retain terms only up to linear powers of $\mathbf{H}$ in the evaluation of the required integrals. A general

[55] H. F. Hameka, *Rev. Mod. Phys.* **34**, 87 (1962).
[56] F. London, *J. Phys. Radium* **8**, 397 (1937).
[57] R. Ditchfield, *J. Chem. Phys.* **56**, 5688 (1972).
[58] R. Ditchfield, *Chem. Phys. Lett.* **15**, 203 (1972).

integral of the form $\langle u|\hat{O}|v\rangle$ is expanded as follows[57]:

$$\int \xi_\mu^* \hat{O} \xi_\nu \, d\tau = \int \exp[(i/c)(\mathbf{A}_\mu - \mathbf{A}_\nu) \cdot \mathbf{r}] \, \chi_\mu \hat{O} \chi_\nu \, d\tau$$

$$= \int \chi_\mu \hat{O} \chi_\nu \, d\tau - (1/2c) \mathbf{H} \times (\mathbf{R}_\mu - \mathbf{R}_\nu) \cdot \int \mathbf{r} \chi_\mu \hat{O} \chi_\nu \, d\tau. \quad (2.35)$$

Thus the integrals to be evaluated contain an additional factor of **r** in the integrand.

The results using GIAO theory have been extremely good. Ditchfield[58] has shown that even using a minimal Gaussian-type orbital (GTO) basis set of GIAO one obtains results which are in better overall agreement with experiment than the moderately extended to extended sets of DMP, Kern *et al.*, and Arrighini *et al.*

In addition the minimal set calculations have been moderately successful in reproducing experimental proton shifts.[58] Other *ab initio* methods have yielded quite poor results for proton shielding and also usually contain large gauge dependences; Ditchfield's method has been able not only to reproduce the observed trends of relative proton shifts, but to do so nearly quantitatively for a fairly large number of compounds.

In Table VI are shown some of the most recent GIAO results for heavy atom shielding using the slightly extended GTO[17] basis set along with the available experimental quantities.

For the ^{13}C shielding the agreement between theory and experiment is remarkably good. The theoretical values are within the estimated experimental uncertainties with the single exception of CH_3F. We feel that the extremely successful predictions of the isotropic shielding constants for this ensemble of systems indicate that the shielding tensor components and hence the anisotropies are quite reliable.

The few ^{19}F, ^{17}O, and ^{15}N shielding constants which have been evaluated using this method also agree very well with experimental shifts. It should be noted that the fluorine shielding calculated from the previously used methods has been extremely unreliable for CH_3F.[33] The ^{17}O experimental values for H_2O and CH_3OH are based upon the observed shifts[59] relative to CO for which $\sigma = -16$ ppm. (See Table V.)

It is fair to conclude that there is now available an *ab initio* method for calculating reliable absolute chemical shielding tensors for moderately small molecules of first and second period atoms. If the theory of chemical shielding were capable solely of reproducing observed shifts and of confirming our understanding of the factors which contribute to these quantities

[59] A. Velenik and R. M. Lynden-Bell, *Mol. Phys.* **19**, 371 (1970).

and our confidence in *ab initio* methods, it would be extremely worthwhile. However this method gives in many cases information beyond that which can ordinarily be measured by experimental methods—namely, the tensor components—and it can be applied to molecules inaccessible to experimental techniques. Further it is anticipated that the most successful form of the theory may in the future be extended to third period atoms and larger, more complex systems.

TABLE VI

THEORETICAL SHIELDING TENSORS AND ANISOTROPIES[a] USING GAUGE-INVARIANT ATOMIC ORBITALS[b]

	Nucleus	σ_{xx}[c]	σ_{yy}[c]	σ_{zz}[c]	$\Delta\sigma$[d]	σ_{iso}	σ_{iso}(obs.)[e]
CH_4	C	204.8	204.8	204.8	0	204.8	196 ± 10
CH_3F	C	112.4	112.4	196.5	84.1	140.4	122 ± 15
CH_3OH	C	137.0	139.6	201.2	62.9	159.3	146 ± 10
CH_3NH_2	C	164.6	167.2	203.1	37.2	178.3	181 ± 10
C_2H_6	C	193.9	193.9	204.5	10.6	197.4	188 ± 10
HCN	C	−17.3	−17.3	278.4	295.7	81.3	76 ± 10
HC≡CH	C	54.8	54.8	279.2	224.4	129.6	120 ± 10
CHF_3	C	94.4	94.4	79.0	−15.4	89.3	—
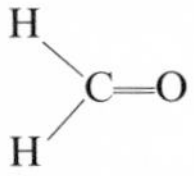	C	−7.9	−115.4	139.0	200.7	5.2	-1 ± 10
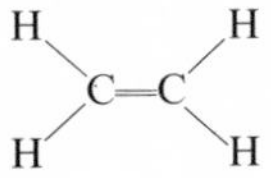	C	96.3	−66.0	191.7	176.6	74.0	70 ± 10
HF	F	376.2	376.2	482.8	106.6	411.7	411 ± 20
F_2	F	−504.7	−504.7	488.4	993.1	−173.7	-210 ± 25
CH_3F	F	509.5	509.5	438.0	−71.5	485.7	468 ± 15
H_2O	O	320.1	355.4	308.8	—	328.1	334 ± 15[f]
CH_3OH	O	400.3	318.5	303.0	—	340.6	371 ± 15[f]
HCN	N	−230.3	−230.3	338.9	569.2	−40.6	-37 ± 20
NH_3	N	281.9	281.9	239.4	−42.5	267.7	$\left\{\begin{matrix}260\pm20\\175\pm20\end{matrix}\right\}$

[a] Units ppm.

[b] Calculations from Robert Ditchfield (personal communication) (1972).

[c] For geometrics see Tables I–V.

[d] $\Delta\sigma \equiv \sigma_{zz} - \frac{1}{2}(\sigma_{xx} + \sigma_{yy})$.

[e] See Tables I–V.

[f] Obtained using $\sigma(CO) = -16$ from Gierke and Flygare[50] and relative shifts from Velenik and Lynden–Bell.[59]

D. Magnetic Susceptibility Calculations

It is interesting to observe that although the theoretical formulations for magnetic shielding and susceptibility are fundamentally the same, the methods of approach and the kinds of success for the calculations of these properties have been quite different. Semiempirical and empirical schemes for breaking down the total susceptibility into atomic or bond contributions such as Pascal Constants[60,61] have given very reasonable results, while this technique has not worked well at all for the shielding. For this review article only *ab initio* type susceptibility calculations will be considered in detail. A recent review which covers in more detail the general problem of susceptibility calculations is that of Ditchfield.[17] (See also Section V,C.)

The variational-perturbation method of Karplus and Kolker[36] was used to calculate the magnetic susceptibilities for several diatomic molecules. Their results were fairly accurate for χ but gave large errors for the magnetic anisotropies as shown in Table VII. Although this method was not extended to larger molecules, it was instrumental in demonstrating the possibility of calculating reasonably accurate second-order properties given a sufficiently flexible first-order wavefunction.

The magnetic susceptibility tensors are shown in Table VII. The perturbed Hartree–Fock theory has been used to calculate these properties by the same groups (Ditchfield *et al.*, Stevens *et al.*, Arrighini *et al.*, and Kern *et al.*) which studied shielding. The general result has been to show that the use of limited basis sets is quite inadequate for the calculation of accurate susceptibilities. By using very large basis sets Lipscomb and co-workers[15] were able to obtain susceptibilities which were very accurate and gauge-independent. The anisotropies were also found to be in good agreement with experiment when these were available.

The attempt to extend these results to polyatomic molecules by Arrighini *et al.*[29] demonstrated that for nearly all but the most extensive wavefunctions employed the theoretical values of χ^{p} were significantly in error. Since these basis sets were 3 or 4 times minimal basis-set size, the method has been limited to a few molecules. In addition, the Arrighini *et al.* results showed a discernible gauge dependence. Ditchfield[62] has suggested that it might be possible to obtain reasonably good results by including 2P orbitals on the protons and fewer extra orbitals on the heavy atoms.

The more limited basis set susceptibility calculations of DMP[28] and Kern and co-workers[32,34] gave rather poor results for almost all the systems

[60] P. W. Selwood, "Magnetochemistry." Wiley (Interscience), New York, 1943.

[61] A. Pacault, *Rev. Sci.* **86**, 38 (1948).

[62] R. Ditchfield, personal communication (1972).

[63] A. J. Sadlej, *Mol. Phys.* **20**, 593 (1971).

TABLE VII

THEORETICAL MAGNETIC SUSCEPTIBILITY TENSORS AND ANISTROPIES[a]

	Ref.	χ_{xx}[b]	χ_{yy}[b]	χ_{zz}[b]	$\Delta\chi$[c]	χ_{av}	$\chi_{expt.}$[d]
Li*H	16	−6.71	−6.71	−9.50	−2.79	−7.63	
LiH*	16	−6.74	−6.74	−9.50	−2.76	−7.66	−7.64[e]
LiH*	36	−8.99	−8.99	−9.44	−0.45	−9.13	
HF*	15	−10.56	−10.56	−10.07	0.49	−10.40	
H*F	15	−11.32	−11.32	−10.07	1.25	−10.90	−8.6±0.1
H*F	36	−8.79	−8.79	−8.15	0.64	−8.58	
F_2	38	(−7)[f]	(−7)[f]	(−17)[f]	(−10±2)[f]	−10.6	−9.7[e]
F_2	36	−21.1	−21.1	−13.7	7.4	−18.6	
B*H	15	34.06	34.06	−11.88	−45.96	18.75	—
BH*	15	33.72	33.72	−11.88	−45.63	18.52	—
B*F	15	−12.37	−12.37	−18.21	−5.84	−14.32	—
BF*	15	−12.21	−12.21	−18.21	−6.00	−14.21	—
CO	20	−10.93	−10.93	−17.91	−6.98	−13.26	−9.8
C*O	36	−15.40	−15.40	−16.14	−0.74	−15.65	−15.2±1[e]
Li_2^*	40	(−28)[f]	(−28)[f]	(−31)[f]	(−3)[f]	−28.94	—
Li_2	36	−34.06	−34.06	−30.81	3.25	−32.97	—
N_2	49	−10.98	−10.98	−18.37	−7.39	−13.44	−12.6[e]
N_2	36	−17.85	−17.85	−16.28	1.57	−17.3	−13.3±3
AlH	49	8.88	8.88	−21.87	−30.75	−1.37	—
CO_2	28	−91.7	−91.7	−25.9	65.8	−69.8	−21.8[e]
C_2H_2	28	—	—	—	—	−34.7	−20.8±0.8
	34	−41.7	−41.7	−25.3	16.4	−36.2	
HC≡N	28	—	—	—	—	−24.7	−14±4[g]
HC≡CF	28	−111.5	−111.5	−29.5	82.0	−84.2	−26.0[e]
FC≡CF	28	−269.4	−269.4	−35.3	234.1	−191.4	−33±4[g]
FC≡N	28	−95.5	−95.5	−26.5	69.0	−72.5	−20±2[g]
H_2O	28	—	—	—	—	−14.1	−13.0±0.1
	34	−14.35	−13.80	−15.73	—	−14.63	
	29	−14.55	−14.22	−14.35	—	−14.37	
NH_3	28	—	—	—	—	−18.90	−16.3±0.8
	34	−19.0	−19.0	−20.6	−1.6	−19.5	
	29	−16.83	−16.83	−16.18	0.65	−16.61	
N^*H_3	63	−17.37	−17.37	−19.34	−1.97	−18.03	
CH_4	28	−24.4	−24.4	−24.4	0	−24.4	−17.4±0.8[h]
	34	−26.7	−26.7	−26.7	0	−26.7	
	31	−19.0	−19.0	−19.0	0	−19.0	
C_2H_6	28	—	—	—	—	−58.6	−26.8±0.8[h]
	34	−77.2	−77.2	−50.2	27.0	−68.2	

continued

TABLE VII—continued

	Ref.	$\chi_{xx}{}^{b}$	$\chi_{yy}{}^{b}$	$\chi_{zz}{}^{b}$	$\Delta\chi^{c}$	χ_{av}	$\chi_{expt.}{}^{d}$
$CH_3C{\equiv}N$	28	−100.3	−100.3	−40.8	59.5	−80.5	-27.6 ± 0.3
$CH_3C{\equiv}CH$	28	−119	−119	−44	75	−93.5	-32.4 ± 2.0^{e}
	28	—	—	—	—	−43.7	
CH_3F	33	−53.7	−53.7	−32.8	20.9	−46.8	-17.8 ± 0.8
	31	−29.1	−29.1	−24.0	5.1	−27.4	
CH_2F_2	28	−106.4	−115.6	−41.1	−69.9	−87.7	-24.0^{e}
CHF_3	28	−114.4	−114.4	−189.1	−74.7	−139.3	-30 ± 4^{g}
CF_4	28	−193.5	−193.5	−193.5	0	−193.5	-36 ± 5^{g}
CH_3CF_3	28	−195.3	−195.3	−208.0	−12.7	−199.5	—
CH_3OH	34	$(-59.7)^{f,i}$		−36.8	22.9	−51.9	-21.4^{j}
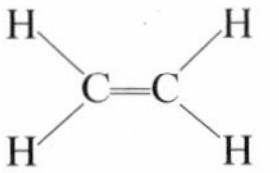	28	—	—	—	—	−40.4	-18.8 ± 0.8^{h}
	34	−27.8	−46.2	−61.3	−24.3	−45.1	
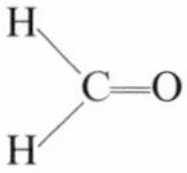	28	—	—	—	—	−22.9	-6.9 ± 0.3
	32	−2.3	−28.2	−43.7	−28.4	−24.7	
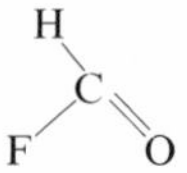	28	−22.8	−86.5	−100.9	−46.2	−70.1	-13.7 ± 4.0^{g}
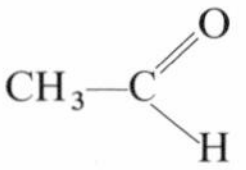	28	−33.3	−90.7	−111.2	−49.2	−78.4	-22.7^{k}
$CH_2{=}C{=}CH_2$	28	−111.3	−111.3	−33.5	77.8	−85.3	-25.3 ± 0.8^{e}
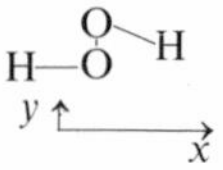	28	$(-30.4)^{i}$		−36.5	−6.1	−32.4	-17.7^{l}

[a] Units 10^{-6} erg gauss^{-2} mole^{-1} (cgs–ppm). Origin of vector potential at nuclear center of mass except where asterisk indicates a nucleus as origin.

[b] See Tables I–V for approximate geometries; see also footnote *c*, Table II.

[c] $\Delta\chi \equiv \chi_{zz} - \frac{1}{2}(\chi_{xx} + \chi_{yy})$.

[d] χ(experimental) values taken from Tables XVII–XX unless otherwise indicated.

[e] Based on theoretical value for χ^{d}.

[f] Approximate.

[g] Estimated from data on similar compounds.

[h] Barter *et al.*[64]

[i] Average of χ_{xx} and χ_{yy}.

[j] Broersma.[65]

[k] "Handbook of Chemistry and Physics."[66]

[l] Foex.[67]

considered. Specifically these groups found that the paramagnetic susceptibilities, computed at the center of mass, were much too small, usually by a factor of 2 or greater; the resulting values of χ were then considerably larger than the observed values. It can be inferred that the limited basis set provides an inadequate representation of the perturbed wavefunction.

It may be concluded that for limited basis sets the principal theoretical model considered here, the perturbed Hartree–Fock theory, has been much less successful in calculating magnetic susceptibilities than it was for shielding.

The Karplus–Kolker variational-perturbation method, on the other hand, although applied only to diatomics gave reasonably accurate results for χ but was considerably worse for σ calculations. The ultimate success of a particular model depends on how well the first-order wavefunction describes the magnetic perturbations. It is possible that in using a limited basis set within the perturbed Hartree–Fock model, the localized region near the nucleus, as sampled by the operator M_α/r^3, is more adequately described.

It should be kept in mind that the perturbed Hartree–Fock shielding constants obtained with the use of simple Slater- or Gaussian-type atomic orbitals were reliable only for a limited number of systems, but that with GIAO the theory has given outstanding agreement with experiment. The GIAO method has not yet been used to calculate magnetic susceptibilities although it is anticipated for the near future. It should at the very least provide more information about the relative difficulties of calculating shielding and susceptibility using *ab initio* procedures.

III. Nuclear Magnetic Shielding Anisotropies

A. Experimental Methods for Shielding Anisotropies

The nuclear shielding of a molecule is defined with respect to a secondary magnetic field. Since each of these fields can have components in any of three directions, $\boldsymbol{\sigma}$ is a second-rank tensor with nine elements, which can always be reduced to three principal tensor components. The anisotropy of the shielding is the difference between two particular components. In principle it can be experimentally determined by observing the response of a molecule to magnetic fields in the appropriate directions. In liquids

---

[64] C. Barter, R. G. Meisenheimer, and D. P. Stevenson, *J. Phys. Chem.* **64**, 1312 (1960).

[65] S. Broersma, *J. Chem. Phys.* **17**, 873 (1949).

[66] "Handbook of Chemistry and Physics" (C. D. Hodgman, ed.), 44th Ed., pp. 2731–2761. Chem. Rubber Publ. Co., Cleveland, Ohio, 1962.

[67] G. Foex, "Constantes Selectionees, Diamagnetism et Paramagnetism: Tables de Constants et Donnees Numeriques," Ch. 1. Masson, Paris, 1957.

and gases the rapid molecular reorientation processes allow one to observe only the rotationally-averaged isotropic value of the chemical shielding tensor. For methods which would enable the determination of anisotropies or individual components, one requires a state in which there is some preferred orientation.

1. *Solid State Methods*

In the solid state the molecular motion is greatly reduced, and, on an NMR time scale, one observes a spectrum which is a superposition rather than an average of the shielding tensor elements. However the solid state NMR spectra are normally very broad due to the dipolar coupling terms, which are not averaged to zero as in the liquid phase NMR. The interpretation of solid state spectra is often a formidable task even for relatively simple systems. There have been developed several techniques for handling such data.

One method is the use of measured values of first and second moments.[68] This method makes use of the known dependence of the moments upon the magnetic field,[69] Eq. (3.1),

$$(\Delta\omega)^2 = (\Delta\omega)^2_{\text{dipolar}} + (\Delta\omega)^2_{\text{anisotropy}}, \tag{3.1a}$$

$$(\Delta\omega)^2_{\text{anisotropy}} = K(\sigma_{\parallel} - \sigma_{\perp})^2 H_0^2, \tag{3.1b}$$

to obtain the isotropic and anisotropic contributions to the shielding. The accuracy of this method has been questioned[70] due to the difficulty of obtaining reliable moments at different fields and the reliability of the assumption of axial symmetry.

A method has been used which employs coherent averaging[70,71] techniques to effectively narrow the dipolar-broadened lines of powdered solids. The observed spectrum is curve-fitted with a broadened theoretical chemical shift distribution to give the principal components of the shielding tensor in the form of the individual anisotropies $(\sigma_{ii} - \frac{1}{3}\mathrm{Tr}\,\sigma)$. This information can sometimes be combined with other experimental data to relate the principal axes to the molecular axes. The method has been used by Mehring *et al.*[70] to investigate the fluorine shielding in several molecules.

A method has recently been developed for observing the magnetically dilute nuclei in a solid by cross-polarizing these dilute spins, e.g., ^{13}C, with the abundant spins, e.g., ^{1}H.[72] This nuclear enhancement technique

[68] D. L. VanderHart and H. S. Gutowsky, *J. Chem. Phys.* **49**, 261 (1968).
[69] N. Bloembergen and T. J. Rowland, *Phys. Rev.* **97**, 1679 (1955).
[70] M. Mehring, R. G. Griffin, and J. S. Waugh, *J. Chem. Phys.* **55**, 746 (1971).
[71] U. Haeberlen and J. S. Waugh, *Phys. Rev.* **175**, 453 (1968).
[72] A. Pines, M. G. Gibby, and J. S. Waugh, *J. Chem. Phys.* **56**, 1776 (1972).

has yielded data on the shielding tensors of a variety of molecules including some having several differently shielded carbons in the same molecule.[73]

In a few cases the experimental shielding tensors of single crystals[74] have been studied. The procedure consists of observing the positions and shapes of the lines at different crystal orientations. However, one often requires certain assumptions about the symmetry of the system[75] in order to extract from the data the complete shielding tensors. Various other techniques[76–78] have been used to obtain shielding tensors or anisotropies for solids, but are applicable only to a few systems.

2. *Methods Using Partially Oriented Systems*

There are systems, other than pure solids, in which some orientations are preferred to others. An important example is that of certain liquid crystals[79] which, in the presence of a magnetic or electric field, orient themselves so that there is a net alignment along the field axis. The liquid crystal molecules themselves are complex and give rise to extremely complicated, largely indecipherable NMR spectra due to the large number of dipolar-broadened peaks. However it is found that certain molecules when dissolved in liquid crystals will also become partially oriented. The solute molecules are not held rigidly, but are able to translate relatively freely in one direction. This motion causes the intermolecular dipole–dipole interaction, but not the intramolecular interaction, to be averaged to zero. In addition since the solute molecules cannot reorient equally in all directions, there will be an anisotropic contribution to the observed shielding.

The orientation of the solute molecules is described by the tensor **S** defined by[80]

$$S_{ij} = \tfrac{1}{2}\langle 3\cos\theta_i \cos\theta_j - \delta_{ij}\rangle \qquad i,j = 1,2,3, \tag{3.2}$$

where θ_i and θ_j are the angles between the axes of the solute molecule and the applied magnetic field. The diagonal elements of this tensor in its principal axis system are called the degrees of order, with S varying between 1 and $-\frac{1}{2}$, and being equal to 1, $-\frac{1}{2}$, and 0 corresponding to alignment always parallel to the field axis, always perpendicular to the field axis, and randomly oriented with respect to the field axis respectively.[81]

[73] A. Pines, M. G. Gibby, and J. S. Waugh, *Chem. Phys. Lett.* **15**, 373 (1972).
[74] P. C. Lauterbur, *Phys. Rev. Lett.* **1**, 343 (1958).
[75] W. Derbyshire, J. P. Stuart, and D. Warner, *Mol. Phys.* **17**, 449 (1969).
[76] H. M. McConnell and C. H. Holm, *J. Chem. Phys.* **25**, 1289 (1956).
[77] W. H. Jones, Jr., T. P. Graham, and R. G. Barnes, *Phys. Rev.* **132**, 1898 (1963).
[78] J. F. Baugher, P. C. Taylor, T. Oja, and P. J. Bray, *J. Chem. Phys.* **50**, 4914 (1969).
[79] A. D. Buckingham and K. A. McLauchlan, *Progr. NMR Spectrosc.* **2**, 63 (1967).
[80] A. Saupe, *Z. Naturforsch. A* **19**, 161 (1964).
[81] G. Englert and A. Saupe, *Mol. Cryst.* **1**, 503 (1966).

The observed chemical shift in the nematic phase is given by

$$\sigma_{\text{nem}} = \sigma_{\text{iso}} + f(S_{11}, S_{22}, S_{33}; \boldsymbol{\sigma}). \tag{3.3}$$

In molecules with linear or C_{3v} symmetry there is only one independent S parameter and the above expression reduces to

$$\sigma_{\text{nem}} = \sigma_{\text{iso}} + \tfrac{2}{3}S(\sigma_{\parallel} - \sigma_{\perp}). \tag{3.4}$$

A plot of σ_{nem} versus S will yield $\Delta\sigma$ from the slope. If one assumes that medium effects can be neglected, it is also possible to obtain $\Delta\sigma$ from the nematic to isotropic phase shift at the transition temperature. The latter method, though simpler, is usually less reliable; however these two methods have been found to give entirely equivalent results for experimental ^{13}C anisotropies.[82] For fluorine shielding studies there are often large uncertainties[3] in the value of S derived from the observed splittings of the NMR line patterns. The splittings depend upon the indirect coupling constants, J_{ij}'s, their anisotropies, the direct dipolar coupling terms, and hence the internuclear distances, as well as upon S.[83] There are also rather large medium effects associated with fluorine and proton liquid crystal NMR studies.[3] Some recent work in our laboratory has indicated that there are significant temperature dependences of fluorine shifts in both isotropic and nematic phases.[82] Therefore the anisotropies determined from liquid crystal measurements, particularly using the single-point method, may contain large errors.

It is also possible to use clathrate crystal lattices[84,85] as the orienting media which are required for deriving anisotropies, since small molecules can sometimes be trapped in the cavities of such crystals. In the β-quinol clathrates,[86] at liquid helium temperatures, gas molecules are frozen into the ellipsoidal cages of the lattice.

The nearly 100% alignment (effective S value = 1) allows the chemical shift anisotropy to be determined directly from the spectra at 0° and 90° orientation angles. However this particular technique is confined to very small molecules (CF_4 was too large). The overall utility of the crystal lattice method is limited because of the lack of suitable hosts[79] and the possibility of strong guest host interactions; consequently relatively few anisotropies have been measured using it.

[82] P. Bhattacharyya, personal communication (1972).
[83] L. C. Snyder, *J. Chem. Phys.* **43**, 4041 (1965).
[84] E. Hunt and H. Meyer, *J. Chem. Phys.* **41**, 353 (1964).
[85] A. B. Harris, E. Hunt, and H. Meyer, *J. Chem. Phys.* **42**, 2851 (1965).
[86] J. H. van der Waals and J. C. Platteeuw, *Advan. Chem. Phys.* **2**, 1 (1959).

3. *Spin-Rotation Methods*

The spin-rotation interaction is the interaction of the nuclear magnetic dipole moment[87] with the magnetic field produced by the end-over-end rotation of the molecule as a whole. It is an additional second-order energy term which is of importance because it can sometimes be obtained from the analysis of the high resolution rotational spectrum of a molecule.

The parameter describing the interaction is the spin-rotation tensor **C**, which is defined by the following term of the Hamiltonian:[88]

$$\mathscr{H}_{\mathrm{SR}} = \mathbf{I}\cdot\mathbf{C}\cdot\mathbf{J}. \tag{3.5}$$

I and **J** are the nuclear spin and rotational angular momentum operators respectively.

The spin-rotation and nuclear magnetic shielding tensors are two closely related quantities which both describe essentially the same magnetic field, i.e. the one produced at the nucleus by electronic motion. They are distinguishable by the manner in which they are observed experimentally. The relationship between the quantities **C** and **σ** has been discussed in detail by Flygare.[88,25] The diagonal components of the paramagnetic shielding [see Eq. (2.10)] and the spin-rotation tensors in the principal inertial system can be expressed in terms of the same quantities and hence in terms of each other.[25] Thus,

$$C_{xx} = (2e\mu_0 g_I G_x/\hbar c)\Bigg[\sum_l Z_l(z_l^2 + y_l^2)/r_l^3 - (1/m)\sum_{p>0}(E_p - E_0)^{-1}\langle 0|L_x|p\rangle\langle p|L_x/r^3|0\rangle + \text{c.c.}\Bigg] \tag{3.6}$$

$$\sigma_{xx}^{\mathrm{p}} = -(e^2/2m^2c^2)\sum_p (E_p - E_0)^{-1}\langle 0|L_x|p\rangle\langle p|L_x/r^3|0\rangle + \text{c.c.} \tag{3.7}$$

Here μ_0 is the nuclear magneton, g_I the nuclear g-factor, G_x the rotational constant of the x-axis, and L_x the angular momentum operator. The resulting expression for σ_{xx}^{p} is

$$\sigma_{xx}^{\mathrm{p}} = -(e^2/2mc^2)\sum_l Z_l(z_l^2 + y_l^2)/r_l^3 + \tfrac{1}{2}(M_p/mg_I)(C_{xx}/G_x), \tag{3.8}$$

with similar expressions for σ_{yy}^{p} and σ_{zz}^{p}.

It should be noted that the infinite sum of excited state contributions has been eliminated, and the paramagnetic shielding components can be expressed in terms of experimental quantities. Thus from a knowledge of

[87] G. C. Wick, *Phys. Rev.* **73**, 51 (1948).
[88] W. H. Flygare, *J. Chem. Phys.* **41**, 793 (1964).

the moments of inertia, the internuclear distances and the spin-rotation constants, one can obtain the σ^p components directly.

Most of the experimental work on spin-rotation interactions has been carried out in the gas phase on small molecules with a fairly high degree of symmetry. In linear molecules there is only one spin-rotation constant, while for symmetric top molecules there are two independent components $C_{\parallel}$ and $C_{\perp}$. Experimentally one sometimes obtains the two equivalent parameters C_{av} and $\Delta C = C_{\parallel} - C_{\perp}$, although the latter quantity frequently has a much larger uncertainty.

One advantage the molecular beam method of determining spin-rotation constants has over other methods of measuring σ is that the measurements are done in the gas phase or "free molecule" state. Thus they are not subject to medium effects and should correspond most closely to the "true" values of the shielding. In addition the anisotropies derived from spin-rotation constants can be used as a check on the reliability of other methods.

The diamagnetic shielding components required to convert the σ^p of spin-rotation measurements into total shielding components are discussed in a later section.

4. *Relaxation Methods*[89]

An environment whose response to a magnetic field changes with orientation—i.e. an anisotropic one—can provide a mechanism for spin-lattice relaxation via the fluctuating field.[89] The expression for the contribution to the spin-lattice relaxation time of the anisotropic shielding for a linear or symmetric top molecule developed by Abragam[89] has been discussed by Huntress.[90]

$$1/T_1(\Delta\sigma) = \tfrac{8}{15}\pi^2(\sigma_{\parallel} - \sigma_{\perp})^2 H_0^2 \tau, \tag{3.9}$$

where τ is the correlation time of molecular reorientation. There are a couple of difficulties in using this expression.

The experimentally measured spin-lattice relaxation time T_1 is comprised of a number of contributing terms including dipolar terms $T_{1\mathrm{D}}$, spin-rotation terms $T_{1\mathrm{SR}}$, scalar coupling terms $T_{1\mathrm{SC}}$, and quadrupolar terms $T_{1\mathrm{QC}}$, as well as the shielding anisotropy terms, which are related as follows.[89]

$$\frac{1}{T_1} = \frac{1}{T_{1\Delta\sigma}} + \frac{1}{T_{1\mathrm{D}}} + \frac{1}{T_{1\mathrm{SR}}} + \frac{1}{T_{1\mathrm{SC}}} + \frac{1}{T_{1\mathrm{QC}}}. \tag{3.10}$$

[89] A. Abragam, "The Principles of Nuclear Magnetism," Ch. VIII. Oxford Univ. Press, London and New York, 1961.

[90] W. T. Huntress, Jr., *J. Chem. Phys.* **48**, 3524 (1968).

In principle it is possible to pick out the anisotropy term by varying the observing field since the $1/T_{1\Delta\sigma}$ is the only term with a direct field dependence. However in practice it often turns out that at the usual laboratory fields this term is much smaller than the other terms and cannot be detected, although the situation might be improved with the use of superconducting magnets. Nevertheless there have been several molecules for which the anisotropic shielding term has been dominant or has made a sufficiently significant contribution to the overall relaxation, and $1/T_{1\Delta\sigma}$ has been determined.[91]

In order to obtain $\Delta\sigma$ it is still necessary to estimate or otherwise derive a value for the correlation time. It can sometimes be determined from the dipolar contribution to the spin-lattice relaxation time[89] or estimated from theoretical expressions.[92] There are often large uncertainties associated with this method. In many cases a known anisotropy is used to provide an estimate for the correlation time of Eq. (3.9).

As noted earlier the spin-rotation interaction also provides a mechanism for spin-lattice relaxation. The expression corresponding to Eq. (3.9) is[92]

$$1/T_{1\mathrm{SR}} = (2IkT/3\hbar^2)(2C_\perp^2 + C_\parallel^2)\,\tau_{\mathrm{SR}}. \tag{3.11}$$

In this equation I is the mean moment of inertia, T is the absolute temperature, and τ_{SR} the rotational correlation time. It was shown in the preceding section that from a knowledge of the spin-rotation constants it is often possible to derive the shielding anisotropies.

However from the above expression it is seen that the relaxation method gives only an average or effective mean-squared value for the spin-rotation constants, which is suitable for obtaining isotropic, but not anisotropic shielding. There are, though, a couple of indirect methods which do provide estimates of $\Delta\sigma$, one of which is discussed in the section on phosphorus anisotropies, and the other described in the latter part of this section. Deverell[93] compares the average spin-rotation constants obtained by several methods and shows that in many cases the relaxation methods are the least reliable.

Using the expression for the spin-rotation contribution to $1/T_1$, Eq. (3.11), and the relationship between the angular velocity and molecular reorientation correlation times,[92] Spiess *et al.*[94] have derived an expression for the product of $1/T_{1\mathrm{SR}}$ and $1/T_{1\Delta\sigma}$ for a linear molecule:

$$(1/T_{1\mathrm{SR}})(1/T_{1\Delta\sigma}) = (4/135)(\omega_0^2\, I^2/\hbar^2)\, C_k^2\,(\Delta\sigma)^2, \tag{3.12}$$

[91] P. Rigny and J. Virlet, *J. Chem. Phys.* **47**, 4645 (1967).

[92] P. S. Hubbard, *Phys. Rev.* **131**, 1155 (1963).

[93] C. Deverell, *Mol. Phys.* **18**, 319 (1970).

[94] H. W. Spiess, D. Schweitzer, U. Haeberlen, and K. H. Hausser, *J. Magn. Resonance* **5**, 101 (1971).

which does not contain any correlation times. Here ω_0 is the resonance frequency and C_k the spin-rotation constant. By varying the temperature and magnetic field independently, it is possible to obtain experimental values for C_k and $\Delta\sigma$.

5. *Methods for Estimating* $\Delta\sigma^{d}$

It has been demonstrated that most *ab initio* methods give quite accurate values for the diamagnetic shielding tensors. Since most of the smaller molecules of second row atoms have been subjected to these methods, the quantities are often available. However for larger molecules and molecules containing third row atoms it may be necessary to use a semiempirical approach for estimating the diamagnetic components which are needed to complement spin-rotation techniques.

One of the first and most useful methods of estimating diamagnetic shielding constants was the Ramsey hypothesis.[95] This hypothesis states that the potential, V_N, of a nucleus, given by Eq. (3.13), is constant for the same nucleus in different molecules.

$$V_N = -(3mc^2/e)\sigma^{d} + \sum_{\alpha} (Z_\alpha e/R_\alpha), \tag{3.13}$$

where σ^{d} is the isotropic diamagnetic shielding and the summation runs over the other nuclei in the molecule. Since the constant V_N can be obtained from accurate calculations on atoms,[96] a knowledge of the molecular geometry is sufficient to provide values for σ^{d}. The hypothesis has been used in various forms by Flygare and Goodisman,[97] Chan and Dubin,[98] and Deverell,[93,99] often in conjunction with spin-rotation constants and generally with good results.

However, when the summation term (also called σ^{NUC}) was split into terms corresponding to individual diamagnetic shielding components the results were unreliable.[97] Recently Gierke and Flygare[50] have introduced a new method, the atom-dipole method,[100] which is quite reliable for a large number of molecules of known σ^{d} components. The atom-dipole method, which closely resembles the Ramsey hypothesis for σ^{d}_{iso}, has also been utilized to estimate $\Delta\sigma^{d}$ of molecules for which only $\Delta\sigma^{p}$ data were available previously. One point in favor of this procedure is that the paramagnetic shielding components derived from spin-rotation constants usually have an uncertainty of several ppm, so the diamagnetic components may contain a certain amount of error and still remain useful.

[95] N. F. Ramsey, *Amer. Sci.* **49**, 509 (1961).
[96] G. Malli and C. Froese, *Int. J. Quantum Chem.* **1**, Suppl. 95 (1967).
[97] W. H. Flygare, and J. Goodisman, *J. Chem. Phys.* **49**, 3122 (1968).
[98] S. I. Chan and A. S. Dubin, *J. Chem. Phys.* **46**, 1745 (1967).
[99] C. Deverell, *Mol. Phys.* **17**, 551 (1969).
[100] T. D. Gierke, H. L. Tigelaar, and W. H. Flygare, *J. Amer. Chem. Soc.* **94**, 330 (1972).

B. Experimental Results for Shielding Anisotropies

It was noted previously that all *ab initio* theoretical models evaluate the anisotropies as secondary quantities in that the theories themselves provide the individual components. We therefore applied the criterion that a theory need at least give accurate values for σ_{iso} in order to be considered reliable for shielding anisotropies. Experimentally it is often the case that the anisotropy is determined directly from a measured quantity. In such a case the parallel and perpendicular shielding components can be derived indirectly from the assumed absolute isotropic shift, and only if the anisotropy is uniquely defined. These methods of measuring anisotropy, which include liquid crystal, second-moment, and relaxation methods are referred to in the tables to follow as type A methods. There are also experimental methods which, like the theoretical methods, determine the shielding components first, and from them derive the anisotropies. These type B methods include some of the powder pattern and single crystal methods and the method of spin-rotation constants. In some cases there may be some ambiguity in this distinction since some methods, depending upon how detailed an analysis is performed, can give either the anisotropies or the components. Our classification relies upon the manner in which the data is presented by the authors of the work reported.

1. ^{13}C *Shielding Anisotropies*

The experimental ^{13}C anisotropies are given in Table VIII for linear and symmetric top molecules and in Table IX for asymmetrical molecules. The absolute isotropic shifts shown in these tables are based upon the theoretical calculation of ^{13}C shielding in CO by Stevens and Karplus,[20] which has been found to be in very good agreement with experimental methods of fixing the ^{13}C shielding scale.[101]

The values of $\sigma_{\parallel}$ and $\sigma_{\perp}$ of Table VII are either direct experimental quantities (type B) or derived from σ_{iso} and $\Delta\sigma$ (type A). The shielding components in Table IX were all determined experimentally by Pines *et al.*[73] using the cross-relaxation technique.

For linear molecules the anisotropies are, as expected, very large. This results from the fact that the electron distribution is cylindrically symmetric about the molecular axis. Thus the precession of the electron cloud about the axis is unhindered, i.e. no anti-shielding factors, whereas in the perpendicular direction there is in general a large antishielding factor, particularly since linear molecules usually have multiple bonds. From the theoretical considerations given earlier, the phenomenon of zero parallel anti-shielding is of course due to the fact that the angular momentum operator

[101] Survey of five ^{13}C absolute scales. See Section II,B,2.

TABLE VIII: EXPERIMENTAL ^{13}C SHIELDING TENSORS AND ANISOTROPIES FOR LINEAR AND SYMMETRIC TOP MOLECULES[a]

	σ_{η}	$\sigma_{\perp}$	σ_{iso}[b]	$\Delta\sigma$[c]	Method[d]	Type[e]	Ref.
HCN	264	−18	76 ± 10[b]	282 ± 20	LC	A	21,82
C_2H_2	283	38	120 ± 10[b]	245 ± 20	LC	A	22
CO	268	−116	12 ± 10[f]	384 ± 10	MB	B	19
					$+\sigma^d$(T)		20
OCS	283	−89	35 ± 20[g]	372 ± 42	MB	B	102
CS_2	288	−143	1 ± 10[f]	438 ± 44	T_1	A	94
				425 ± 16	S—PP	A	23
				431 ± 30	[avg]		
CH_3F	167	99	122 ± 15[h]	68 ± 15	LC	A	82
CH_3Cl	189	161	170 ± 10[h]	28 ± 10	LC	A	82
CH_3Br	185	185	185 ± 10[h]	0 ± 5	LC	A	82
CH_3I	169	244	219 ± 10[h]	-75 ± 20	LC	A	103
				-77 ± 20	LC	A	82
				-30 ± 5	LC	A	104
CH_3OH[i]	190	124	146 ± 10	70 ± 20	LC	A	82
				63 ± 10	S—DR	B	73
C^*H_3CN	196	191	193 ± 10[h]	5 ± 10	LC	A	103
CH_3C^*N	283	−28	76 ± 10	311 ± 30	LC	A	82
$C^*H_3C{\equiv}CCH_3$	199	185	190 ± 10	14 ± 10	S—DR	B	73
$CH_3C^*{\equiv}CCH_3$	251	50	117 ± 10	201 ± 10	S—DR	B	73
$(C_6H_5{-}C{\equiv}C^*{-})_2$[j]	300	30	120 ± 10[k]	270 ± 30	T_1	A	105
$CHCl_3$	101	125	117 ± 10[h]	-24 ± 5	LC	A	82
C_6H_6[l]	188	5	66 ± 10[h]	180 ± 10	S—DR	A	106
				190 ± 4	LC	A	107
				180 ± 5	S—DR	B	73
				183 ± 6	[avg]		
C_6F_6[l]	86	46	59 ± 10	40 ± 10	S—DR	B	73
$C^*_6(CH_3)_6$[l]	173	5	61 ± 10	168 ± 10	S—DR	B	73
$C_6(C^*H_3)_6$[l]	173	173	173 ± 10	0 ± 5	S—DR	B	73
$CaCO_3$[i]	75	0	25 ± 10	75	SC	A	74
				76 ± 5	S—PP	A	23

[a] Units are ppm.

[b] Based on absolute scale of $\sigma_{CO} = 12\pm10$ and relative shielding—see text and Table I.

[c] $\Delta\sigma \equiv \sigma_{\parallel} - \sigma_{\perp}$ ($\sigma_{\parallel}$ along symmetry axis).

[d] LC—liquid crystal, MB—molecular beam, S—DR—solid-double resonance, T_1—spin-lattice relaxation time, σ^d(T)—theoretical diamagnetic shielding, SC—single crystal, S—PP—solid-powder pattern, avg—average value.

[e] Type A: $\Delta\sigma$ observed directly, $\sigma_{\parallel}$, $\sigma_{\perp}$ derived from $\Delta\sigma$. Type B: $\Delta\sigma$ derived from reported values of $\sigma_{\parallel}$, $\sigma_{\perp}$, see text.

[f] See footnote *e*, Table I.

[g] See Gierke and Flygare.[50]

[h] See Emsley *et al.*[1] (Ch. 12).

[i] Assumed C_{3v} symmetry.

[j] Approximate axial symmetry assumed.

[k] See Levy and Nelson.[108]

[l] $\sigma_{\parallel}$ is along C_6 axis.

[102] F. H. De Leeuw and A. Dymanus, *Chem. Phys. Lett.* **7**, 288 (1970).

[103] I. Morishima, A. Mizuno, and T. Yonezawa, *Chem. Phys. Lett.* **7**, 633 (1970).

[104] C. S. Yannoni and E. B. Whipple, *J. Chem. Phys.* **47**, 2508 (1967).

[105] G. C. Levy, D. M. White, and F. A. L. Anet, *J. Magn. Resonance* **6**, 453 (1972).

[106] C. S. Yannoni and H. E. Bleich, *J. Chem. Phys.* **55**, 5406 (1971).

[107] G. Englert, *Z. Naturforsch. A* **27**, 715 (1972).

TABLE IX

EXPERIMENTAL ^{13}C SHIELDING TENSORS AND ANISOTROPIES FOR ASYMMETRIC MOLECULES[a,b]

	σ_{11}[c]	σ_{22}[c]	σ_{33}[c]	σ_{iso}[d]	$\Delta\sigma$[e]
$\overset{*}{C}H_3CHO$[f]	141	153	193	162 ± 10	46 ± 10
$CH_3\overset{*}{C}HO$	-83	-41	106	-6 ± 10	168 ± 10
$\overset{*}{C}H_3CH_2OH$	165	174	189	176 ± 10	19 ± 10
$CH_3\overset{*}{C}H_2OH$	118	118	175	137 ± 10	57 ± 10
$(CH_3)_2SO$	129	151	177	152 ± 10	37 ± 10
$(\overset{*}{C}H_3)_2CO$	147	147	197	164 ± 10	50 ± 10
$(CH_3)_2\overset{*}{C}O$	-83	-69	117	-12 ± 10	193 ± 10
$(C_6H_5)_2\overset{*}{C}O$[g]	-78[h]	-35[h]	95[h]	-6 ± 10	151 ± 10
$(\overset{*}{C}H_3CH_2)_2O$	168	178	192	179 ± 10	19 ± 10
$(CH_3\overset{*}{C}H_2)_2O$	99	111	178	129 ± 10	73 ± 10
$(\overset{*}{C}H_3CO)_2O$	159	162	197	173 ± 10	36 ± 10
$(CH_3\overset{*}{C}O)_2O$	83	83	-82[i]	28 ± 10	-165 ± 10
$(\overset{*}{C}H_3O)_2CO$	119	125	190	145 ± 10	68 ± 10
$(CH_3O)_2\overset{*}{C}O$	77	77	-44[i]	37 ± 10	-121 ± 10[i]
$(CH_3S)_2$	151	181	193	175 ± 10	27 ± 10
$\overset{*}{C}H_3CO_2H$	162	162	197	174 ± 10	35 ± 10
$CH_3\overset{*}{C}O_2H$	-64	21	95	17 ± 10	117 ± 10
$\overset{*}{C}_6H_5CH_3$	-35	52	190	69 ± 10	182 ± 10
$C_6H_5\overset{*}{C}H_3$	160	171	188	173 ± 10	22 ± 10
C_6H_6[j]	6	6	186	66 ± 10	180 ± 5

[a] Units ppm.

[b] All data in this table unless otherwise noted are obtained from solid state double resonance techniques given in Pines *et al.*[73]

[c] Shielding components are along principal carbon shielding axes ordered by $\sigma_{11} \leqslant \sigma_{22} \leqslant \sigma_{33}$.

[d] See Table VIII, footnote *b*.

[e] $\Delta\sigma \equiv \sigma_{33} - \frac{1}{2}(\sigma_{11} + \sigma_{22})$.

[f] Asterisk indicates nucleus of interest.

[g] Data for this compound were derived from single crystal measurements, see Kempf *et al.*[109]

[h] For direction of crystal axes see Kempf *et al.*[109]

[i] σ's arranged so that σ_{33} is unique.

[j] Included for sake of comparison.

[108] G. C. Levy and G. L. Nelson, "Carbon-13 NMR for Organic Chemists." Wiley (Interscience), New York, 1972.

[109] J. Kempf, H. W. Spiess, U. Haeberlen, and H. Zimmerman, *Chem. Phys. Lett.* **17**, 39 (1972).

about the parallel axis does not mix ground and excited states. Also of interest is the fact that $\sigma_{//}$ is nearly the same for each of these linear molecules. Since the charge distribution for these molecules can be expected to vary significantly, it can be inferred that the shielding is rather insensitive to the distribution of charge along the symmetry axis. The $\sigma_{\perp}$ values of these same molecules vary from -140 ppm for CS_2 to $+116$ ppm for CO, indicating that there are probably significant differences in the electronic environments of the carbon nuclei.

The trend discussed above is also noticed in molecules which have approximately cylindrical symmetry. The substituted acetylenes and cyanides are linear with respect to their nearest neighbors. Accordingly it is seen that for the molecules acetonitrile ($CH_3C{\equiv}N$), dimethylacetylene, and phenylacetylene the parallel shielding components of the designated carbons are 283, 251, and 248 ppm respectively, compared with the 260–290 ppm range of $\sigma_{//}$ for linear molecules.

Another symmetry which has some similarity to cylindrical symmetry is that of a three-fold axis of symmetry. The three-fold axis, particularly in the case of a CH_3 group, can be envisaged as offering a fairly small amount of hindrance to the free precession of the electrons about the axis.

Pines *et al.*[73] have pointed out that the shielding components along the C-3 axes (σ_{33} of Table IX, $\sigma_{//}$ of Table VIII) were fairly constant for nearly all the CH_3-containing systems they studied. The result has also been observed in a recent study by Bhattacharyya[82] of the methyl halides. Furthermore, when these and other results are placed on a single scale (Tables VIII, IX), the hypothesis is further confirmed. The σ_{zz} (σ_{33}) components in the large majority of these compounds lie in the range of 180–210 ppm.

The methyl halides are an interesting group in that the observed anisotropies decrease regularly from F to I. For CH_3Br the anisotropy is zero within experimental error. Spiesecke and Schneider[110] have shown that the experimental isotropic ^{13}C shifts in methyl monohalides correspond fairly closely with the electronegativities of the halogens. The observed shifts for the polyhalomethanes, however, could not be explained by halogen electronegativities. For the monohalides, since the parallel shielding stayed relatively constant, the variation in σ_{iso} was due to changes in $\sigma_{\perp}$ which turned out to correlate with the electronegativities. For the polyhalides all the shielding components were less clearly identifiable in relation to the molecular axes, and they depended upon the charge distribution and electronegativities in a more complicated manner.

The effect of the benzene ring upon the ^{13}C shielding of a substituted CH_3 group such as in toluene or hexamethyl benzene is very small.

[110] H. Spiesecke and W. G. Schneider, *J. Chem. Phys.* **35**, 722 (1961).

$\sigma_{//}$ is 173 ppm for this carbon in both these molecules, and the anisotropies are 22 and 0 ppm respectively. There is thus little contribution from the ring current to the shielding even when the magnetic field is $\perp$ to the ring.

The shielding of the aromatic carbons shows a very large anisotropy, with the most shielded component again being that along the nearly axially symmetric hexad axis. The in-plane shielding for benzene is close to zero indicating that the planar ring effectively prevents the induction of a substantial secondary field at the corners of the ring.

The anisotropy is greatly reduced for C_6F_6 relative to benzene (39 versus 185) resulting from a slightly increased shielding perpendicular to the symmetry axis and a greatly decreased shielding along the axis. The effect of the fluorines is most likely an inductive one which decreases the net electron density of the ring.

The correspondence between a threefold axis and a cylindrically symmetric axis should be considered in more detail. For a methyl group it has been observed that the parallel shielding is approximately 190–200 ppm. This value is still significantly different from the 280 ppm for $\sigma_{//}$ of a linear molecule. In applying the approximation of axial symmetry one is in effect stating that the paramagnetic contribution to the parallel shielding is zero, and the net $\sigma_{//}$ is entirely accounted for by the diamagnetic circulation of electrons about the symmetry axis. *Ab initio* shielding calculations[34] show that for CH_3X compounds, the paramagnetic contribution, $\sigma^{p}_{//}$, is relatively small, i.e. 60–100 ppm, while the diamagnetic contribution is about 300 ppm, or just slightly more than for a linear molecule.

The axial symmetry assumption, which is only qualitatively valid for CH_3 groups, is probably almost totally inadequate to describe the shielding in CX_3 groups. To support the contention, examine the shielding of $CHCl_3$ with respect to that of CCl_4. From the anisotropic and isotropic shielding of $CHCl_3$, it is deduced that $\sigma_{//} = 101$ ppm. For CCl_4 it is assumed that $\Delta\sigma = 0$, $\sigma_{//} = \sigma_{iso} = 98$ ppm;[110] thus the shielding along the CCl_3 axis is 98 ppm. An axially symmetric distribution of Cl's would be expected to yield larger values of $\sigma_{//}$ for these two molecules. It is expected that the value of $\sigma^{d}_{//}$ will be at least around 300 ppm, indicating that there is a large paramagnetic contribution to $\sigma_{//}$. For CF_4 the theoretical $\sigma^{d}_{//}$ of 502 ppm[62] combined with the isotropic shift of 75 ppm[44] yields a paramagnetic value of $\sigma^{p}_{//} = 437$ ppm. The conclusion is that a threefold axis is frequently a poor approximation to axial symmetry. For CH_3 groups the contributions of the protons are relatively small, and the proton bond charge density does not interact significantly with the rest of the molecule, resulting in a small constant contribution of the CH_3 group to shielding along the C_3 axis.

For molecules without linear or C_3 symmetry, it is usually not possible to define a unique anisotropy. The shielding tensors and molecules in this category which are presented in Table IX are referred to the principal axes of the carbon nuclear shielding. The convention used is $\sigma_{11} \leqslant \sigma_{22} \leqslant \sigma_{33}$. The anisotropy is defined as $\Delta\sigma = \sigma_{33} - \frac{1}{2}(\sigma_{11} + \sigma_{22})$ since the most shielded axis is usually the closest to a unique axis, although in a few cases the ordering of the components is reversed.

Note has already been made of the relative constancy of the σ_{33} shielding components of the CH_3 groups which is observed even for molecules with a molecular C_3 axis.

Pines *et al.*[73] have also grouped together a series of compounds containing a CH_2 group singly bonded to an oxygen atom. These carbons are shown to have shielding tensors which are similar in most aspects, including σ_{iso}. In addition they point out that some of the shielding components for carbonyl carbons are very sensitive to substitution. A more detailed analysis of these similarities and differences is expected in the near future. It is noted that with the use of the absolute shielding scale, it can be seen that for carbonyl or carboxy carbons at least one component of the shielding is negative with respect to a bare C nucleus.

2. ^{19}F *Shielding Anisotropies*

In the area of fluorine shielding anisotropies there is a great deal of confusing and conflicting data. We will try to interpret these results in a consistent manner. For some systems the total evidence strongly suggests one set of data over another. For others it is possible only to point out the discrepancies and await future developments to resolve them.

The task of comparing fluorine anisotropies and tensors among molecules and even among different studies of a given molecule is complicated by the fact that for fluorine there are usually a few sets of axes for which the components of the shielding tensor can be defined.[70] Although it is in principle possible to transform from one set to another, in practice there is usually not enough information available; the axis system used is determined primarily by the type of data obtained from the experiment. In liquid crystal studies the anisotropy is obtained in terms of the axis system of the ordering matrix, *S*, which is usually very closely related to the molecular axis system for molecules of planar or C_3 symmetry, although in general the observed anisotropic shift depends on all the components of *S*. The anisotropies obtained from clathrates are defined with respect to the laboratory axes, but since the method has only been used for symmetric molecules in a rigid lattice, the fluorine bond axes coincide with or can be related to the laboratory axes. Methods which rely upon second moment analyses or curve-fitting of solid state spectra normally provide information of anisotropies or shielding components in terms of the principal axes of the fluorine nuclear shielding. The spin-rotation constants and the resulting shielding components, derived from molecular beam experiments, are also defined with respect to the principal axes of the nucleus. However for certain very high resolution molecular beam work[102,111] the anisotropies observed are in terms of the principal inertial system. Finally the relaxation methods, which derive the anisotropies from T_1 measurements, give these quantities in the molecular axis system.

As already stated the transformations between these axis systems are usually not available. However the transformation formulas[70] become simplified for certain types of symmetry. For symmetric-top systems, the approximate transformation formula, Eq. (3.14), has been useful, although it is strictly valid only when the shielding is axially symmetric about the bond.

$$\Delta\sigma_{\mathrm{bond}} = \Delta\sigma_{\mathrm{axis}}/P_2(\cos\theta), \tag{3.14}$$

where θ is the angle between the axis of symmetry and the set of symmetric bonds. This formula has been used and discussed by Yannoni *et al.*[112]

The experimental fluorine anisotropies are given in Tables X–XIII. The entries are classified as type A or B as explained in the preceding section on ^{13}C anisotropies, depending on whether or not the anisotropy was a direct experimental quantity. The absolute scale is that of Hindermann and Cornwell.[26] In Tables X and XI the shielding tensors are defined with respect to the molecular and C–F bond axis systems, respectively. For many of the molecules in the other tables the molecular and C–F bond axis systems coincide. If the axis system is unspecified the molecular axis one is to be assumed.

The diatomic molecules, F_2 and HF, show widely differing anisotropies of 1050 and 100 ppm, respectively. As expected the parallel shielding is the same for both molecules. From the $\sigma_\perp$ terms it is apparent that there is a very large antishielding factor in F_2 ($\sim$1000 ppm) whereas the result in HF reflects the dominance of the fluorine over the total electron density.

There has been a considerable amount of interest in the fluoromethanes, as the effects of fluorine substitution are often quite dramatic. The first reported study of the fluorine anisotropy of CH_3F was the liquid crystal work of Bernheim *et al.*[3] which gave a value of -157 ppm. This value depended strongly upon some uncertain liquid crystals parameters such as J anisotropy and ordering parameter S, and disagreed in sign with the predictions of semiempirical theory. The molecule was reinvestigated by

[111] S. G. Kukolich and A. C. Nelson, *J. Chem. Phys.* **56**, 4446 (1972).
[112] C. S. Yannoni, B. P. Dailey, and G. P. Ceasar, *J. Chem. Phys.* **54**, 4020 (1971).

TABLE X

EXPERIMENTAL ^{19}F SHIELDING TENSORS AND ANISOTROPIES OF SMALL MOLECULES IN MOLECULAR AXIS[a] SYSTEMS[b]

	$\sigma_{\perp}$	$\sigma_{\parallel}$[c]	σ_{iso}[d]	$\Delta\sigma$[e]	Method[f]	Type[g]	Ref.
F_2	-560	490	$-210 \pm 25^{h,i}$	1050 ± 50	Solid[j]	A	113
			-233 ± 25^{k}	1055^{l}	MB	B	114
HF	380	482	$414 \pm 15^{h,k}$	102	MB	B	115
					$+\sigma^d$(T)		35
LiF	321	480	374 ± 20^{h}	159 ± 20	MB	B	116
			369 ± 20^{m}		$+\sigma^d$(T)		35
KF	156	479	264 ± 10^{h}	323 ± 10	MB	B	117
					$+\sigma^d$(SE)		50
RbF	167	486	273 ± 10^{h}	319	MB	B	118
					$+\sigma^d$(SE)		50
	221	481	308 ± 30^{n}	260 ± 30	MB	A	119
CsF	-22	491	149 ± 15^{h}	513	MB	B	120
					$+\sigma^d$(SE)		50
			308 ± 30^{n}	502 ± 20	MB	A	121
ClF	702	488	631 ± 20^{h}	-214 ± 30	MB	B	122
			637 ± 20^{k}		$+\sigma^d$(SE)		50
				-157 ± 10	LC	A	3
CH_3F	489	425	468 ± 20	-66 ± 8	Clath.	A	84
				-61 ± 15	MB	B	123
CH_2F_2	$288^{o,p}$	$447^{o,p}$		$159 \pm 10^{o,p}$	LC	A	3
	$(229, 380)^{o}$	415°	341 ± 10^{q}	110 ± 15°	MB	B	111
	$(393, 402)^{r}$	229^{r}		-168 ± 18^{r}			
CHF_3	301	221	274 ± 10^{k}	-80 ± 3	LC	A	3
	286	251		-35 ± 20	Clath.	A	85
CH_3CF_3	276	212	255 ± 10^{q}	-64 ± 5	LC	A	124
$CF_3{\equiv}CH$	248	206	234 ± 10^{q}	-42 ± 6	LC	A	125
$CF_3C{\equiv}CF_3$	266	218	250 ± 10	-48	LC	A	126
CF_3I	197	189	194 ± 10^{q}	-8 ± 3	LC	A	112
CF_3CCl_3	296	220	271 ± 10^{q}	-76 ± 3	LC	A	112
	246	322		76 ± 3			
$CDFCl_2$[s]	340°	130°	270 ± 10^{q}	-210 ± 30°	—	A	127
CF_2BrCF_2Br	165°	425°	252 ± 10^{t}	$260 \pm 40^{o,p}$	PC	A	128
$CFCl_2CFCl_2$	178°	418°	258 ± 10^{t}	$240 \pm 40^{o,p}$	PC	A	128
CF_3CO_2H	293	215	267 ± 10	-78 ± 1^{u}	LC	A	129
$CF_3CO_2^-Ag^+$	278	204	253 ± 10	-74^{u}	CA	B	70
CF_3Cl	231	205	222 ± 10	-26 ± 2	LC	A	130

Footnotes to TABLE X

[a] Except where noted otherwise.

[b] Units ppm.

[c] Parallel axis is symmetry axis.

[d] Based on absolute shielding scale of Hindermann and Cornwell,[26] Relative shifts are from references on right-hand side of this table unless otherwise noted.

[e] $\Delta\sigma = \sigma_{||} - \langle\sigma_{\perp}\rangle_{av}$.

[f] MB—molecular beam, LC—liquid crystal, Clath.—clathrate, PC—polycrystalline, CA—coherent-averaging, σ^d(T)—theoretical diamagnetic shielding, σ^d(SE)—semiempirical σ^d, S—M2—solid state second moment.

[g] Type A: $\Delta\sigma$ a primary quantity; Type B: $\Delta\sigma$ a secondary quantity; see also text.

[h] Absolute shift determined directly from spin-rotation constants and σ^d theoretical.

[i] See Gierke and Flygare.[50]

[j] Five different methods averaged.

[k] See Hindermann and Cornwell.[26]

[l] $\Delta\sigma^p$ from spin-rotation constants, $\Delta\sigma^d$ from Kolker and Karplus.[35]

[m] See Cornwell[131] for relative shift with respect to F_2.

[n] See Deverell *et al.*[132] for relative shift with respect to CF_3COOH.

[o] C–F bond axis system used.

[p] Axial symmetry about C–F bond assumed.

[q] See Dungan and Van Wazer[46] for relative shifts with respect to CCl_3F.

[r] Principal inertial axis system used.

[s] Broad-line NMR of glassy and single crystal states.

[t] See Dungan and Van Wazer[46] for relative shift with respect to CCl_3F.

[u] Axial symmetry about C_3 axis assumed.

---

[113] D. E. O'Reilly, E. M. Peterson, Z. M. El Saffar, and C. E. Scheie, *Chem. Phys. Lett.* **8**, 470 (1971).

[114] I. Ozier, L. M. Crapo, J. W. Cederberg, and N. F. Ramsey, *Phys. Rev. Lett.* **13**, 482 (1964).

[115] M. R. Baker, H. M. Nelson, J. A. Leavitt, and N. F. Ramsey, *Phys. Rev.* **121**, 807 (1961).

[116] R. Braunstein and J. W. Trischka, *Phys. Rev.* **98**, 1092 (1955).

[117] R. van Wachem and A. Dymanus, *J. Chem. Phys.* **46**, 3749 (1967).

[118] J. C. Zorn, T. C. English, J. T. Dickinson, and D. A. Stephenson, *J. Chem. Phys.* **45**, 3731 (1966).

[119] G. Graff and O. Runolfsson, *Z. Phys.* **187**, 140 (1965).

[120] T. C. English and J. C. Zorn, *J. Chem. Phys.* **47**, 3896 (1967).

[121] G. Graff, R. Schonwasser, and M. Tonutti, *Z. Phys.* **199**, 157 (1967).

[122] R. E. Davis and J. S. Muenter, *J. Chem. Phys.* **57**, 2836 (1972).

[123] S. C. Wofsky, J. S. Muenter, and W. Klemperer, *J. Chem. Phys.* **55**, 2014 (1971).

[124] D. N. Silverman and B. P. Dailey, *J. Chem. Phys.* **51**, 655 (1969).

[125] A. D. Buckingham, E. E. Burnell, C. A. de Lange, and A. J. Rest, *Mol. Phys.* **14**, 105 (1968).

[126] A. D. Buckingham, E. E. Burnell and C. A. de Lange, *Mol. Phys.* **15**, 285 (1968).

[127] C. MacLean and E. L. Mackor, *Proc. Phys. Soc., London* **88**, 341 (1966).

[128] E. R. Andrew and D. P. Turnstall, *Proc. Phys. Soc., London* **81**, 986 (1963).

[129] M. B. Dunn, *Mol. Phys.* **15**, 433 (1968).

[130] N. Zumbulyadis, personal communication (1972).

[131] C. D. Cornwell, *J. Chem. Phys.* **44**, 874 (1966).

[132] C. Deverell, K. Schaumburg, and H. J. Bernstein, *J. Chem. Phys.* **49**, 1276 (1968).

TABLE XI

EXPERIMENTAL ^{19}F SHIELDING TENSORS AND ANISOTROPIES OF C–F BONDS OF CF_3 SYSTEMS[a]

	$\langle\sigma_\perp\rangle^b$	$\sigma_{\parallel}{}^c$	$\sigma_{iso}{}^d$	$\Delta\sigma^e$	Method[f]	Ref.
CHF_3	199	424	274 ± 10	225 ± 15	LC	3
	239	344		105 ± 20	Clath.	85
ClF_3	186	210	194 ± 10	24 ± 9	LC	112
CH_3CF_3	191	383	255 ± 10	192 ± 15	LC	124
CCl_3CF_3	196	423	271 ± 10	228 ± 9	LC	112
$CF_3C{\equiv}CH$	186	331	234 ± 10	145 ± 15	LC	125
$CF_3C{\equiv}CCF_3$	202	346	250 ± 10	144	LC	126
CF_3CO_2H	189	424	267 ± 10	235 ± 3^g	LC	129
$CF_3CO_2^-Ag^+$	179	401	253 ± 10	222 ± 10^g	CA	70
$(CF_3CO)_2O$	189	399	259 ± 10	210 ± 10^g	CA	70

[a] Units ppm.
[b] $\langle\sigma_\perp\rangle$ = average value of shielding perpendicular to C–F bond.
[c] Along C–F bond.
[d] See footnote *d*, Table X.
[e] Assuming tetrahedral geometry and relation $\Delta\sigma(\text{C–F bond}) = -3\Delta\sigma(C_{3v}\text{ axis})$; see discussion of Eq. (3.14).
[f] See footnote *f*, Table X.
[g] Rapid molecular motion assumed to produce effective three-fold symmetry.

clathrate[84] and molecular beam techniques.[123] These gave results of -66 ± 8 and -61 ± 15 ppm, respectively, and also agreed with the most reliable *ab initio* calculations.[62] There is considerably more ambiguity in the results for CF_2H_2 and CF_3H for each of which two apparently conflicting sets of data have been reported. The molecular beam study of CF_2H_2[111] provides the shielding in terms of the inertial axes and in terms of the fluorine axes. The shielding component along the C–F bond agrees with the parallel shielding of CH_3F which is also along the C–F bond (415 versus 425 ppm). The liquid crystal results[3] for CF_2H_2, in addition to assumptions concerning the sign of J_{HF} and the direction of orientation, are also based upon an approximation of axial symmetry about the C–F bond. This assumption is obviously not borne out by the molecular beam study. For CHF_3 both the clathrate[85] and liquid crystal[3,130] studies give the anisotropy with respect to the molecular C_3 axis but the results do not agree (-80 ± 3 and -35 ± 20). Using the transformation formula, the shielding along the C–F bond becomes 424 for the liquid crystal result and 344 for the clathrate result (see Table XI). The liquid crystal result thus corresponds more closely with the previously discussed C–F bond shielding of CH_3F and CF_2H_2. In addition the $\sigma_\perp$ value of 194 ppm of the liquid crystal study is more in line with the other values of this quantity as shown in Table XI. However as discussed earlier the liquid crystal results for fluorine are subject to large temperature effects. A study of CF_4, which appears to have a measurable anisotropy,[133,134] would be very useful in evaluating these trends for fluoromethanes, particularly that of C–F bond shielding. Also of interest is the investigation of possible double-bond character, the occurrence of which has been postulated theoretically by Ditchfield[62] and Karplus and Das[135] and also discussed by Dailey and co-workers.[112,124]

There is a substantial number of molecules with CF_3 groups (besides CF_3H) whose fluorine anisotropies have been studied. These include CH_3CF_3, CF_3I, CF_3CCl_3, $CF_3C{\equiv}CH$, $CF_3C{\equiv}CCF_3$, CF_3CO_2H, CF_3CO_2Ag, and $(CF_3CO_2)_2O$. For most of these molecules the anisotropy has been measured along the C_3 axes as shown in Table X. The corresponding anisotropies in Table XI have been derived from those in Table X by assuming trigonal molecular symmetry with a tetrahedral CF_3X configuration. An interesting feature of these quantities is that the fluorine shielding along the C_3 axis and the fluorine shielding perpendicular to the CF bond are both more uniform than the shielding along the CF bond. Using this trend it appears that the correct value for $\Delta\sigma(C_{3V})$ of CF_3CCl_3, whose sign was in doubt, is -76 ppm. These patterns also hold for the molecules CF_3CO_2Ag, CF_3CO_2H, and $(CF_3CO)_2O$ which do not have true threefold axes.

Some of the less symmetric fluorine-substituted organic compounds given in Table X are $CDFCl_2$, CF_2BrCF_2Br, and $CFCl_2CFCl_2$. The C–F bond anisotropies of these three were determined from second moment analyses of polycrystalline solids under the assumptions of rapid molecular motion about the C–F bonds. For CF_2BrCF_2Br and $CFCl_2CFCl_2$ the anisotropies of 260 and 240 ppm are not too different from the values obtained for some of the CF_3-containing molecules under the assumption of axial symmetry. However the assumption of axial symmetry has been shown to lead to an unreasonable value of $\Delta\sigma = \pm 2000$ ppm for C_2H_5F.[136] The C–F bond anisotropy of $CDFCl_2$ of -210 is of the same order of magnitude as the others but differs in sign. Davies[137] points out that the anisotropy of $CDFCl_2$ leads to an unreasonable value of $\Delta\sigma^d = -737$ ppm when use is made of the Ramsey hypothesis. A change of the sign of $\Delta\sigma$ gives better agreement with other available data in Tables X and XI.

For the fluorine shielding of the aromatic fluorocarbons shown in

[133] I. Ozier, L. M. Crapo, and S. S. Lee, *Phys. Rev.* **172**, 63 (1968).

[134] R. Dong and M. Bloom, *Can. J. Phys.* **48**, 793 (1970).

[135] M. Karplus and T. P. Das, *J. Chem. Phys.* **34**, 1683 (1961).

[136] A. D. Buckingham, E. E. Burnell, and C. A. de Lange, *Mol. Phys.* **16**, 191 (1969).

[137] D. W. Davies, *Mol. Phys.* **15**, 587 (1968).

Table XII, the coordinate system has been chosen as follows: The 3 direction is perpendicular to the ring, the 1 direction is in the plane of the ring and perpendicular to the C–F bond, and the 2 direction is parallel to the C–F bond. The anisotropy has been defined with respect to the 3 axis ($\Delta\sigma = \sigma_{33} - \frac{1}{2}[\sigma_{11} + \sigma_{22}]$) for a couple of reasons. For most of the fluorobenzenes the axis of orientation in liquid crystals appears to be the axis perpendicular to the plane. In addition this direction is normally the most highly shielded one.

Hexafluorobenzene has been the subject of several investigations, two in liquid crystals and one in the solid state. It has now been well established that σ_{33} is the most shielded component and that $\Delta\sigma$ is 155 ppm. Mehring *et al.*[70] have also been able to demonstrate that above 84°K in the solid state there is a rapid rotation about the C_6 axis and that the shielding tensor is axially symmetric about this axis. Similar behavior has been observed for C-13 of benzene above 40°K.[138]

For the mono-, di-, and tri-substituted fluorobenzenes Nehring and Saupe[140] have assumed that, as in hexafluorobenzene, the molecular planes are aligned perpendicular to the magnetic field, and that consequently the perpendicular axis is the most shielded axis. Further they have been able to deduce the fluorine shielding components for C_6H_5F, 1,3 $C_6H_4F_2$, and 1,3,5 $C_6H_3F_3$ by assuming that the effect of *meta* substituents upon this quantity is negligible. They also discuss the more significant *ortho* effects and show that they cannot be used to predict the shielding components.

Nehring and Saupe's results for fluorobenzene are disputed by a molecular beam study of Chan and Dubin,[98] which indicates that the most shielded component is along the C–F bond and not perpendicular to the plane. Therefore by our definition their anisotropy is negative. However their method required, besides the spin-rotation constants, the assumption that the sum of $(\sigma^{d}_{\alpha\alpha} + \sigma^{NUC}_{\alpha\alpha})$ be constant for $\alpha = x, y, z$. This assumption has been shown to be generally unreliable by Flygare and co-workers.[50,97]

A study has also been made of 1,2,4 $C_6H_3F_3$ in the polycrystalline state, but the interpretation of the results is complicated by the presence of three chemically different fluorines. Mehring *et al.*'s[70] study using coherent-averaging techniques included the double-ring systems perfluoronaphthalene, perfluorobiphenyl, and perfluorobenzophenone. They show that the former displays an anisotropy similar to that of C_6F_6 and that it too is able to rotate about an axis perpendicular to the molecular plane. Other interesting features of these molecules can be found in the original paper. The molecules tetrafluorohydroquinone and trifluoracetanilide were investigated by Derbyshire *et al.*[75] The interpretations of the spectra of these molecules, as for some of the preceding ones, are complicated by occurrences of multiple fluorine sites.

TABLE XII

EXPERIMENTAL ^{19}F SHIELDING TENSORS AND ANISOTROPIES OF SOME AROMATIC MOLECULES[a]

		σ_{11}[b]	σ_{22}[b]	σ_{33}[b]	σ_{iso}[c]	$\Delta\sigma$[d]	Method[e]	Type[e]	Ref.
						154 ± 5	LC	A	139
C_6F_6		300	300	456	352 ± 10[c]	158 ± 4	LC	A	140
						155 ± 6	CA	B	70
C_6H_5F		285	441	276	334 ± 10[f]	-87 ± 30[f]	MB	B	98
		206	330	370	302 ± 10[c]	102 ± 10	LC	A	83
1,2-$C_6H_4F_2$					328 ± 10[c]	117 ± 5[g]	LC	A	140
1,3-$C_6H_4F_2$		206	326	367	299 ± 10[c]	101 ± 6	LC	A	140
1,4-$C_6H_4F_2$					309 ± 10[c]	23 ± 5	LC	A	140
1,3,5-$C_6H_3F_3$		206	322	364	297 ± 10[c]	101 ± 5	LC	A	140
		(263)[h]		367		104 ± 2	LC	A	141
1,2,4-$C_6H_3F_3$						250[g]	PC	A	128
1,2,4,5-$C_6H_2F_4$					329 ± 10[c]	122 ± 4[g]	LC	A	140
$C_{10}F_8$[i]		280	280	439	333 ± 10[j]	159 ± 10	CA	B	70
$C_{12}F_{10}$[k]	meta	310	310	441	354 ± 10	131 ± 10	CA	B	70
	ortho	287	287	436	336 ± 10	149 ± 10			

[a] Units ppm.
[b] $_{\parallel}$ and $_{\perp}$ defined with respect to ring axis 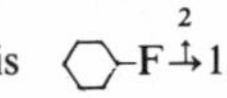 1, $3_{\perp}$ to plane.
[c] See footnotes *d*, *n*, Table X.
[d] $\Delta\sigma \equiv \sigma_{33} - \frac{1}{2}(\sigma_{11} + \sigma_{22})$.
[e] See footnotes *f*, *g*, Table X.
[f] Assumed $(\sigma^{d}_{\alpha\alpha} + \sigma^{NUC}_{\alpha\alpha})$ is a constant. See Chan and Dubin[98] and also discussion on semiempirical methods of σ^{d}.
[g] See original references for definition of this quantity.
[h] Average of σ_{11} and σ_{22}.
[i] Perfluoronaphthalene.
[j] Average of several F shifts.
[k] Perfluorobiphenyl.

The anisotropies of a series of inorganic fluorides are given in Table XIII. Most of these anisotropies are quite large, with the uncertainties also usually large. Notice that among the heavy metal fluorides $\Delta\sigma_F$ is negative for MoF_6, WF_6, and UF_6 but positive for XeF_4. The isotropic shifts for the first three have been obtained using the Ramsey hypothesis in conjunction with spin-rotation constants obtained from liquid state spin-lattice relaxation techniques, while the anisotropies are obtained from solid state T_1 measurements. The derived shielding components show an extremely large range and should be considered only rough estimates. The clathrate

[138] M. G. Gibby, A. Pines, and J. S. Waugh, *Chem. Phys. Lett.* **16**, 296 (1972).
[139] L. C. Snyder and E. W. Anderson, *J. Chem. Phys.* **42**, 3336 (1965).
[140] J. Nehring and A. Saupe, *J. Chem. Phys.* **52**, 1307 (1970).
[141] C. T. Yim and D. F. R. Gilson, *Can. J. Chem.* **46**, 2783 (1968).

TABLE XIII

EXPERIMENTAL ^{19}F SHIELDING TENSORS AND ANISOTROPIES OF SOME INORGANIC FLUORIDES AND OTHER SYSTEMS[a]

	$\sigma_{\perp}$[b]	$\sigma_{\parallel}$[b]	σ_{iso}[c]	$\Delta\sigma$[d]	Method[e]		Ref.
NF_3	$(-155-15)^f$	305	45 ± 10^g 75 ± 10^h	390 ± 60^f	Clath.	A	85
SF_6	107^f	207^f	140 ± 20^g	100^f	Clath.	A	142
XeF_4	28	598	218 ± 20^g 235 ± 20^i	570 ± 40	S–M2	A	143
MoF_6	11	-289	-89 ± 30^g	-300	T_1	A	91
WF_6	190	-310	23 ± 30^g	-500	T_1	A	91
UF_6	-402	-1052	-619 ± 100 -540 ± 30^h	-650	T_1	A	91 144
PtF_6	—	—	—	1300 ±200	T_1	A	144
XeF_2	—	—	413 ± 30^i	105 ± 10	S–M2	A	145
$Ca_5F(PO_4)_3$ (Fluorapatite)	319	235	291 ± 20	-84	SC	B	146
$BaFPO_3$	—	—	—	182 ± 22^j	S–M2	A	147
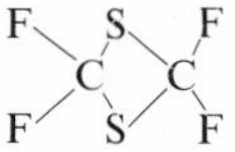	(295, 232)	198	242 ± 20	-66	LC	A	148
$(C_2F_4)_n^k$	—	—	—	-171 ± 15	S–M2	A	149
$C_6F_4O_2$ (Fluoranil)	$(251, 290)^f$	438^f	326 ± 10	167 ± 10^f	CA	B	150
PF_3	241	192	225 ± 10^g	-49 ± 5	LC	A	130

[a] Units ppm.

[b] Parallel and perpendicular shielding defined with respect to molecular axis system unless noted otherwise.

[c] Based on absolute F–19 shielding scale of Hinderman and Cornwell,[26] using standard of $\sigma(CC1_3F) = 189$ ppm. Relative shifts are from references on right-hand side of this table unless noted otherwise.

[d] $\Delta\sigma = \sigma_{\parallel} - \langle\sigma_{\perp}\rangle_{avg.}$

[e] See footnotes *f* and *g* of Table X.

[f] With respect to F-bond axis system.

[g] See Dungan and Van Wazer[46] and Emsley and Phillips[47] for relative shifts with respect to $CC1_3F$.

[h] See Baker *et al.*[27]

[i] See Brown and Verdier.[151]

[j] Axial symmetry about P–F bond assumed.

[k] Polytetrafluoroethylene.

[142] M. B. Dunn and C. A. McDowell, *Chem. Phys. Lett.* **13**, 268 (1972).

[143] R. Blinc, I. Zupancic, S. Maricic, and Z. Veksli, *J. Chem. Phys.* **39**, 2109 (1963); **40**, 3739 (1964).

[144] R. Blinc, E. Pirkmajer, J. Slivnik, and I. Zupancic, *J. Chem. Phys.* **45**, 1488 (1966).

studies of NF_3 and SF_6 and liquid crystal study of PF_3, on the other hand, should provide reasonably accurate predictions of the shielding tensors and anisotropies.

Several other studies of shielding components of some fairly complex molecules in the solid state are also presented to indicate the capabilities of some of the various methods of analysis.

3. ^{31}P *Shielding Anisotropies*

The experimental results for phosphorus shielding tensors and anisotropies are given in Table XIV. As for the C and F shielding studies, the attempt has been made to place all the shielding quantities (except anisotropies) on a single absolute scale. According to the usual convention ^{31}P shifts are given with respect to an 85% solution of H_3PO_4. Deverell[99] has arrived at the value of σ (H_3PO_4) = 320 ppm by assigning values to ($\sigma^p - \sigma^{NUC}$), by a ratio of spin rotation constants, and to ($\sigma^d + \sigma^{NUC}$), presumably using the Ramsey hypotheses. It is felt that a more reliable value for phosphorus shielding is that derived using σ^p of PH_3 from spin-rotation measurements of Davies *et al.*[152] and σ^d from *ab initio* calculations of Rothenberg *et al.*[153] This procedure yields a value of $\sigma(PH_3)$ = 595 ppm (very close to that derived by Gierke and Flygare)[50] and $\sigma(H_3PO_4)$ = 356 ppm.

The study of phosphorus shielding has been carried out primarily from solid state experiments with some data also available from molecular beam, T_1, and liquid crystal studies. Gibby *et al.*[154] have pointed out that for nuclei such as ^{31}P the large chemical shifts and relatively small dipolar broadening enable the solid state spectra to be more easily analyzed without using special techniques. They investigated the dependence of the shielding upon complete rotations about a set of laboratory axes for a single crystal P_4S_3. From this experiment they were able to deduce the principal shielding tensor components for the apical and basal phosphorus atoms and to demonstrate the occurrence of a rotation about a threefold axis which passes through the apical P and the center of the plane of the basal P's.

---

[145] D. K. Hindermann and W. E. Falconer, *J. Chem. Phys.* **50**, 1203 (1969).
[146] J. L. Carolan, *Chem. Phys. Lett.* **12**, 389 (1971).
[147] D. L. VanderHart, H. S. Gutowsky, and T. C. Farrar, *J. Chem. Phys.* **50**, 1058 (1969).
[148] R. C. Long, Jr., and J. H. Goldstein, *J. Chem. Phys.* **54**, 1563 (1971).
[149] C. W. Wilson, III, *J. Polym. Sci.* **61**, 403 (1962).
[150] M. Mehring, R. G. Griffin, and J. S. Waugh, *J. Amer. Chem. Soc.* **92**, 7222 (1970).
[151] T. H. Brown and P. H. Verdier, *J. Chem. Phys.* **40**, 2057 (1964).
[152] P. B. Davies, R. M. Neumann, S. C. Wofsky, and W. Klemperer, *J. Chem. Phys.* **55**, 3564 (1971).
[153] S. Rothenberg, R. H. Young, and H. F. Schaefer, *J. Amer. Chem. Soc.* **92**, 3243 (1970).
[154] M. G. Gibby, A. Pines, W. K. Rhim, and J. S. Waugh, *J. Chem. Phys.* **56**, 991 (1972).

TABLE XIV

EXPERIMENTAL ^{31}P SHIELDING TENSORS AND ANISOTROPIES[a]

	$\sigma_{\perp}$	$\sigma_{\parallel}$	σ_{iso}[b]	$\Delta\sigma$	Method[c]	Type[d]	Ref.
PN	−406	970	53±25	1376±30	MB + σ^d(SE)	B	155, 50
P_4[g]	711	997	806±20[b]	286±15	T_1(liq + cryst)	A	156
PH_3	612	562	595±10[b]	−50±15	MB + σ^d(SE)	B	152, 153
				−55±50	LC	A	130
$P(CN)_3$	449	557 (454)[e]	492±20[f]	128±30	S–M2	A	157
P_4O_{10}[g,h]	297	623 (327)[e]	406±20[f]	326±60	S–M2	A	157
P_4S_{10}[g,h]	260	436 (219)[e]	311±20[f]	188±30	S–M2	A	157
P_4S_3[g]							
Apical	331±10, 353±10	118±12	267±20[i]	−224	SC	B	154
Basal	255±15, 312±14	762±20	443±20[i]	479			
$Na_4P_2O_7$[h] $[O_3P—O—PO_3]^{-4}$	297	497	364±15[b]	200±100	S–M2	A	75
$BaFPO_3$	—	—	—	−145±20	S–M2	A	147
PCl_3	(321)	(−230)	137±20[b]	(−551)[j]	T_1(liq) + σ^d(SE)	B	158, 50
PBr_3	(297)	(−202)	131±20[b]	(−499)[j]	T_1(liq) + σ^d(SE)	B	158, 50
$POCl_3$	(493)	(63)	350±20[b]	(−430)[j]	T_1(liq) + σ^d(SE)	B	158, 50
Zn_3P_2[g]	—	—	—	120	S–PP	B	154
PF_3	183	411	259±20[b]	228±2	LC	A	130

[a] Units ppm, parentheses indicate highly uncertain quantities.

[b] Based on absolute shielding of PH_3 of 595 ppm, from which is obtained $\sigma(H_3PO_4) = 356$ ppm; relative shifts are obtained from ref. 1, Vol. 2, Section 12.6, unless otherwise noted.

[c] MB—molecular beam, T_1—spin-lattice relaxation time, S–M2—solid state second moment, S–PP—solid state powder pattern, σ^d(SE)—semiempirical σ^d, SC—single crystal.

[d] See footnote *g*, Table X.

[e] Based on values given for $\sigma_{\parallel} - \sigma(H_3PO_4)$ in Lucken and Williams[157] for which designated component ($\sigma_{\parallel}$) may be in error; see footnote.[160]

[f] Relative shifts with respect to H_3PO_4 found in Purdela.[159]

[g] See original paper for geometry.

[h] Axial symmetry assumed.

[i] See Gibby *et al.*[54] for relative shift with respect to H_3PO_4.

[j] See text for method used to obtain these quantities.

Several other groups, using polycrystalline powders rather than single crystals, have had to assume axial symmetry in order to derive the shielding tensors shown in Table XIV. In addition to the reported value of $\Delta\sigma$ of anyhydrous pyrophosphate, Derbyshire *et al.*[75] attempted a complete tensor analysis by the second moment method of sodium pyrophosphate decahydrate, but found two of the off-diagonal elements to be indeterminate.

For those molecules in which there is a unique P on a threefold axis, it is seen that in most cases, e.g. PH_3, $P(CN)_3$, P_4O_{10}. P_4S_{10}, $P_2O_7^{-4}$, the parallel axis is the most shielded[160] axis, but P_4S_3 and $BaFPO_3$ are exceptions—i.e. $\Delta\sigma$ is negative. It is also noted that phosphorus, being a second period atom, has greater electron density near the nucleus. Consequently the shielding components and the isotropic shifts, in addition to showing a large total range, are all fairly large positive numbers.

The molecular beam experiment on PH_3[152] which was used to establish the absolute shielding scale shows that phosphine, like ammonia, has a small anisotropy. The molecule PN is interesting as it is the only diatomic molecule of phosphorus which has been studied.[155]

The anisotropies of PCl_3, PBr_3, and $POCl_3$ were evaluated using the following approximate procedure. From the observed liquid state spin-lattice relaxation times[158] is derived an ellipsoidal expression for $C_\perp^2$ and $C_\parallel^2$. See Eq. (3.11). Combining this with the quantity σ^1 $(=\sigma^p-\sigma^{NUC})$ gives two possible sets of values for $C_\perp$ and $C_\parallel$ of which one set is assumed to be more reasonable. The paramagnetic shielding components obtained in this manner can be combined with diamagnetic components obtained using the atom-dipole method[50] to give the total shielding components and the anisotropies.

4. *^{14}N and ^{15}N Shielding Anisotropies*

There are two isotopes of nitrogen, ^{14}N and ^{15}N, whose shielding anisotropies have been investigated. The ^{15}N isotope is difficult to study by conventional NMR techniques because of its low natural abundance (0.365%), and most of the anisotropy data are from molecular beam experiments. The experimental results are shown in Table XV in which the absolute shielding scale of Baker *et al.*[27] and Chan *et al.*,[161] based on N_2, is

---

[155] J. Raymonda and W. Klemperer, *J. Chem. Phys.* **55**, 232 (1971).

[156] N. Boden and R. Folland, *Mol. Phys.* **21**, 1123 (1971).

[157] E. A. C. Lucken and D. F. Williams, *Mol. Phys.* **16**, 17 (1969).

[158] K. T. Gillen, *J. Chem. Phys.* **56**, 1573 (1972).

[159] D. Purdela, *J. Magn. Resonance* **5**, 23 (1971).

[160] The values given for $\sigma_\parallel-\sigma(H_3PO_4)$ in Ref. 157 are in strong disagreement with those determined from $\Delta\sigma$ and σ_{iso} in Table XIV, possibly due to a reporting error in Ref. 157.

[161] S. I. Chan, M. R. Baker, and N. F. Ramsey, *Phys. Rev. A* **136**, 1224 (1964).

used as a reference. However, for each of these molecules an absolute scale can be derived by combining the σ^p from spin-rotation constants with *ab initio* or empirical values of σ^d. The isotropic shift of $CH_3{}^{14}NC$, studied in the liquid crystal phase by Yannoni,[162] has been determined from the experimental shift with respect to the known ^1H shift of CH_4 and also using the N_2-based scale. There is a 33 ppm discrepancy between these two methods.

TABLE XV

EXPERIMENTAL ^{14}N AND ^{15}N SHIELDING CONSTANTS AND ANISOTROPIES[a]

	$\sigma_\perp$	$\sigma_\parallel{}^b$	$\sigma_{iso}{}^c$	$\Delta\sigma^d$	Method[e]	Type[f]	Ref.
$^{15}N_2$	−319	338	-100 ± 20^g	657 ± 20	MB + σ^d(T)	B	161, 49
$P^{14}N$	−698	350	-349 ± 20^g	1047 ± 20	MB + σ^d(SE)	B	155, 50
$CH_3{}^{14}NC$	10 43	370 403	130 ± 20^h 163 ± 20^i	360 ± 73	LC	A	162
$^{14}NH_3$	273	234	$260\pm20^{g,j}$ $(175\pm20)^c$	-39 ± 10	MB + σ^d(T)	B	164, 30
$H—C\equiv{}^{14}N$	−229	348	$-37\pm20^{g,j}$	577 ± 20	MB + σ^d(SE)	B	165, 50
$Cl—C\equiv{}^{14}N$	−635	350	$-306\pm20^{g,j}$	985 ± 20	MB + σ^d(SE)	B	165, 50
$NH_4^+{}^{15}NO_3^-$	[−198, −172]	25	$(-115\pm20)^k$	210 ± 5	DR–PP	B	163

[a] Units ppm.
[b] Parallel axis is symmetry or molecular axis.
[c] Absolute shielding scale based on N_2 molecular beam data, see Baker *et al.*[27]
[d] $\Delta\sigma \equiv \sigma_\parallel - \sigma_\perp$.
[e] MB—molecular beam, LC—liquid crystal, σ^d(T)—theoretical σ^d, σ^d(SE)—semiempirical σ^d, DR—double resonance, PP—powder pattern.
[f] Type A: $\Delta\sigma$ a primary quantity; Type B: $\Delta\sigma$ a secondary quantity.
[g] Isotropic shift determined directly from spin-rotation constants and diamagnetic shielding.
[h] See Chan *et al.*[161] for relative shifts with respect to NH_3.
[i] See Yannoni[162] for shift relative to that of ^1H of CH_4.
[j] See Gierke and Flygare.[50]
[k] Absolute value of $\sigma(NO_3^-)$ from Baker *et al.*[27] used.

[162] C. S. Yannoni, *J. Chem. Phys.* **52**, 2005 (1970).
[163] M. G. Gibby, R. G. Griffin, A. Pines, and J. S. Waugh, *Chem. Phys. Lett.* **17**, 80 (1972).
[164] S. G. Kukolich and S. C. Wofsky, *J. Chem. Phys.* **52**, 5477 (1970).
[165] C. H. Townes and A. L. Schawlow, "Microwave Spectroscopy." McGraw-Hill, New York, 1955.

There are four linear molecules shown in Table XV and the parallel shielding is almost constant for each of them, a result also observed in ^{13}C and ^{19}F shielding. The linear molecules all contain a triply-bonded terminal N, and the resulting anisotropies are extremely large, 600–900 ppm. The corresponding anisotropies for triply-bonded carbons were only about 300 ppm, which is similar to that of N of CH_3NC. The isotropic shifts for the linear molecules are dominated by the antishielding perpendicular components. The small anisotropy of NH_3 ($\sigma_{//} = 234$, $\sigma_{\perp} = 273$, $\Delta\sigma = -39$ ppm) is fairly similar to CH_4 ($\sigma_{//} = \sigma_{\perp}$ 196, $\Delta\sigma = 0$) (see Table I) in its responses to differently oriented magnetic fields. The powder pattern study of NH_4NO_3[163] revealed that the nitrate ion does not possess a threefold axis as would be expected for the free ion.

5. *Miscellaneous Shielding Anisotropies*

The shielding anisotropies and tensors of several additional nuclei are shown in Table XVI. Much of this data has been derived from a combination of spin-rotation constants and semiempirical diamagnetic shielding constants, a procedure which is probably most reliable for linear molecules. A few noteworthy aspects of the table are discussed below.

The chlorine shielding in ClF has a large antishielding in the perpendicular direction which results in a huge anisotropy of 2389 ppm. The fluorine shielding is also anomalous as it has a negative anisotropy (Table X). Among the hydrogen halides the ratio of $\sigma_{//}$ to $\sigma_{\perp}$ is approximately constant for the halogen shielding, i.e. HF, 1.27; HCl, 1.34; HBr, 1.32; and HI, 1.37.

The perpendicular components of the ^{17}O shielding are negative for each of the three molecules listed—CO, OCS, and H_2CO—in all of which O is doubly bound to its neighbor. Interestingly the perpendicular shielding components of each of their neighboring carbons (as well as the sulfur in OCS and several other carbonyl carbons, Table X) are also all negative quantities, indicating that this feature may be characteristic of certain types of double bonds.

The determination of the shielding anisotropies in RbF and CsF by direct observation[119] and spin-rotation constants[118,120] affords an opportunity to assess the reliability of the empirical diamagnetic shielding scheme for those heavy metals. The results are consistent within the limits of uncertainty of the high resolution molecular beam studies. The anisotropy studies of $^{51}V_2O_5$[167] and $^{207}PbMoO_4$[169] are important in that they were

[166] R. Fusaro and J. W. Doane, *J. Chem. Phys.* **47**, 5446 (1967).
[167] J. L. Ragle, *J. Chem. Phys.* **35**, 753 (1961).
[168] S. D. Gornostansky and C. V. Stager, *J. Chem. Phys.* **46**, 4959 (1967).
[169] P. C. Lauterbur and J. J. Burke, *J. Chem. Phys.* **42**, 439 (1965).

TABLE XVI

EXPERIMENTAL SHIELDING ANISOTROPIES OF MISCELLANEOUS NUCLEI[a]

	Nucleus	$\sigma_{\parallel}$[b]	$\sigma_{\perp}$	σ_{iso}[c]	$\Delta\sigma$[c]	Method[d]	Type[e]	Ref.
H–Cl	^{35}Cl	1146	854	952	292	MB, σ^d(SE)	B	50
F–Cl	^{35}Cl	1149	−1240	−444	2389	MB, σ^d(SE)	B	50
Cl–C≡N	^{35}Cl	1152	293	579	859	MB, σ^d(SE)	B	50
H–Br	^{79}Br	3124	2364	2617	760	MB, σ^d(SE)	B	50
$NaBrO_3$	^{81}Br	—	—	—	|90|	SC	A	166
H–I	^{127}I	5503	4015	4510	1488	MB, σ^d(SE)	B	50
CO	^{17}O	411	−229	−15	640	{MB σ^d(T)	B	52 54
OCS	^{17}O	421	−440	−153	861	MB, σ^d(SE)	B	50
H_2CO	^{17}O	465 ± 100	[−1185 ± 50[f] −400 ± 300[g]]	−373	1258	{MB σ^d(T)	B	53 54
OCS	^{33}S	1060	331	574	729	MB, σ^d(SE)	B	50
LiF	^{7}Li	106	78	87	28	{MB σ^d(T)	B	116 35
KF	^{39}K	1331	1124	1193	207	{MB σ^d(SE)	B	117 50
RbF	^{85}Rb	3369	3224	3272	145	{MB σ^d(SE)	B	118 50
		—	—	—	380 ± 210	MB—HR	A	119
CsF	^{133}Cs	5782	5622	5675	160	{MB σ^d(SE)	B	120 50
		—	—	—	171 ± 21	MB–HR	A	119
V_2O_5	^{51}V	—	—	—	−550 ± 60	S–M2	A	167
		—	—	—	−535 ± 30	SC	A	168
$PbMoO_4$	^{207}Pb	—	—	—	−189 ± 10	SC	A	169
$(CH_3)_4Si$	^{29}Si	0	0	0	0	S–DR	B	170
$(CH_3)_3SiOCH_3$	^{29}Si	10	−30	−17[h]	40	S–DR	B	170
$(CH_3)_2Si(OCH_3)_2$	^{29}Si	[34][i]	[−13][i]	3[h]	—	S–DR	B	170
$CH_3Si(OCH_3)_3$	^{29}Si	68	29	42[h]	39	S–DR	B	170
$Si(OCH_3)_4$	^{29}Si	80	80	80[h]	0	S–DR	B	170
$(CH_3)_3SiC_6H_5$	^{29}Si	24	−4	5[h]	28	S–DR	B	170

[a] Units ppm.
[b] Parallel axis is molecular or symmetry axis.
[c] $\sigma_{iso} = \frac{1}{3}(\sigma_{\parallel} + 2\sigma_{\perp})$; $\Delta\sigma = \sigma_{\parallel} - \sigma_{\perp}$.
[d] MB—molecular beam, SC—single crystal, HR—high resolution, S–M2—solid second moment, S–DR—solid double resonance, σ^d(SE)—semiempirical σ^d, σ^d(T)—theoretical σ^d.
[e] Type A: $\Delta\sigma$ a primary quantity: Type B: $\Delta\sigma$ a secondary quantity.
[f] Direction C–O bond.
[g] Direction $\perp$ to C–O bond in molecular plane.
[h] Liquid state value with respect to TMS.
[i] $\sigma_{33} = 34$, $\sigma_{11} = \sigma_{22} = -13$.

[170] M. G. Gibby, A. Pines, and J. S. Waugh, *J. Amer. Chem. Soc.* **94**, 6231 (1972).

among the first examples of the use of second moment analyses for the derivation of anisotropies of powders and single crystals, respectively.

The reported ^{29}Si shielding anisotropies[170] are a further application of the cross-polarization technique. Since there is not yet available an absolute shielding scale for silicon, the quantities have been reported with reference to the TMS shift. The magnitudes of these anisotropies are small for a third period atom, which indicates the small effect of replacing a methyl with a methoxy group.

IV. Magnetic Susceptibility Anisotropies

A. Experimental Methods for Susceptibility Anisotropies

The magnetic susceptibility anisotropy is, like the shielding anisotropy, a property which can be determined either directly from experimental measurements or indirectly as the difference between the appropriate tensor components. The distinction between these two methods is not trivial since one of the most widely used methods for deriving the $\chi_{\alpha\alpha}$ components is that of molecular g-value measurements, which itself provides $\chi^{\mathrm{p}}_{\alpha\alpha}$ and requires a theoretical calculation or estimation of $\chi^{\mathrm{d}}_{\alpha\alpha}$ in order to yield values for $\chi_{\alpha\alpha}$'s and hence $\Delta\chi$'s. Fortunately there are available a number of accurate *ab initio* calculations and other reliable methods for predicting these quantities, thereby lessening the uncertainty of the indirectly determined anisotropies. In addition, several of the more direct methods of evaluating χ are dependent upon parameters such as polarizability anisotropies, medium effects and bulk magnetic susceptibility measurements, which are often not very reliable.

The following discussion of the various methods used for determining $\Delta\chi$ will include such factors as scope of applicability, state of matter, need for outside parameters, and the axis system in which the susceptibilities are defined. The discussion of the methods will be followed by a short review of some of the applications of these techniques.

1. *Crystallographic Methods*

The original methods of determining the isotropic or mean molar magnetic susceptibility consisted of measuring the forces experienced by bulk samples in the presence of magnetic fields. Although in principle it is possible to determine the anisotropy likewise, i.e. by measuring these forces on crystals of known orientation, the method is not very accurate.[60]

An alternate method for measuring the magnetic anisotropies of crystals has been developed by Krishnan *et al.*[171–173] It is based on the principle that an anisotropic sample of arbitrary shape in a uniform magnetic field will experience a force tending to align the axis of the algebraically largest susceptibility parallel to the magnetic field. A sample, suspended by a thin fiber, is rotated until this alignment is obtained, i.e. no net force upon the sample. If the crystal is then given a small displacement, it oscillates with a period which can be related directly to the anisotropy. Also needed is the torsion constant, which can be calibrated using a substance of known moment of inertia.[60]

A slight modification of the above procedure, known as the flip-angle method,[173] has also been widely used. One determines the number of radians of rotation of the torsion head required to rotate the crystal by an angle of 45°, at which point the crystal suddenly flips around to a new equilibrium position. This method gives the difference between the minimum and maximum molar susceptibilities of the axes in the plane of oscillation. In general it should be possible to determine three separate anisotropies, which provide a check on the accuracy since their sum is zero.

To obtain the absolute values of the principal axis susceptibilities from the measured anisotropies, it is necessary to know the mean susceptibilities or one of the principal susceptibilities. Selwood[60] has described a method due to Rabi,[174] in which the principal susceptibilities can be measured directly by placing a crystal of known orientation in a solvent of known and variable susceptibility in a nonuniform magnetic field.

A major difficulty encountered in using crystal methods is that of the relationship between the principal axes of the crystal susceptibility and the principal axes of the molecular susceptibility. This relationship, which has been treated in detail by Lonsdale and Krishnan,[175] depends a great deal upon the symmetry of the crystals. They have shown that as the crystal symmetry decreases the amount of obtainable information about the molecular susceptibility increases although the analysis may become more complicated. This situation exists because compounds of low crystal symmetry contain only a few centrosymmetrically arranged molecules. Thus for a triclinic system one obtains directly the numerical values of the three principal molecular susceptibilities, while for a tetragonal crystal only under certain special conditions can even two of these quantities be determined. For a crystal of cubic symmetry only χ_{average} is obtainable.

[171] K. S. Krishnan, B. C. Guha, and S. Banerjee, *Phil. Trans. Roy. Soc. London, Ser. A* **231**, 235 (1933).

[172] K. S. Krishnan and S. Banerjee, *Phil. Trans. Roy. Soc. London, Ser. A* **234**, 265 (1935).

[173] P. W. Selwood, *in* "Physical Methods in Organic Chemistry" (A. Weissberger, ed.), 3rd Ed., Ch. 43. Wiley (Interscience), New York, 1960.

[174] I. I. Rabi, *Phys. Rev.* **29**, 174 (1927).

[175] K. Lonsdale and K. S. Krishnan, *Proc. Roy. Soc., Ser. A* **156**, 597 (1936).

Most of the applications of these crystallographic methods were made for large molecules, particularly organic ring systems. These have especially large anisotropies between the axis perpendicular to the planar ring and the axes within the plane. Lonsdale[176] used the Krishnan oscillation method in conjunction with mean powder susceptibilities to evaluate principal susceptibilities of some long-chained and layered aliphatics in addition to many aromatic systems.

In spite of the added uncertainties of this procedure due to the required knowledge of the crystal structure, the crystallographic method has been found to be fairly reliable. In some molecules the crystal structure is known quite accurately. Comparisons with anisotropies by other methods have shown good agreement in most cases.[177] Le Fevre *et al.*[178–180] have tentatively concluded that the magnetic anisotropy is a property which is independent of the state of matter. A review of much of the work on anisotropies of crystals is given by Bothner-By and Pople.[181]

2. *Magnetic Birefringence Methods*

Birefringence is the property whereby an isotropic substance exhibits double refraction when light is passed through it at right angles to a strong electric or magnetic field. The phenomenon of magnetic double refraction, known as the Cotton–Mouton effect, is analagous to the Kerr electro-optical birefringence.[182–184]

The Cotton–Mouton effect is usually quite small and one must be careful to eliminate or take account of the much larger Faraday effect in which there is a rotation of the plane of polarization of a beam of light parallel to the magnetic field.[183] Experimentally the Cotton–Mouton effect is observed as a phase difference, D, of components of light vibrating parallel and perpendicular to the lines of force in a magnetic field. The expression is[182]

$$D = 2\pi l(\eta_{\parallel} - \eta_{\perp})/\lambda = 2\pi C l H^2, \tag{4.1}$$

where $\eta_{\parallel}$ and $\eta_{\perp}$ are the two indices of refraction, l the distance traversed by the light, H the magnetic field strength, C the Cotton–Mouton constant, and λ the wavelength.

[176] K. Lonsdale, *Proc. Roy. Soc., Ser. A* **171**, 541 (1939).
[177] See Table XX.
[178] R. J. W. Le Fevre, P. H. Williams, and J. M. Eckert, *Aust. J. Chem.* **18**, 1133 (1965).
[179] R. J. W. Le Fevre and D. S. N. Murthy, *Aust. J. Chem.* **22**, 1415 (1969).
[180] R. J. W. Le Fevre and D. S. N. Murthy, *Aust. J. Chem.* **23**, 193 (1970).
[181] A. A. Bothner-By and J. A. Pople, *Annu. Rev. Phys. Chem.* **16**, 43 (1965).
[182] J. W. Beams, *Rev. Mod. Phys.* **4**, 133 (1932).
[183] A. D. Buckingham and J. A. Pople, *Proc. Phys. Soc., London, Sect. B* **69**, 1133 (1956).
[184] A. D. Buckingham and J. A. Pople, *Proc. Phys. Soc., London, Sect. A* **68**, 905 (1955).

The theory of the Cotton–Mouton effect, based on the Langevin–Born model, is discussed by Beams,[182] Raman and Krishnan,[185] and more recently by Buckingham and Pople.[183] The expression for C is a function of the molecular polarizability and susceptibility tensors, $\boldsymbol{\alpha}$ and $\boldsymbol{\chi}$, respectively. In the case of actual or assumed axial symmetry the expression reduces to[173]

$$C = (2/135)\,\pi N_0[\eta + (2/3kT)\,\Delta\alpha\Delta\chi]. \tag{4.2}$$

The first term in the brackets, η, to be discussed later, is often assumed negligible; thus from a knowledge of $\Delta\alpha$ and C at a given temperature it is possible to deduce a value for the magnetic anisotropy, $\Delta\chi$.

Le Fevre *et al.*[186–188] have investigated in great detail the determination of the principal polarizabilities, α_1, α_2, α_3, as well as $\Delta\alpha$. They have shown how the measured molar Kerr constants[187,189] can be used to deduce polarizability ellipsoids and calculate molecular and bond polarizabilities.[190] From these and other studies values have become available for the polarizabilities required in Eq. (4.2). The reliability of these quantities has been questioned, but Le Fevre and co-workers[178–180] have shown that they lead to values of $\Delta\chi$ which are in substantial agreement with $\Delta\chi$'s by other methods. Alternative methods of determining the polarizability anisotropies are discussed by Bridge and Buckingham[191] and Scharpen *et al.*[192] Attempts to obtain this quantity theoretically have had little success thus far.[38,193–195]

Buckingham and Pople[183] have discussed the temperature-independent term, η, in Eq. (4.2), which is the electric anisotropy induced directly by the magnetic field. For atoms or spherical molecules it is the only term. A theoretical calculation for atomic hydrogen indicates that it is probably considerably smaller than the temperature-dependent term.[196] Buckingham *et al.*[196] have measured the Cotton–Mouton constants for some nearly spherical molecules. They found them to be about an order of magnitude smaller for argon and CH_4 but larger for SF_6. There is also some evidence that η might be important for conjugated molecules with large

[185] C. V. Raman and K. S. Krishnan, *Proc. Roy. Soc., Ser. A* **117**, 1 (1927).
[186] R. J. W. Le Fevre, *Advan. Phys. Org. Chem.* **3**, 1 (1965).
[187] R. J. W. Le Fevre and C. G. Le Fevre, *J. Chem. Soc.* p. 1577 (1954).
[188] M. Aroney and R. J. W. Le Fevre, *J. Chem. Soc.* p. 3600 (1960).
[189] R. J. W. Le Fevre and C. G. Le Fevre, *J. Chem. Soc.* p. 4041 (1953).
[190] R. J. W. Le Fevre and L. Radom, *J. Chem. Soc. B* p. 1295 (1967).
[191] N. J. Bridge and A. D. Buckingham, *Proc. Roy. Soc., Ser. A* **295**, 334 (1966).
[192] L. H. Scharpen, J. S. Muenter, and V. W. Laurie, *J. Chem. Phys.* **53**, 2513 (1970).
[193] H. J. Kolker and M. Karplus, *J. Chem. Phys.* **39**, 2011 (1963).
[194] G. P. Arrighini, M. Maestro, and R. Moccia, *Chem. Phys. Lett.* **1**, 242 (1967).
[195] D. W. Davies, *Mol. Phys.* **20**, 605 (1971).
[196] A. D. Buckingham, W. H. Prichard, and D. H. Whiffen, *Trans. Faraday Soc.* **63**, 1057 (1967).

Cotton–Mouton constants.[197] The neglect of this term is an approximation which has not been entirely justified.

The most extensive series of experimental determinations of $\Delta\chi$ from Cotton–Mouton effects are those of Le Fevre and Murthy.[179,198–200] In order to minimize intermolecular and solvent effects they have used very dilute solutions in nonpolar solvents such as CCl_4 to obtain the molar Cotton–Mouton and Kerr constants required to deduce the anisotropies. The previously mentioned agreement between their $\Delta\chi$ results and those of Lonsdale[176] indicates the apparent validity of the assumptions used. Earlier data of Ramanadham[201] is less reliable because of lesser sensitivity.

Buckingham *et al.*[196] have undertaken a study of the magnetic birefringence of a number of gases of varying symmetry and size in order to investigate the temperature-independent term explicitly (see above). Using measured values of polarizability anisotropy they also derived $\Delta\chi$'s for some diatomic and polyatomic molecules of less than tetrahedral symmetry. Use of the experimental values of χ in conjunction with the mean susceptibilities allows a determination of the principal susceptibility components. In addition to the molecular susceptibility work, a large amount of effort has gone into the task of computing and correlating the bond susceptibilities and anisotropies[17,181,202–210] of various molecules. Zurcher[211] has extended these methods to include bond polarizabilities as well as susceptibilities.

The field of bond susceptibilities and anisotropies, although very closely related to the methods and purposes of the determination of molecular susceptibilities and anisotropies, is a very large topic in itself, the details of which are beyond the scope of this review. This section is confined to methods of determining χ which are either experimental measurements of

[197] K. H. Groddie, *Phys. Z.* **39**, 772 (1938).
[198] C. L. Cheng, D. S. N. Murthy, and G. L. D. Ritchie, *Mol. Phys.* **22**, 1137 (1971).
[199] R. J. W. Le Fevre and D. S. N. Murthy, *Aust. J. Chem.* **19**, 179 (1966).
[200] R. J. W. Le Fevre, D. S. N. Murthy, and P. J. Stiles, *Aust. J. Chem.* **21**, 3059 (1968).
[201] M. Ramanadham, *Indian J. Phys.* **4**, 15, 109 (1929).
[202] R. L. Mital and R. R. Gupta, *J. Chem. Phys.* **54**, 3230 (1971).
[203] H. A. Allen and N. Muller, *J. Chem. Phys.* **48**, 1626 (1968).
[204] A. Pacault, J. Hoarau, and A. Marchand, *Advan. Chem. Phys.* **3**, 171 (1961).
[205] H. F. Hameka, *J. Chem. Phys.* **34**, 1996 (1961).
[206] D. W. Davies, *Mol. Phys.* **6**, 489 (1963).
[207] L. Salem, "The Molecular Orbital Theory of Conjugated Systems." Benjamin, New York, 1966.
[208] L. M. Jackman, "Nuclear Magnetic Resonance Spectroscopy." Pergamon, Oxford, 1959.
[209] W. Weltner, Jr., *J. Chem. Phys.* **28**, 477 (1958).
[210] W. T. Raynes, *Mol. Phys.* **20**, 321 (1971).
[211] R. F. Zurcher, *J. Chem. Phys.* **37**, 2421 (1962).

the primary quantities or *ab initio* theoretical evaluations, with the understanding that experimental methods often rely on estimated or semiempirical type quantities. The reader is referred to the above references for discussions of semiempirical and bond susceptibility methods. (See also Section V, C.)

3. *Microwave Spectroscopy Methods*

Within the last several years microwave spectroscopy has become the most important and extensive method for determining magnetic susceptibility tensors and anisotropies. Flygare's group[25,212] has been primarily responsible for this trend.

One of the earliest uses of microwave spectroscopy was in the determination of rotational magnetic moments[213] (also known as molecular g-values) of $^1\Sigma$ molecules. These moments are directly related to the paramagnetic susceptibility components. The theory of rotational magnetic moments is given by Strandberg *et al.*[214,215] who showed that the small rotation-induced electronic angular momentum can be treated as a perturbation to the rigid rotor Hamiltonian. The expression for g_{xx} is, in Flygare's notation,[25]

$$g_{xx} = (4\pi G_x M_p/\hbar m)\left[m \sum_n Z_n(y_n^2 + z_n^2) + 2 \sum_{p>0}^{\infty} |\langle 0|M_x|p\rangle|^2/(E_o - E_p) \right], \tag{4.3}$$

G_x is the rotational constant, $\mathbf{M}$ the electronic angular momentum, M_p the proton mass, Z_n the nuclear charge, and the n and p summations are over nuclei and electronic energy states respectively.

From Section II, the diagonal component of the paramagnetic susceptibility is

$$\chi_{xx}^{p} = (-e^2/2m^2c^2) \sum_{p>0}^{\infty} |\langle 0|M_x|p\rangle|^2/(E_o - E_p), \tag{4.4}$$

whence

$$\chi_{xx}^{p} = (e^2/4mc^2) \sum_n Z_n(y_n^2 + z_n^2) - (e^2/16\pi mc^2) \cdot g_{xx}/G_x M_p. \tag{4.5}$$

Therefore from a knowledge of internuclear distances, the rotational constant, and the molecular g-value, the paramagnetic susceptibility can be obtained directly. The matter of obtaining the corresponding diamagnetic susceptibility is considered separately in a later section (Section IV, A,4).

[212] W. H. Flygare and R. C. Benson, *Mol. Phys.* **20**, 225 (1971).
[213] C. K. Jen, *Phys. Rev.* **74**, 1396 (1948); **81**, 197 (1951).
[214] J. R. Eshbach and M. W. P. Strandberg, *Phys. Rev.* **85**, 24 (1952).
[215] B. F. Burke and M. W. P. Strandberg, *Phys. Rev.* **90**, 303 (1953).

With the development of more sensitive and high powered apparatus a more general theory including higher order effects of rotational magnetism was required. Huttner and Flygare[25] considered in detail the Hamiltonian and energy expressions for a rotating molecule in a magnetic field, including effects due to spin-rotation coupling, induced magnetic moments, and molecular and shielded nuclear Zeeman effects. There are two terms in the first-order energy which depend on g or χ. In the uncoupled case (spin-rotation interaction $H_{SR} = 0$) these are[25]

$$E^1(g, \chi) = -\mu_0 [M_j/J(J+1)] \sum_{\alpha} g_{\alpha\alpha} \langle J_\alpha^2 \rangle H_z$$

$$- \{\tfrac{1}{2}\chi + \{[3M_j^2 - J(J+1)]/(2J-1)(2J+3)\} \cdot [1/J(J+1)]$$

$$\times \sum_{\alpha} (\chi_{\alpha\alpha} - \chi) \langle J_\alpha^2 \rangle\} H_z^2, \tag{4.6}$$

where J and M_j are rotational quantum numbers and $\langle J_\alpha^2 \rangle$ is the reduced matrix element of J_α^2. A similar but more complicated expression is obtained in the coupled basis.

Huttner *et al.*[216] have shown in detail how the parameters $g_{\alpha\alpha}$ and $[(\chi_{\alpha\alpha} - \chi), (\alpha = x, y, z)]$ of Eq. (4.6) can be derived from an analysis of the $\Delta J = 0$ transitions of the $J = 1$, 2, and 3 states of H_2CO by measuring the first- and second-order splittings. It should be noted that the results of carrying out such an analysis are three anisotropies (two are independent) and three g-values, or a total of five independent parameters.

In a recent review article Flygare and Benson[212] have described the origin and significance of the terms in the magnetic field, rigid rotor Hamiltonian including the approximations involved and possible refinements of the theory. They discuss also the possible sources of error and the experimental limitations of the methods compared with other methods. In addition they show which secondary experimental quantities can be derived from the ones already mentioned, and finally give an extensive review and interpretation of most of the measurements carried out to that date ($\sim$1970).

4. *Methods for Estimating Diamagnetic Susceptibilities*

Although it should be clear from the context, it is pointed out that the diamagnetic susceptibilities referred to are the χ^d's or *ground state parts* of the total magnetic susceptibility. The latter quantity χ_{average} is, for most closed-shell molecules, a negative quantity and hence gives rise to a diamagnetic effect, i.e. one opposed to the applied magnetic field. The

[216] W. Huttner, M.-K. Lo, and W. H. Flygare, *J. Chem. Phys.* **48**, 1206 (1968).

need to obtain reliable χ^d's was seen to arise because of the availability of the experimental g-values, which lead directly to χ^p. Thus to obtain $\chi\,(=\chi^p+\chi^d)$ and its components and anisotropies one needs χ^d's.

The Van Vleck expression for χ^d_{xx},

$$\chi^d_{xx} = -(e^2/4mc^2)\left\langle 0\left|\sum_i^{\text{el.}} (y_i^2+z_i^2)\right|0\right\rangle$$
$$= -(e^2/4mc^2)[\langle y^2\rangle + \langle z^2\rangle], \tag{4.7}$$

shows a direct relationship to the electronic second moments. *Ab initio* SCF methods[217] and some approximate SCF methods have been able to calculate the second moments with reasonable to very high accuracy. For most common small molecules these quantities are available theoretically and in some cases experimentally.

The semiempirical methods based on bond parameters have also been moderately successful in predicting values for second moments and χ^d's. This result is not very surprising since the magnetic susceptibilities themselves have been known since Pascal's work[61,218] to obey additivity rules fairly well.

Among the recent methods[100,219,220] for evaluating second moments perhaps the most reliable and extensive one is that of Gierke, Tigelaar, and Flygare.[100] Their method, known as the atom-dipole method, expresses the electronic second moments in terms of sums over empirical atomic-dipole and quadrupole moments. The results for molecular dipole moments, quadrupole moments, and diamagnetic susceptibilities are in extremely good agreement with the corresponding experimental or accurate theoretical quantities. One limitation of this method, common to many empirical schemes, is that an unknown system must have bonds similar to those of a known system in order that the empirical dipoles and quadrupoles be established. Thus, for example, a molecule such as ClF_3 is currently beyond the scope of this method.

5. *Molecular Beam Methods*

The molecular beam techniques can be used to evaluate molecular g-values from microwave or radio-frequency splittings and in some cases to observe the anisotropies directly from the quadratic Zeeman effect. However it appears to be less well-suited to either of these quantities

[217] See Sect. II.
[218] See Ref. 60, Ch. III, Sect. 6.
[219] R. Rein, G. R. Pack, and J. R. Rabinowitz, *J. Magn. Resonance* **6**, 360 (1972).
[220] Z. B. Maksic and J. E. Bloor, *Chem. Phys. Lett.* **13**, 571 (1972).

than is conventional high resolution microwave spectroscopy. It is more useful in determining spin-rotation constants since the beam resonance methods often detect nuclear spin transitions rather than pure rotational transitions. It has been pointed out that, although providing high resolution, the molecular beam magnetic resonance technique does not select rotational states, and in heavier molecules the spectra become increasingly complex.

There are some molecules for which the molecular beam method has been used to good advantage.[212] These include many small-to-moderate sized diatomics such as H_2, D_2, Li_2, Na_2, LiF, and F_2 and some larger diatomics such as Rb_2, RbF, CsF, TiF, and RbCl.[221] For the latter group Graff *et al.* [119,121] used a combination of electric and magnetic resonance in order to detect some low-lying rotational and vibrational states.

The technique of focusing or state selection has been used in MASER molecular beam resonance experiments.[222,223] Several groups[224–228] have shown that the beam-maser method allows an extremely high degree of resolution and has provided a very precise value for the small magnetic anisotropy of NH_3.[229] The beam-maser technique though is very limited in scope. Some of the problems of sensitivity, selectivity, and time are discussed by Verhoeven and Dymanus.[226]

In conclusion the use of molecular beams for susceptibility parameters is not nearly as generally applicable as microwave spectroscopy. It is however an important complementary technique since there are a number of systems including diatomics, NH_3, and CH_4 (along with other molecules without a permanent dipole moment) for which molecular beams provide information unavailable by other methods.

B. TABLES OF MAGNETIC SUSCEPTIBILITY ANISOTROPIES

The experimental susceptibility tensors and anisotropies are presented in Tables XVII through XX corresponding to linear, symmetric top, asymmetric top, and aromatic molecules respectively. For the linear and symmetric top molecules, there are four quantities given $\chi_{\perp}$, $\chi_{\parallel}$, χ_{avg}, and $\Delta\chi$, any two

[221] See references cited in Ref. 212.
[222] J. P. Gordon, H. J. Zeiger, and C. H. Townes, *Phys. Rev.* **99**, 1264 (1955).
[223] P. Thaddeus and L. C. Krisher, *Rev. Sci. Instrum.* **32**, 1083 (1961).
[224] P. Thaddeus, L. C. Krisher, and J. H. N. Loubser, *J. Chem. Phys.* **40**, 257 (1964).
[225] J. Verhoeven, A. Dymanus, and H. Bluyssen, *J. Chem. Phys.* **50**, 3330 (1969).
[226] J. Verhoeven and A. Dymanus, *J. Chem. Phys.* **52**, 3222 (1970).
[227] S. G. Kukolich, *J. Chem. Phys.* **54**, 8 (1971).
[228] S. L. Rock and W. H. Flygare, *J. Chem. Phys.* **56**, 4723 (1972).
[229] S. G. Kukolich, *Chem. Phys. Lett.* **5**, 401 (1970).

TABLE XVII: EXPERIMENTAL MAGNETIC SUSCEPTIBILITY TENSORS AND ANISOTROPIES[a] OF SOME LINEAR MOLECULES

	$\chi_\perp$	$\chi_{\|\|}$	χ_{av}[b]	$\Delta\chi$[b]	Method[c]	Ref.
N_2	−9.7	−18.4	−12.6[d]	−8.7[d]	g−MB	161
					+ χ^d(T)	49
	−9.2	−17.7	−12.0±0.1[e]	−8.4	CM gas	196
				−8.56±0.34	MB	212
CO	−10.0	−18.0	−12.7[d]	−8.0[d]	g−MB + χ^d(T)	20
			−9.8[e]	−8.2±0.9	MS	212
				−10.2	CM gas	196
CS	−9.1	−33.1[f]	−17.1[d]	−24.0±3.0	MS	212
LiH	−6.69	−9.55	−7.64[d]	−2.86[d]	g−MB	230
					χ^d(T)	16
HF	−10.4	−10.1	−10.3[d]	0.3[d]	g−MB + χ^d(T)	231
			−8.6±0.1		Liq.	231
F_2	−6.0	−17.0[f]	−9.7[d]	−11.0[d]	g−MB	144
					+ χ^d(T)	38
ClF	−14.7	−30.6[f]	−20.0[d]	−15.9±0.7[g]	MS	232
BrF	−18.0	−39.0[f]	−25.0	−21.0±1.0[g]	MS	232
CO_2	−19.8	−25.9[h]	−21.8[d]	−6.0	CM gas	196
			−24[e]	−6.23±0.18	Ref. 233	—
OCS	−29.7	−37.9	−32.4[i]	−8.0±1.0	MS	212
				−8.35±0.28	MS	234
	−29.3	−38.6	−32.4	−9.27±0.10	MS	233
				−9.43±0.01	MB	102
CS_2	−42.5	−58.5[j]	−47.8[d]	−14.0	CM soln.	200
			−40[e]			
N_2O	−15.4	−25.8	−18.9[i]	−10.8	CM gas	196
				−10.1±0.2	MS	235
H—C≡C—H	−17.5	−25.2	−20.1	−7.7[d]	g−MB	236
					+ χ^d(T)	34
	−20.1	−24.6[f]	−21.6[d]	−4.5	Estimate	237
			−20.8±0.8[e]			
F—C≡C—H	−24.3	−29.5[h]	−26.0[d]	−5.19±0.15	MS	212
Cl—C≡CH	−35.3	−44.6[f]	−38.4[d]	−9.3±0.5	MS	237
Br—C≡CH	−43.6	−53.1[f]	−46.8[d]	−9.5±0.9	MS	237
F—C≡N	−20.8	−28.0[f]	−23.2[d]	−7.2±0.8	MS	241
Cl—C≡N	−28.8	−39.6[f]	−32.4[d]	−10.8±0.5[g]	MS	232
Br—C≡N	−37.9	−50.1[f]	−42.0[d]	−12.2±0.7[g]	MS	232
I—C≡N	−50.9	−67.1[f]	−56.3[d]	−16.2±0.5[g]	MS	232
H—C≡N	−14.4	−21.6[f]	−16.8[d]	−7.2±0.4	MS	237
H—C≡P	−27.7	−36.1[f]	−30.5[d]	−8.4±0.9	MS	237

[a] Units 10^{-6} erg gauss^{-2} mole^{-1}.

[b] $\chi_{av} \equiv \frac{1}{3}(2\chi_\perp + \chi_{||})$, $\Delta\chi \equiv \chi_{||} - \chi_\perp$. Unless otherwise indicated, these are the two primary experimental quantities.

of which yield the others. Normally the primary quantities determined experimentally are $\Delta\chi$ and χ_{av}. In cases in which the molecular g-values and theoretical values for χ^d are used the primary experimental quantities are $\chi_\perp$ and $\chi_\parallel$. In these instances $\Delta\chi$ and χ_{av} are secondary quantities, a fact which will be indicated in the tables. A third possibility, particularly for linear molecules, is that in which $\Delta\chi$ is determined experimentally and $\chi_\parallel = \chi_\parallel^d$ can be easily estimated from empirical parameters. When there are several independent determinations of $\Delta\chi$ which are very close, the average is usually used to compute the secondary quantities. Otherwise the tables are self-explanatory.

The discussion accompanying these tables will be brief for the following reasons.

1. The trends for susceptibility tensors and anisotropies are less revealing than for shielding since the former depend more strongly upon molecular size than anything else. Consequently there are very regular increments in susceptibilities, e.g. see Table XX, and these effects have been well-studied and analyzed.

---

[c] MS—microwave spectroscopy, g—molecular g-value, MB—molecular beam, CM—Cotton–Mouton effect, χ^d(T)—theoretical diamagnetic susceptibility.

[d] Secondary quantity, obtained as sum of χ^p(expt.) and $\chi_\parallel{}^d$(theor.), e.g., Ewing *et al.*[232] or from $\Delta\chi$(expt.) and $\chi_\parallel{}^d$(theor.).

[e] See "Handbook of Chemistry and Physics"[66] (N_2), Bitter[238] (CO_2), Foex[67] (CO, CS_2), and Barter *et al.*[64] (C_2H_2).

[f] χ^d obtained from empirical values of $\langle a^2\rangle = \langle r_\perp^2\rangle$. See Gierke *et al.*[100]

[g] Obtained as average of two isotopes of halogen.

[h] R. Ditchfield (personal communication) (1972).

[i] See Landolt-Bornstein.[239]

[j] See Fischer and Kemmey.[240]

[230] T. R. Lawrence, C. H. Anderson, and N. F. Ramsey, *Phys. Rev.* **130**, 1865 (1963).

[231] R. M. Stevens and W. N. Lipscomb, *J. Chem. Phys.* **41**, 184 (1964).

[232] J. J. Ewing, H. L. Tigelaar, and W. H. Flygare, *J. Chem. Phys.* **56**, 1957 (1972).

[233] W. H. Flygare, W. Huttner, R. L. Shoemaker, and P. D. Foster, *J. Chem. Phys.* **50**, 1714 (1969).

[234] H. Taft and B. P. Dailey, *J. Chem. Phys.* **48**, 597 (1968).

[235] W. H. Flygare, R. L. Shoemaker, and W. Huttner, *J. Chem. Phys.* **50**, 2414 (1969).

[236] J. W. Cederberg, C. H. Anderson, and N. F. Ramsey, *Phys. Rev. A* **136**, 960 (1964).

[237] S. L. Hartford, W. C. Allen, C. L. Norris, E. F. Pearson, and W. H. Flygare, *Chem. Phys. Lett.* **18**, 153 (1973).

[238] F. Bitter, *Phys. Rev.* **36**, 1648 (1930).

[239] H. Landolt-Bornstein, "Zahlenwerte und Funktionen," Vol. II, Part 10. Springer-Verlag, Berlin and New York, 1951.

[240] C. R. Fischer and P. J. Kemmey, *Mol. Phys.* **22**, 1133 (1971).

[241] S. L. Rock, J. C. McGurk, and W. H. Flygare, *Chem. Phys. Lett.* **19**, 153 (1973).

TABLE XVIII

MAGNETIC SUSCEPTIBILITY TENSORS AND ANISOTROPIES[a] OF SOME SYMMETRIC TOP MOLECULES

	$\chi_\perp$	$\chi_\parallel$	χ_{av}[b]	$\Delta\chi$[b]	Method[c]	Ref.
CH_3F	−15.1	−23.3	-17.8 ± 0.8[d]	-8.2 ± 0.8	MS	212
CH_3Cl	−29.4	−37.3	-32.0 ± 0.60[e]	-7.95 ± 0.50	MS	242
CH_3Br	−40.0	−48.5	-42.8 ± 0.80[e]	-8.50 ± 0.40	MS	242
CH_3I	−53.5	−64.5	-57.2 ± 1.2[e]	-10.98 ± 0.45	MS	242
$CH_3C{\equiv}N$	−24.1	−34.6	-27.6 ± 0.3[d]	-10.5 ± 0.5	MS	243
$CH_3N{\equiv}C$	−23.1	−36.6	-27.6 ± 1.0[f]	-13.5 ± 0.7	MS	243
$CH_3C{\equiv}CH$	−29.5	−37.2	-32.1 ± 2.0[g]	-7.70 ± 0.14	g−MS χ^d(T)	224 220
C_2H_6	−25.6	−29.2	-26.8 ± 0.8[d]	−3.6	CM gas	210
NH_3	−16.4	−16.1	-16.3 ± 0.8[d]	0.37 ± 0.04	MB	229
PH_3	−27.1	−24.4	-26.2 ± 0.8[d]	2.7 ± 0.8	MS	245
PF_3	−36.9	−38.3	−37.4[h]	−1.4	MS	246
CHF_3	—	—	—	1.2 ± 0.6	MS	212
NF_3	—	—	—	3.0 ± 1.5	MS	246

[a] Units 10^{-6} erg gauss^{-2} mole^{-1}.
[b] $\chi_{av} \equiv \frac{1}{3}(2\chi_\perp + \chi_\parallel)$; $\Delta\chi \equiv \chi_\parallel - \chi_\perp$. Unless otherwise indicated, $\Delta\chi$ and χ_{av} are obtained from references given on rhs of table.
[c] See footnote *c*, Table XVII.
[d] See Barter *et al.*[64] (CH_3F, C_2H_6, NH_3, PH_3) and Foex[67] (CH_3CN).
[e] Pascal.[247]
[f] Assumes $\chi_{av}(CH_3NC) = \chi_{av}(CH_3CN)$. See also Rein *et al.*[219]
[g] $\chi_{av} = \chi^{p}_{av}$(expt.) + χ^{d}_{av}(theor.).
[h] Estimated, see Stone *et al.*[246]

2. Many of the most interesting aspects associated with susceptibilities are the related parameters such as g-values and quadrupole moments, discussions of which are beyond the scope of this review.

3. The principal use of the properties tabulated in this section will be for comparison with the quantities discussed in other sections. Thus it is very

[242] D. L. VanderHart and W. H. Flygare, *Mol. Phys.* **18**, 77 (1970).
[243] J. M. Pochan, R. L. Shoemaker, R. G. Stone, and W. H. Flygare, *J. Chem. Phys.* **52**, 2478 (1970).
[244] R. L. Shoemaker and W. H. Flygare, *J. Amer. Chem. Soc.* **91**, 5417 (1969).
[245] S. G. Kukolich and W. H. Flygare, *Chem. Phys. Lett.* **7**, 43 (1970).
[246] R. G. Stone, J. M. Pochan, and W. H. Flygare, *Inorg. Chem.* **8**, 2647 (1969).
[247] M. P. Pascal, *Ann. Chim. Phys.* **19**, 5 (1910).

useful in evaluating the reliability of theoretical methods to have as much detailed experimental data as possible, especially the tensor components. In addition, the susceptibility anisotropies will be analyzed in relation to the shielding anisotropies to try to determine from theoretical and physical considerations the causes of the differences between them.

For the linear molecules, (Table XVII) note is made of the effect—applicable to almost all susceptibilities—that, within a particular series, as the molecular size increases, all the susceptibility terms likewise increase. Examples are the series F_2, ClF, BrF; CO_2, OCS, CS_2; HC≡CH, FC≡CH, ClC≡CH, BrC≡CH; and HCN, FCN, ClCN, BrCN, ICN. Also it is found that for isoelectronic molecules, e.g. N_2 and CO, or CO_2 and N_2O, the susceptibility terms remain approximately constant.

Turning now to the symmetric top molecules of Table XVIII it is noted that the quantities $\chi_\perp$, $\chi_\parallel$, and χ_{av} all increase regularly from CH_3F to CH_3I but the anisotropy stays relatively constant, changing only by 25% while χ_{av} more than triples. Flygare and Benson[212] have shown that for CH_3-containing molecules the value of $g_\parallel$ is practically constant (0.310 nm). This effect is reasonable since for a linear molecule $g_\parallel = 0$ and for the symmetric top systems only the CH_3 group is likely to contribute to $g_\parallel$. Recall that the ^{13}C shielding of a methyl carbon along the C_3 axis was also very nearly constant for almost all cases investigated. Thus the changes of $\chi_\parallel$ among these compounds must be caused by changes in $\chi_\parallel^d$ (the radial second moments) since $\chi_\parallel^p$ is directly related to $g_\parallel$ and should be reasonably constant.[242] This contention is further supported by the observation that for linear molecules $\chi_\parallel^p = 0$, while $\chi_\parallel^d$ changes significantly among different molecules.

In Table XIX the principal susceptibilities for some asymmetric top molecules are given. For these molecules there are three possible anisotropies; the out-of-plane minus average in-plane anisotropy has been reported except where noted otherwise. In order to determine the susceptibility components, one requires the average susceptibility in addition to the anisotropies. Most of the anisotropies were determined by Flygare's group, who also did measurements on additional molecules not included in this table.[212] In several cases the value of χ_{av} was estimated using some of the empirical methods for either χ_{av} itself or χ^d. For H_2O and CF_2H_2 the beam-maser results agree quite well with those of microwave spectroscopy. Several trends including replacement of H by F and O by S are discussed in the original papers and in the general references.

Table XX contains the susceptibility tensors and anisotropies of a large

TABLE XIX

MAGNETIC SUSCEPTIBILITY TENSORS AND ANISOTROPIES[a] OF SOME ASYMMETRIC TOP MOLECULES

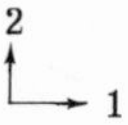	χ_1	χ_2	χ_3	χ_{av}[b]	$\Delta\chi$[b]	Method[c]	Ref.
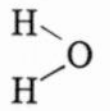	−13.5	−12.1	−13.7	[−13.1 ± 0.1][d]	−0.9 ± 0.6	MS	248
	−13.5	−12.3	−13.5		−0.6 ± 0.6	MS	249
D_2O	−13.2	−12.9	−13.2		−0.1	MS	226
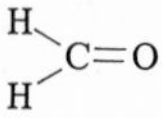	1.57	−8.17	−13.96	−6.85 ± 0.3[e]	[−10.8 ± 0.2]	MS	227
					[−10.7 ± 0.4]	MS	216
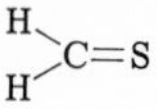	−1.7	−20.8	−35.0	−19.2 ± 1.8[f]	−23.7 ± 0.8	MS	228
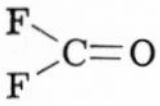	−25.9	−24.6	−28.7	−26.4[g]	−3.5 ± 0.6	MS	250
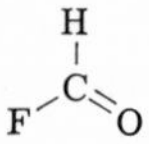	−11.7	−11.7	−17.7	−13.7 ± 4.0[h]	−6.0 ± 0.3	MS	251
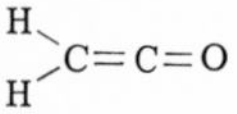	−25.9	−24.5	−22.8	−24.4[f]	2.6 ± 0.5	[g − MS	252
						+ χ^d(T)]	253
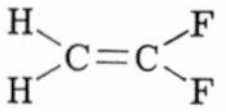	−29.3	−26.0	−30.4	−28.6[g]	−2.7 ± 0.4	MS	250
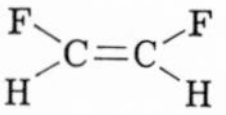	−25.3	−27.7	−28.6	−27.2[g]	−2.1 ± 0.3	MS	250
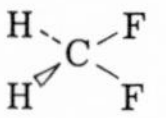	−23.7	−25.3	−23.0	[−24.0][g]	1.6 ± 0.4	MS	250
	−23.8	−25.4	−22.8		1.79 ± 0.01	MB	111
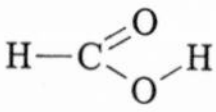	−16.8	−18.8	−24.1	−19.90[i]	−6.4 ± 0.3	MS	212

number of aromatic compounds. These results show clearly the additive effects of the substituents F, Cl, Br, and CH_3 upon the average susceptibilities. The principal susceptibilities and anisotropies also show some additivity effects, although the out-of-plane components, χ_3, show some reversals in the series C_6H_5F, 1,4 $C_6H_4F_2$, and 1,3,5 $C_6H_3F_3$. The trend among the monohalides from F to I is that the average in-plane susceptibilities increase in magnitude more rapidly than the out-of-plane susceptibilities; thus the anisotropy decreases in magnitude from -58 cgs-ppm in C_6H_5F to -36 cgs-ppm in C_6H_5I. Cheng *et al.*[198] have pointed out that the introduction of fluorine atoms has little effect on $\chi_{\parallel}$ but causes a large decrease in $\chi_{\perp}$.

The data in the table permit a comparison of the values obtained using three different methods, and in particular enable an evaluation of some of the approximations and uncertainties involved. In using the Cotton–Mouton effect Le Fevre and Murthy have assumed the temperature-independent term to be negligible. From the closeness of the microwave spectroscopy and the Cotton–Mouton effect results for C_6H_6, C_6H_5F, and pyridine the assumption is apparently justified. As previously stated, the direct observation of the principal susceptibilities in crystals is limited by the uncertain relation between the crystal axes and the molecular axes. Thus discrepancies of up to 15 cgs-ppm are observed in the anisotropies determined by crystal and Cotton–Mouton studies. From the rather good agreement between some of the crystal and other method results (e.g. for C_6H_6), it has been inferred that the above discrepancies are due to uncertainties in the crystal axes and not to a difference in $\Delta\chi$ in different states of matter.

Footnotes to TABLE XIX.

[a] Units 10^{-6} erg $gauss^{-2}$ $mole^{-1}$.
[b] See footnote *b*, Table XVIII.
[c] See footnote *c*, Table XVII.
[d] Foex.[67]
[e] Huttner *et al.*[216]
[f] Obtained as sum of χ^p(expt.) and χ^d(theor.).
[g] χ^d's estimated from experimental and empirical second moments.
[h] Estimated from Pascal's constants, see Rock *et al.*[251]
[i] Broersma.[65]

[248] H. Taft and B. P. Dailey, *J. Chem. Phys.* **51**, 1002 (1969).
[249] S. G. Kukolich, *J. Chem. Phys.* **50**, 3751 (1969).
[250] R. P. Blickensderfer, J. H. S. Wang, and W. H. Flygare, *J. Chem. Phys.* **51**, 3196 (1969).
[251] S. L. Rock, J. K. Hancock, and W. H. Flygare, *J. Chem. Phys.* **54**, 3450 (1971).
[252] W. Huttner, P. D. Foster, and W. H. Flygare, *J. Chem. Phys.* **50**, 1710 (1969).
[253] J. H. Letcher, M. L. Unland, and J. R. Van Wazer, *J. Chem. Phys.* **50**, 2185 (1969).

TABLE XX

MAGNETIC SUSCEPTIBILITY TENSORS AND ANISOTROPIES[a] OF SOME AROMATIC MOLECULES

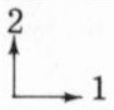	χ_1	χ_2	χ_3	χ_{av}[b]	$\Delta\chi$[b]	Method[c]	Ref.
	−34.9	−34.9	−94.6	[−54.7 ± 0.6][d]	−59.7 ± 0.2	MS	254
					−59.7	Crystal	181
	−36.8	−36.8	−90.4		−57.8 ± 1.8	CM liq.	198
					−53.6 ± 2.4	CM gas	255
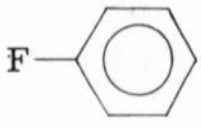	−40.7	−37.1	−97.2	−58.3 ± 0.6[e]	−58.3 ± 1.0	MS	256
					−57.2 ± 1.2	CM liq.	198
	[−45.6][f]		−96.2	−62.5 ± 2.0[g]	−50.6 ± 3.0	CM liq.	198
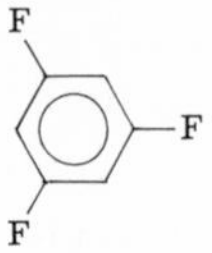	−53.5	−53.5	−92.8	−66.6 ± 3.0[g]	−39.7 ± 1.8	CM liq.	198
					−39.1 ± 1.8	CM gas	255
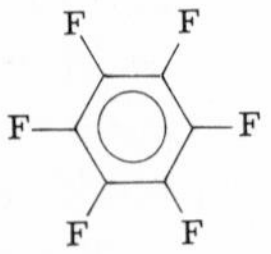	−67.4	−67.4	−102.9	[−79.2 ± 2.4]	−35.5 ± 1.2	CM liq.	198
	−68.6	−68.6	−100.5		−31.9 ± 1.2	CM gas	255
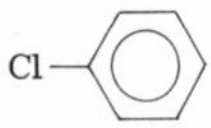	[−54.9][f]		−100.1	−70.0 ± 2.0[h]	−45.3 ± 1.8	CM liq.	198
	[−72.0][f]		−112.3	−85.4 ± 2.0[h]	−40.3 ± 1.2	CM liq.	198
	−78.3	−50.3	−120.2	−82.9	−55.9	Crystal	181
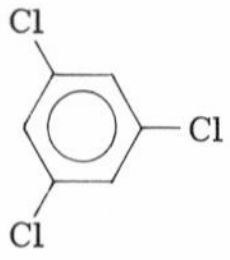	−88.2	−88.2	−123.7	−100.0[g]	−35.5 ± 1.2	CM liq.	198
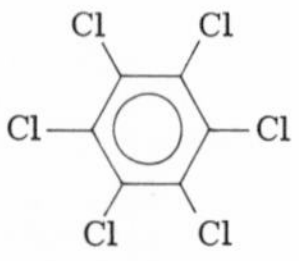	−132.8	−132.8	−171.3	−145.6[h]	−38.5 ± 1.8	CM liq.	198
	−128.0	−128.0	−182.0	−146.0	−54.0	Crystal	181
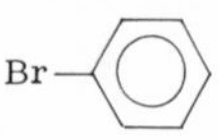	[−63.4][f]		−110.0	−78.9 ± 0.3	−46.6	CM liq.	199
	−97.1	−70.5	−136.7	−101.4	−52.9	Crystal	181
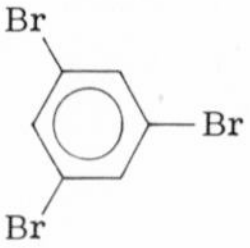	−122.5	−122.5	−170.0	−138.3	−47.5	Crystal	181

TABLE XX—continued

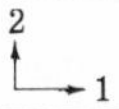	χ_1	χ_2	χ_3	χ_{av}[b]	$\Delta\chi$[b]	Method[c]	Ref.
	−87.6	−59.9	−129.0	−92.2	−55.2	Crystal	181
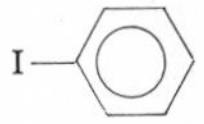	[−79.5][f]		−115.1	−91.3	−35.6 ± 2.0	CM liq.	199
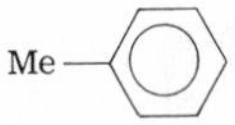	[−45.5][f]		−105.7	−65.6[i]	−60.2 ± 5.4	CM liq.	198
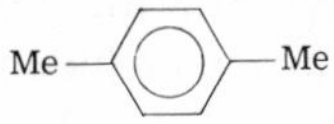	[−54.3][f]		−122.3	−77.0[i]	−68.0 ± 9.6	CM liq.	198
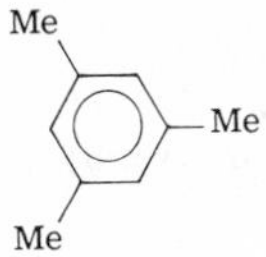	−60.5	−60.5	−124.9	−82.0[i]	−64.4 ± 0.9	CM liq.	198
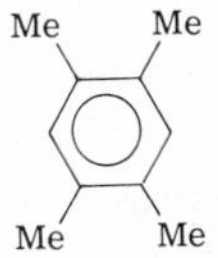	[−78.6][f]		−146.5	−101.2[i]	−67.9 ± 2.0	CM liq.	199
	−77.3	−82.4	−143.9	−101.2	−64.0	Crystal	181
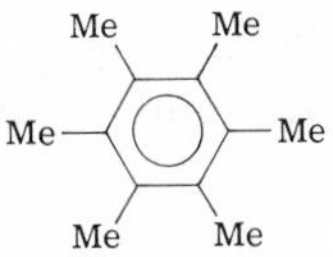	−98.6	−98.6	−170.2	−122.5[i]	−71.6 ± 3.0	CM liq.	198
	−101.1	−102.7	−163.8	−122.5	−61.9	Crystal	181
	[−27.9][f]		−89.3	[48.4 ± 0.1][j]	−61.4 ± 1.8	CM liq.	198
	−30.4	−28.3	−86.8		−57.4 ± 0.7	MS	257

[a] Units 10^{-6} erg gauss^{-2} mole^{-1}.
[b] $\chi_{av} \equiv \frac{1}{3}(\chi_1+\chi_2+\chi_3)$; $\Delta\chi \equiv \chi_3 - \frac{1}{2}(\chi_1+\chi_2)$.
χ_{av} and $\Delta\chi$ are obtained from quoted reference on rhs of table unless indicated otherwise.
[c] MS—microwave spectroscopy, CM—Cotton–Mouton effect, Crystal—Crystallographic method, see text.
[d] Bothner-By and Pople.[181]
[e] Pascal.[258]
[f] Average of χ_1 and χ_2.
[g] Estimated from χ of similar compounds.
[h] Foex.[67]
[i] Le Fevre and Murthy.[199]
[j] Bitter,[238]

[254] R. L. Shoemaker and W. H. Flygare, *J. Chem. Phys.* **51**, 2988 (1969).
[255] M. P. Bogaard, A. D. Buckingham, M. G. Corfield, D. A. Dunmur, and A. H. White, *Chem. Phys. Lett.* **12**, 558 (1972).
[256] W. Huttner and W. H. Flygare, *J. Chem. Phys.* **50**, 2863 (1969).
[257] J. H. S. Wang and W. H. Flygare, *J. Chem. Phys.* **52**, 5636 (1970).
[258] P. Pascal, *C. R. Acad. Sci.* **152**, 1010 (1911).

V. Correlations and Comparisons

Much theoretical and experimental data of the shielding and susceptibility anisotropies have been presented. Included along with the anisotropies have been principal tensor components which were deduced directly from the anisotropic and isotropic quantities or which were obtained directly from the experimental measurements. The theoretical calculations reviewed were *ab initio* type and included molecules containing up to second period atoms, e.g. H, Li, B, C, N, O, F. The discussion of experimental shielding anisotropies was concerned with the data of all available nuclei with the exception of protons, for reasons given elsewhere. These nuclei included C, F, P, N, O, Cl, Br, I, Li, K, Rb, Cs, V, Mo, Pb, S, and Si. The collection of susceptibilities was less complete and concentrated upon molecules of particular or general interest due to symmetry features, similarities to other molecules, or relationships to corresponding shielding anisotropies.

These two magnetic phenomena, shielding and susceptibility, have been studied together for several reasons. In many respects there are great similarities in the mathematical formulations, physical explanations, and the observed trends of these quantities. There are, in addition, several expressions which postulate direct relations between them. In the following sections some of the relationships between shielding and susceptibility will be investigated in a little more detail. To simplify comparison of expressions, atomic units ($\hbar = e = m = 1$) are used in this section.

A. Theoretical Predictions for Shielding and Susceptibility

1. *Perturbed Hartree–Fock Calculations of σ and χ*

The theoretical expressions for σ and χ, Eqs. (2.10) and (2.11) respectively, are directly comparable, the only differences being the appearance of an r^3 factor in the denominators of the operators for shielding. Consequently the major theoretical approaches have been equally applicable to both. However, as shown in Section II, the predictions of *ab initio* models are considerably more accurate for shielding than for susceptibility. For each of these quantities the diamagnetic (ground state) parts of the total quantity are comparatively easy to determine accurately, with the major uncertainty lying in the paramagnetic (high frequency) contribution.

The ability of an *ab initio* method of molecular calculations to reproduce experimental quantities depends primarily upon the type of operators involved and the quality of the wavefunctions. For very accurate wavefunctions, the perturbed Hartree–Fock theory has been able to provide accurate results for both shielding and susceptibility. For moderately extensive basis sets the theory is still sometimes able to give reasonably accurate predictions of σ_{iso}, but it is considerably in error for χ. This difference is formally attributable to the difference in form of the operators required in the integrals. The effect of the $1/r^3$ term in the shielding expressions is to increase the dependence of the paramagnetic and diamagnetic terms on the electron density near the nucleus.[259] The wavefunction is more concentrated and perhaps more accurate in this region since the total wavefunction is comprised of atom-centered LCAO-MOs.

It should be kept in mind that the angular momentum integrals required for the paramagnetic terms are those between the unperturbed ground state and the perturbed first-order wavefunction, which is often expanded in terms of the unperturbed states. The problem can be reformulated as that of how well the coupled excited states reproduce the true perturbed wavefunction, and in particular, following the previous statement, how well it does so in the region of the nucleus.

It can probably be assumed that minimal basis set wavefunctions provide in general a poor overall representation of the first-order wavefunction. For many shielding calculations, it is conceivable that the first few excited states do give a reasonably accurate description of the perturbed electron distribution near the nucleus. However this model breaks down for molecules having highly polar bonds.

The computation of the paramagnetic susceptibility differs from that of shielding in two major respects: (1) the absence of r^3 in the denominator, which results in a comparative delocalization of the contributions over the entire molecule; and (2) the use of the center of mass as the origin for the evaluation of the angular momentum integrals.

As a result of the combined effect of these two features, the susceptibility reflects to a much greater extent the overall electron distribution of the molecule. Thus the magnetic susceptibility contains contributions from distant parts of the molecule, particularly from electrons at distant nuclei. These contributions may be subject to large errors if a procedure is used which does not give a good overall representation of the first-order wavefunction.

The shielding results obtained using the standard perturbed Hartree–Fock theory, though considerably superior to those involving susceptibility, were nevertheless inconsistent and only reliable within certain limitations. For molecules with several fluorine or oxygen substituents the agreement with experiment was poor for ^{13}C or ^{19}F.

[259] See Ref. 32 for the discussion of this point for H_2CO.

2. *Theoretical versus Experimental σ's Using GIAO*

The use of GIAO within the perturbed Hartree–Fock framework was shown in Section II,C to be extremely reliable for predicting isotropic shifts for many molecules, particularly for ^{13}C shielding.

Its reliability for predicting individual shielding components and anisotropies will now be examined by comparing these quantities with experiment (see Table XXI).

The agreement is extremely good with a few exceptions. The theoretical value of N_2 was computed using a large basis set but not with GIAO. The error for the ^{13}C shielding tensor in CH_3F is so far unaccountable. For most of the results presented, the most reliable experimental quantity is $\Delta\sigma$. The approximately 15 ppm errors in $\Delta\sigma$ of ^{13}C in C_2H_2 and HCN and of ^{14}N in HCN are within experimental error and for F_2 the error of 57 ppm is only 6% of the total $\Delta\sigma$ of about 1000 ppm.

TABLE XXI

GIAO: THEORETICAL[a] VERSUS EXPERIMENTAL[b] SHIELDING

	$\sigma_{\parallel}$	$\sigma_{\perp}$	$\Delta\sigma$	σ_{iso}	
C*O[c]	271.2	−118.4	389.5	11.5	
	268	−116	384	12±10	Expt.
HC*N	278.4	−17.3	295.7	81.3	
	264	−18	282	76±10	Expt.
$C_2^*H_2$	279.2	54.8	224.4	129.6	
	283	38	245	120±10	Expt.
C^*H_3F	196.5	112.4	84.1	140.4	
	167	99	68	122±15	Expt.
C^*H_3OH	201.2	138.3	62.9	159.3	
	190	124	66	146±10	Expt.
HF*	482.8	376.2	106.6	411.7	
	480	376	104	410±20	Expt.
F_2	488.4	−504.7	993.1	−173.7	
	490	−560	1050	−210±25	Expt.
CH_3F^*	438.0	509.5	−71.5	485.7	
	425	489	−64	470±20	Expt.
N_2[c]	338.4	−198.9	537.3	−19.8	
	338	−319	657	−101±20	Expt.
HCN*	338.9	−230.3	569.2	−40.6	
	348	−229	577	−37±20	Expt.

[a] All theoretical values except for CO and N_2 are from R. Ditchfield (personal communication) (1972).

[b] Experimental values taken from Tables VIII, X, and XV.

[c] Theoretical results from Stevens *et al.*; see Tables VIII, XV.

B. Comparison of Experimental Shielding and Susceptibility Anisotropies

1. *Basis Trends of $\Delta\sigma$ and $\Delta\chi$*

One of the most critical differences between shielding and susceptibility, evident by considering the methods used to measure them, is that the susceptibility is a bulk property and the shielding is not. The total magnetic moment (susceptibility) of a bulk sample is a sum of the individual moments. On the other hand the magnetic field induced at a nuclear site (i.e. the shielding) in a bulk sample is the same (neglecting medium effects) as that of an isolated molecule.

It has been noted previously that the magnetic susceptibility is a property which can very conveniently be subdivided into atomic or bond contributions, whereas this procedure has been successful for shielding only in a few cases of limited application. There is very little correlation between the nuclear shielding of a molecule and its size. This fact is made even more obvious when one considers that there is often a very large change in the shielding of two similar or different nuclei within a given molecule, e.g. the two carbons in CH_3CN. The magnetic susceptibility is a property of the molecule as a whole. It has significant contributions from the electrons of each of the atoms. The shielding is a property of the various points in the molecules on which nuclei are situated, and it usually contains predominant contributions from local terms.

Examination of the diamagnetic shielding and susceptibility terms shows this difference clearly.

$$\sigma^{\mathrm{d}}_{\mathrm{avg}} = (1/3c^2)\langle 1/r \rangle, \tag{5.1}$$

$$\chi^{\mathrm{d}}_{\mathrm{avg}} = (1/4c^2)\langle r^2 \rangle. \tag{5.2}$$

The average value of $1/r$ is a quantity which normally decreases whereas $\langle r^2 \rangle_{\mathrm{avg}}$ increases at large distances. It should be kept in mind, though, that the paramagnetic susceptibility terms are of a form such that they oppose the diamagnetic terms, thereby preventing the total susceptibility from receiving its largest contributions from the electrons of the atoms most distant from the center of mass. The cancellation of the effect of the distant nuclear charges by the closely bound electrons is a factor in both the shielding and the susceptibility. For the former these pairs of cancelling terms are not nearly so large, a situation which may partially explain in a physical sense why the paramagnetic shielding is more readily calculable than the paramagnetic susceptibility. The paramagnetic quantities are examined further in the next section.

2. *Paramagnetic Susceptibility and Shielding*

The cancellation between large paramagnetic and slightly larger in magnitude diamagnetic susceptibility contributions is responsible for the observed regularity in trends of the total molar (or molecular) susceptibility. Since the total susceptibility and the diamagnetic susceptibility obey these additivity rules it follows that the paramagnetic susceptibility must do likewise, although this is not at all evident from the expression for χ^{p} [Eq. (2.11)]. One possible explanation, put forth by Weltner,[209] is that the summation of products of angular momenta integral terms [see Eq. (5.3), below] can be divided into bond contributions with products between bonds being negligible.

An important relationship used previously (Section IV,A,3) exists between the infinite summation of the paramagnetic susceptibility and the experimental parameters of molecular g-values and molecular geometry [Eq. (5.3)]:

$$g_{xx} = (M_{\mathrm{p}}/cI_x) \sum_k Z_k(y_k^2 + z_k^2) - (2M_{\mathrm{p}}/I_x) \sum_p |\langle 0|L_x|p\rangle|^2/(E_p - E_0)$$
$$= g'_{xx} + g''_{xx}, \tag{5.3}$$

where g'_{xx} and g''_{xx} are the nuclear and electronic contributions, respectively. Weltner[209] has discussed the rigid charge approximation[214] in which the electrons are treated as fixed rigidly in the bonds so that

$$g''_{xx} I_x = M_{\mathrm{p}} \sum_i^{\text{bond el.}} (y_i^2 + z_i^2), \tag{5.4}$$

where y_i and z_i are the distances from the bond centers to the x inertial axis. He has calculated g-values for several molecules with fair results. He also points out that for large molecules g'_{xx} is of the order of the charge-to-mass ratio (0.5) and that g''_{xx} is approximately equal to -0.5, thus causing g_{xx} to tend towards zero. From their use of the expression

$$\chi^{\mathrm{p}}_{xx} = (1/4c^2) \sum_k Z_k(y_k^2 + z_k^2) - (1/4c^2 M_p)\, I_x\, g_{xx}. \tag{5.5}$$

Flygare and Lo[260] have noticed that for most molecules of four or more atoms the first term (χ^{NUC}) is dominant.

This tendency is examined more closely in Table XXII, in which contributions to χ^{p}_{xx} and χ_{xx} and the g-values are presented. For comparison, Table XXIII gives the constituents of the paramagnetic shielding in terms of experimental quantities, which are given by a similar expression

[260] W. H. Flygare and M.-K. Lo, *J. Chem. Phys.* **48**, 1196 (1968).

involving the spin-rotation constants:

$$\sigma_{xx}^{\mathrm{p}} = -(1/2c^2) \sum_k Z_k[(y_k^2 + z_k^2)/r_k^3] + \tfrac{1}{2}(M_{\mathrm{p}}/g_{\mathrm{N}})(M_{xx}/G_x)$$

$$= \sigma_{xx}^{\mathrm{NUC}} + \sigma_{xx}^{\mathrm{SR}}, \qquad (5.6)$$

where g_{N}, G_x, and M_{xx} are the nuclear g-value, the rotational constant, and the spin-rotational constant, respectively. The two terms are referred to as the nuclear and spin-rotation contributions to the paramagnetic shielding.

From Table XXII it is seen that with the exception of the CO bond direction in H_2CO the nuclear term is always the larger term, and for most of these cases it is greater than 80 or 90% of the total paramagnetic susceptibility. Of course, when the paramagnetic susceptibility is added to the diamagnetic susceptibility the χ^{g} term is seen to be a much more significant percentage of the total susceptibility component. The expression for χ (total), using the rigid charge model for g_{xx}'' and using Eqs. (5.3), (5.4), and (2.11) is:

$$\chi_{xx} = -(1/4c^2)(\langle y^2\rangle + \langle z^2\rangle) + (1/4c^2) \sum_i^{\text{bond el.}} (y_i^2 + z_i^2) \qquad (5.7)$$

The attempt here is merely to indicate in a general way why the Pascal scheme and its modifications are workable for magnetic susceptibility. Note from this expression the explicit nature of the cancellation between diamagnetic and paramagnetic terms.

For shielding it is seen from Table XXIII that there is no regular trend in the relative magnitudes of the terms. To neglect the $\sigma_{xx}^{\mathrm{SR}}$ term would be completely unjustified even as a very rough approximation. An important feature of the nuclear term of $\sigma_{xx}^{\mathrm{NUC}}$ is that its largest contribution is from the nearest neighbor(s) of the nucleus of interest and is directly proportional to the nuclear charge. Thus in the case of two bonded nuclei the nucleus with the smaller nuclear charge and hence smaller overall shielding range will exhibit a larger σ^{NUC} term. In these situations the σ^{SR} term must compensate for the σ^{NUC} term. An example is PN.

3. *Linear and Symmetric Top Molecules*

It has been noted previously that the shielding along the molecular axis of a linear molecule or along the C_3 axis of a methyl carbon is a constant among many different molecules. Further analysis reveals that if one divides the parallel shielding $\sigma_{\|}$ into a diamagnetic and paramagnetic component, the two terms $\sigma_{\|}^{\mathrm{d}}$ and $\sigma_{\|}^{\mathrm{p}}$ remain individually constant. This is true provided the nucleus or any other point on the symmetry axis is used as origin for the vector potential. For a linear molecule the situation reduces

TABLE XXII

NUCLEAR AND ELECTRONIC CONTRIBUTIONS TO PARAMAGNETIC SUSCEPTIBILITIES[a]

Approx. geometry	$\chi_{aa}^{\mathrm{NUC}\,b}$ χ_{bb}^{NUC} $(\chi_{cc}^{\mathrm{NUC}})$	$\chi_{aa}^{\mathrm{g}\ b}$ χ_{bb}^{g} (χ_{cc}^{g})	$\chi_{aa}^{\mathrm{p}\ b}$ χ_{bb}^{p} (χ_{cc}^{p})	$\chi_{aa}^{\mathrm{d}\ c}$ χ_{bb}^{d} (χ_{cc}^{d})	$\chi_{aa}^{\ d,e}$ χ_{bb} (χ_{cc})	Ref.
F_2 $\xrightarrow{a}$	0	0	0	−17.0	−17.0	114
	38.3	9.8	48.1	−54.1	−6.0	
CO	0	0	0	−18.1	−18.1	[19, 20]
	18.5	9.9	28.4	−38.4	−10.2	
OCS	0	0	0	−38.6	−38.6	[233, 100]
	176	10	186.2	−215.5	−29.3	
NNO	0	0	0	−25.6	−25.6	235
	84.6	15.2	99.8	−115.3	−15.4	
$H_3C{-}F$ $\xrightarrow{a}$	7.1	2.6	9.7	−33.1	−23.3	100
	44.7	9.6	57.3	−71.9	−15.1	
$H_3C{-}Br$	7.0	2.8	9.8	−58.3	−48.5	242
	130.0	8.2	138.2	−178.2	−40.0	
$H_3C{-}CN$	7.0	2.6	9.6	−44.2	−34.6	243
	130.6	14.7	145.3	−169.4	−24.1	
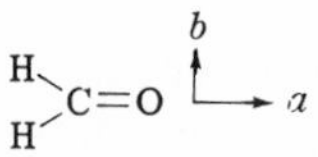	7	23	30	−28.0	1.6	216
	32	14	46	−54.2	−8.2	
	39	9	48	−61.5	−14.0	
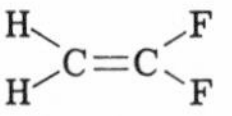	96	8	104.4	−133.8	−29.3	[250, 260]
	118	9	127.4	−153.5	−26.0	
	214	5	218.8	−249.2	−30.4	
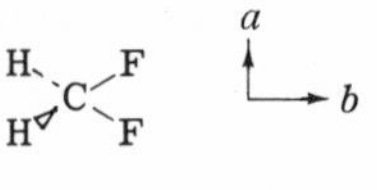	32	2	33.9	−59.2	−25.3	[250, 260]
	100	7	106.9	−130.7	−23.7	
	117	8	115	−147.7	−23.0	
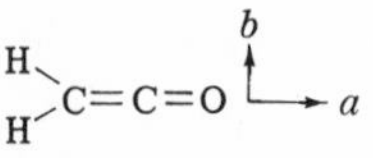	8	3	11	−37	−25.9	[252, 260]
	117	6	123	−148	−24.5	
	125	6	131	−154	−22.8	
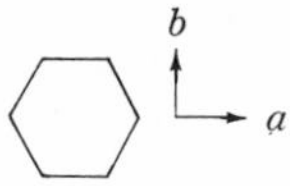	227	24	251	−286	−34.9	[254, 100]
	227	24	251	−286	−34.9	
	454	−39	413	−508	−94.6	
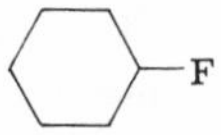	227.5	25.2	252.7	−293.3	−40.6	256
	439.7	32.9	472.6	−509.7	−37.1	
	667.2	−32.0	635.2	−732.4	−97.2	
	214	27	241.5	−271.9	−30.4	257
	210	37	247.4	−275.7	−28.3	
	423	−31	393.9	−480.6	−86.8	

TABLE XXIII

NUCLEAR AND ELECTRONIC CONTRIBUTIONS TO PARAMAGNETIC SHIELDING[a]

	aa (direction)	$\sigma_{aa}^{\mathrm{NUC}}$	$\sigma_{aa}^{\mathrm{SR}}$	σ_{aa}^{P}	Ref.
$^{13}\underline{C}O$	⊥	−100	−385	−485	19
$P^{14}\underline{N}$	⊥	−142	−1015	−1157	155
$\underline{P}N$	⊥	−66	−1363	−1429	
$^{35}Cl\underline{F}$	⊥	−147	256	109	122
$HC^{14}\underline{N}$	⊥	−80	−513	−593	165
$^{17}\underline{O}C^{32}S$	⊥	−156	−824	−980	165
$^{16}OC^{33}\underline{S}$	⊥	−96	−714	−810	
$^{14}\underline{N}H_3$	⊥	−24	−52	−76	164
	∥	−36	−81	−117	
$^{31}\underline{P}H_3$	⊥	−20	−350	−370	152
	∥	−21	−400	−421	
$CH_3\underline{F}$	⊥	−79.0	27.3	−51.7	123
	∥	−5.5	−58.0	−63.5	
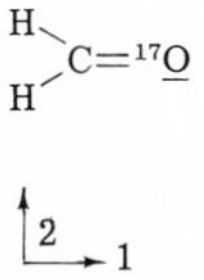	1	−3	−1594	−1597	53
	2	−81	−797	−878	
	3	−84	71	−13	
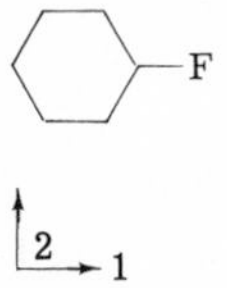	1	−33	−31	−64	98
	2	−189	−184	−373	
	3	−222	−193	−415	

[a] Units ppm; quantities σ^{NUC} and σ^{SR} calculated from Eq. (5.6), see text.

Footnotes to TABLE XXII.

[a] Units 10^{-6} erg gauss^{-2} mole^{-1} or 10^{-6} cm^3 mole^{-1}.
[b] Derived from Eq. (5.5).
[c] See original references given on right-hand side of table.
[d] See Tables XVII–XX.
[e] Since χ^p, χ^d, and χ were evaluated by different methods in some cases, the sum of χ^p and χ^d does not always equal χ.

to the trivial case, $\sigma_{\parallel}^{p} = 0$, $\sigma_{\parallel}^{d} = \sigma_{\parallel}$. In the symmetric top molecules, CH_4, C_2H_6, CH_3F, CH_3OH, CH_3CN, and $CH_3C{\equiv}CH$, we find that $\sigma_{\parallel}^{d} \sim 300$ ppm and $\sigma_{\parallel}^{p} \sim -100$ ppm.

In Section IV it was mentioned that $g_{\parallel}$ is approximately constant for methyl systems. From Eq. (5.5) it can readily be seen that $\chi_{\parallel}^{p}$ will remain constant since $g_{\parallel} \propto I_{\parallel}(=\sum M_k(x_k^2 + y_k^2))$, and $\sum Z_k(x_k^2 + y_k^2)$ will all remain virtually unchanged for systems on which only the three protons are off the symmetry axis. However the $\chi_{\parallel}^{d}$ term, in contrast to the $\sigma_{\parallel}^{d}$ term, changes significantly from one linear molecule to another and from one C_3 system to another.

These data are presented in Table XXIV which shows the g-value, χ's, and $\sigma_{\parallel}$'s for a series of CH_3 molecules. It is interesting that the so-called "distortion terms" $\chi_{\parallel}^{p}$ are quite regular for the series, and that the variations in the quantity $\chi_{\parallel}$ are due to $\chi_{\parallel}^{d}$. The physical explanation of this phenomenon is that the diamagnetic currents (represented by $\chi_{\parallel}^{d}$) about the symmetry axis are influenced strongly by the substituent, e.g. Br, whereas

TABLE XXIV

MOLECULAR g-VALUES, MAGNETIC SUSCEPTIBILITIES, AND PARALLEL SHIELDING FOR SOME CH_3 SYSTEMS[a]

	$\sigma_{\parallel}(^{13}C)$[b]	$g_{\parallel}$	$\chi_{\parallel}^{p}$	$\chi_{\parallel}^{d}$	$\chi_{\parallel}$	$g_{\perp}$	$\chi_{\perp}^{p}$	$\chi_{\perp}^{d}$	$\chi_{\perp}$
CH_4	196	0.3133[c]	9.3[d]	−28.7[e]	−19.4	0.3133[c]	9.3[d]	−28.7[e]	−19.4
CH_3F[f]	167	0.310[c]	9.7	−33.1	−23.3	−0.0612[c]	57.3	−71.9	−15.1
CH_3Cl[g]	189	0.305	9.61	−46.9	−37.3	−0.0161	100.5	−129.8	−29.4
CH_3Br[g]	185	0.294	9.83	−58.3	−48.5	−0.0056	138.2	−178.2	−40.0
CH_3I[g]	169	0.310	9.65	−74.2	−64.5	−0.00655	177.6	−231.2	−53.5
CH_3CN[h]	196	0.310	9.6	−44.2	−34.6	−0.0338	145.3	−169.4	−24.1
CH_3NC[h]	—	0.310	9.6	−46.2	−36.6	−0.0546	137.8	−160.9	−23.1
CH_3CCH[i]	—	0.312	8.6	−46.1	−37.5[k]	0.00350	158.4	−187.9	−29.5[k]
C_2H_6	194[j]	—	27.2	−56.4[l]	−29.2[k]	—	81.5	−107.1[l]	−25.6[k]

[a] Units σ—ppm, g—dimensionless, χ—10^{-6} erg gauss^{-2} mole^{-1}.
[b] See Table VIII.
[c] Flygare and Benson.[212]
[d] Deduced from Equation (5.5); also see footnote *e*.
[e] Arrighini *et al.*[29]
[f] See Table XXII.
[g] VanderHart and Flygare.[242]
[h] Pochan *et al.*[243]
[i] Shoemaker and Flygare.[244]
[j] See Table VI.
[k] See Table XVIII.
[l] Appleman.[34]

TABLE XXV

COMPARISON OF $\Delta\sigma$ AND $\Delta\chi$ FOR LINEAR AND SYMMETRIC TOP MOLECULES[a]

Linear molecules	$\Delta\sigma$	$\Delta\chi$	Sym top molecules	$\Delta\sigma$	$\Delta\chi$
C*O	384	−8.2	C^*H_3F	68	−8.2
CO*	639		CH_3F^*	−63	
C^*O_2	420^b	−6.2	C^*H_3Cl	28	−8.0
OC*S	372		C^*H_3Br	0	−8.5
O*CS	861	−9.3	C^*H_3I	−75	−11.0
OCS*	729				
			C^*H_3CN	5	−10.5
C^*S_2	430	−14.0	CH_3C^*N	311	
HC*N	282	−7.2	CH_3N^*C	360	−13.5
HCN*	577				
HC*CH	245	−7.7	CHF^*_3	-80^c	1.2
F_2	1050	−11.0	N^*H_3	−39	0.37
HF*	102	0.3	NF^*_3	130^c	3.0
ClF*	−214	−15.9	P^*H_3	−50	+2.7
Cl*F	2389				
N_2	657	−8.5			
ClCN*	985	−10.8			
Cl*CN	859				

[a] See Tables VIII–XVI ($\Delta\sigma$) and XVII, XVIII ($\Delta\chi$).
[b] Estimated. See Spiess.[94]
[c] Shielding anisotropy given with respect to C_3 axis.

the distortion of the cylindrical symmetry is effected only by the proton group, which is a structural feature common to each of these systems.

The interpretation of the shielding in terms of the qualitative effects of the electronic currents is similar. However, for shielding, because of the $1/r^3$ term in $\sigma_{\parallel}^{d}$ the largest contributions are from electrons closest to the nucleus. There are small fluctuations in $\sigma_{\parallel}$ in the halogen series with closely bound and possibly partially doubly bound F and the more distant and diffuse I showing the largest deviations. For the other second period substituents, C, N, O, the deviations are smaller.

For linear molecules it is expected and observed that the molecular axis, which has no paramagnetic distortion of its electronic cylindrical symmetry, will be the axis of the largest shielding and susceptibility magnitudes. Consequently these molecules should display relatively large shielding and susceptibility anisotropies. The correspondence between $\Delta\sigma$ and $\Delta\chi$ values for these and symmetric top molecules is shown in Table XXV.

The indication is that the correlation between $\Delta\sigma$ and $\Delta\chi$ is not regular even among molecules of similar structure such as CO_2, OCS, and CS_2. When these quantities are compared for molecules of C_3 symmetry, the correlation between $\Delta\sigma$ and $\Delta\chi$ breaks down entirely. Thus among the methyl halides the ^{13}C shielding anisotropy changes regularly from +68 for $^{13}CH_3F$ to −75 ppm for $^{13}CH_3I$, while the susceptibility anisotropy is practically constant for CH_3F, CH_3Cl, and CH_3Br, and jumps slightly for CH_3I.

The comparison of the shielding anisotropy at different nuclei in the same molecule is also interesting. For most of the linear molecules these anisotropies are fairly comparable, reflecting the size of the nuclear charge, although OCS is an exception since the ^{17}O shielding anisotropy is larger than the ^{33}S anisotropy. The molecule ClF is strongly anomalous with the F shielding anisotropy being negative due to the strong interactions with the Cl. CH_3F is also interesting since the fluorine shielding anisotropy is opposite in sign to that of the carbon shielding, and also to that predicted by semiempirical theory.[135] A very striking example of the localized nature of the shielding is found in CH_3CN, in which the two C atoms have anisotropies of 5 and 311 ppm corresponding to the methyl and cyano carbons respectively. The susceptibility anisotropy shows no unusual features.

4. *Some Molecules of Very Slight Anisotropy*

There is a correspondence between $\Delta\sigma$ and $\Delta\chi$ also in the molecules with small anisotropies in the distribution of electron density. The isoelectronic series of CH_4, NH_3, H_2O, and HF presents in each case a central second period atom surrounded by four pairs of electrons, either as lone pairs or in bonds to hydrogen. The data in Table XXVI show that each of these molecules has relatively small shielding and susceptibility anisotropies. Also included here are the average shielding and susceptibility of the Ne atom which is isoelectronic with the molecules.

An important difference between the shielding and susceptibility of these systems is that the isotropic shielding decreases from Ne to CH_4 while the susceptibility increases. This observation is a further indication of the fact that the susceptibility is a function more of the charge displaced from the nucleus or center of mass while the shielding depends most strongly on the electron density nearest the nucleus. Also of significance is the fact that for the susceptibility the paramagnetic terms increase in magnitude along with the diamagnetic terms, though more slowly, to produce a net increase in the magnitude of χ_{av} from Ne to CH_4. The paramagnetic shielding, on the other hand, increases in magnitude while the diamagnetic shielding decreases from Ne to CH_4, thus strengthening the overall trend of decreased nuclear shielding.

TABLE XXVI

SHIELDING AND SUSCEPTIBILITY OF AN ISOELECTRONIC SERIES

	$\chi_{\parallel}$ $\chi_{\perp}$ χ_{av}	$\chi^{d}_{\parallel}$ $\chi^{d}_{\perp}$ χ^{d}_{av}	$\chi^{p}_{\parallel}$ $\chi^{p}_{\perp}$ χ^{p}_{av}	$\Delta\chi$	$\sigma_{\parallel}$ $\sigma_{\perp}$ σ_{iso}	$\sigma^{d}_{\parallel}$ $\sigma^{d}_{\perp}$ σ^{d}_{iso}	$\sigma^{p}_{\parallel}$ $\sigma^{p}_{\perp}$ σ^{p}_{iso}	$\Delta\sigma$
Ne	−7.46[a]	−7.46	0	0	552.3[b]	552.3	0	0
	−10.1	−10.1	0		480	480.0	0	
HF*[c]	−10.4	−11.3	0.9	0.3	376	482.2	−106	104
	−10.3	−10.9	0.6		411	481.5	−71	
	−13.2[e]	−16.3[f]	3.1		309[g]	415[h]	−106	
H_2O*[d]	(−13.05)	(−15.1)	(2.0)	−0.1	(338)	(415)	−(77)	−29
	−13.1	−15.5	2.4		328	415	−87	
	−16.1[i]	−21.9[f]	5.8		240[j]	351.3[h]	−111	
N*H_3	−16.4	−19.7	3.3	0.4	279	350.3	−71	−39
	−16.3	−20.5	4.2		266	350.6	−85	
C*H_4	−19.4[k]	−28.7[f]	9.3	0	196[k]	296[h]	−100	0

[a] Malli and Froese.[261]
[b] Malli and Froese.[96]
[c] See Tables X and XVII and Stevens and Lipscomb.[231]
[d] Out-of-plane axis is parallel; average of in-plane quantities is perpendicular.
[e] See footnote.[266]
[f] Arrighini *et al.*[29]
[g] Theoretical value, see Table VI.
[h] Arrighini *et al.*[30]
[i] See Table XVIII.
[j] See Table XV.
[k] See Table XXIV.

5. *Relationship Between* $\Delta\chi$ *and* σ *for* ^{13}C *and* ^{19}F *in Benzenes*

The effect of neighboring group anisotropies is normally considered most significant for proton shielding (see Section V,C) whereas our discussion of shielding tensors has excluded those for protons. The magnetic anisotropy effect is typically of the order of a few parts per million, which is significant for proton ranges but smaller than the range of shifts of carbon, fluorine, or other heavy nuclei, and smaller also than the usual uncertainty of the absolute chemical shifts. However there has been some work done in the correlation of relative ^{13}C shifts with magnetic anisotropy in certain systems for which the other factors are assumed invariant.

Spiesecke and Schneider[262] have plotted the observed ^{13}C shifts of CH_3X, CH_3CH_2X, and C_6H_5X versus X-substituent electronegativity and originally[263] attributed the deviation from linearity to magnetic anisotropy

[261] G. Malli and C. Froese, *Int. J. Quantum. Chem.* **1**, Suppl., 99 (1967).
[262] H. Spiesecke and W. G. Schneider, *J. Chem. Phys.* **35**, 73 (1961).
[263] See Ref. 1, p. 997.

effects. Using this procedure he obtained unreasonably large anisotropic effects upon the isotropic shielding, i.e. 30–50 ppm.

One would expect the largest magnetic anisotropy effects for carbon shielding to occur in aromatic systems, particularly for the ring carbons. It has been pointed out that unlike the situation for protons, the isotropic ^{13}C shifts in alkenes are very similar to those in aromatic compounds.[264] Note that for the ring carbons the susceptibility anisotropy changes greatly (see Table XXVII)—(-59.7×10^{-6} cgs in C_6H_6 to -71.6×10^{-6} in $C_6(CH_3)_6$ to -31.9×10^{-6} in C_6F_6—while σ_{iso} remains approximately constant within a total range of 10 ppm. However the behavior of $\sigma_{\parallel}$ and $\sigma_{\perp}$ between C_6H_6 and C_6F_6 is far from constant, a fact reflected also in the shielding anisotropy. The large decrease in $\sigma_{\parallel}$ in C_6F_6 can possibly be explained as an inductive effect, drawing away the ring current density. From the behavior of $\sigma_{\parallel}$ of the ring carbons and of both $\sigma_{\perp}$ and $\sigma_{\parallel}$ of the methyl carbon in CH_3-substituted benzenes, it appears that none of these quantities depends on $\Delta\chi$.

For fluorobenzenes the trends in both $\sigma_{\parallel}$ and $\sigma_{\perp}$ are themselves fairly irregular and not correlated with $\Delta\chi$. For C_6F_6 the shielding anisotropy is largest, 156 ppm, while the magnitude of susceptibility anisotropy is smallest among the fluorobenzenes shown here. Nehring and Saupe[140] have

TABLE XXVII

COMPARISON OF ^{13}C AND ^{19}F SHIELDING AND SUSCEPTIBILITY FOR SOME AROMATIC MOLECULES[a]

	Nucleus	$\sigma_{\parallel}{}^b$	$\langle\sigma_{\perp}\rangle^b$	$\Delta\sigma$	σ_{iso}	$\chi_{\parallel}{}^b$	$\langle\chi_{\perp}\rangle^b$	$\Delta\chi$
C_6H_6	C	188	5	183	66	−94.6	−34.9	−59.7
Toluene[c]	Ring C	190	8	182	69	−105.7	−45.5	−60.2
$C_6(CH_3)_6$	Ring C	173	5	168	61	−170.2	−98.6	−71.6
C_6F_6	C	85	46	39	59	−100.5	−68.6	−31.9
Toluene	CH_3 C	188	166	22	173	−105.7	−45.5	−60.2
$C_6(CH_3)_6$	CH_3 C	173	173	0	173	−170.2	−98.6	−71.6
C_6H_5F	F	370	268	102	302	−97.2	−38.9	−58.3
1,4 $C_6H_4F_2$	F	325	301	24	309	−96.2	−45.7	−50.5
sym-$C_6H_3F_3$	F	367	263	104	297	−92.8	−53.5	−39.3
C_6F_6	F	456	300	156	352	−100.5	−68.6	−31.9

[a] Units σ—ppm, χ—10^{-6} cgs. See Tables VIII, IX (C-13, σ), XII (F-19, σ), and XX (χ) for data.

[b] Parallel direction is parallel to axis of ring plane; perpendicular directions are in ring plane, except for shielding of $^{13}CH_3$ carbons for which parallel is along C_3 axis of methyl group.

[c] Average of ring carbons.

[264] See Ref. 1, p. 999.

pointed out that there is a significant difference between the effects of *ortho*- and *para*-, as opposed to *meta*-, substitution of fluorine; however this trend seems also not to correlate with that of the $\Delta\chi$'s.

C. Neighbor and Bond Anisotropies

1. *Neighbor Anisotropy Effect in Conjugated Rings*

Many of the semiempirical models of shielding are based upon a division of the total isotropic shielding into various types of contributions,[265] particularly local and nonlocal effects. One of the most important nonlocal effects is the neighbor anisotropy effect in which the local electronic currents of nearby atoms or in bonds are replaced by point magnetic dipoles[266] which give rise to anisotropic secondary fields at the nucleus in question. Both classical[9,267,268] and quantum mechanical[8,269,270] models for this quantity were developed in the late fifties with the principal emphasis placed upon the correlation and calculation of proton shifts in aromatic systems. A comparison of these two types of approach is given by Mallion.[271] Other reviews available are those of Ditchfield,[17] Salem,[207] Emsley *et al.*,[1] Jackman,[208] Jones,[272] and Bovey.[273]

The results obtained using these methods have been in good overall qualitative agreement with experiment. However there are some obvious deficiencies in these original models, which have recently been considered in more detail. A major shortcoming of the ring-current model for determining NMR shielding was the tendency to neglect other effects. Thus in order to get some qualitative agreement between the observed and calculated ring-current contributions to proton shielding, it was necessary to use a scale factor relative to benzene.[271] Dailey[274] has shown that the contributions to shielding of local electrons can account for the discrepancy of 40% indicating that the effects other than anisotropy are not negligible. Mallion[271] has discussed this matter and reminds the reader that the observed agreement between the classical model and experiment is strongly dependent on different parameterizations.

[265] A. Saika and C. P. Slichter, *J. Chem. Phys.* **22**, 26 (1954).
[266] See Ref. 2, Sects. 7.4 and 7.5.
[267] J. S. Waugh and R. W. Fessenden, *J. Amer. Chem. Soc.* **79**, 846 (1957).
[268] C. E. Johnson and F. A. Bovey, *J. Chem. Phys.* **29**, 1012 (1958).
[269] J. A. Pople, *Mol. Phys.* **1**, 175 (1958).
[270] R. McWeeny, *Mol. Phys.* **1**, 311 (1958).
[271] R. B. Mallion, *J. Chem. Soc. B* p. 681 (1971).
[272] A. J. Jones, *Rev. Pure Appl. Chem.* **18**, 253 (1968).
[273] F. A. Bovey, "Nuclear Magnetic Resonance Spectroscopy." Academic Press, New York, 1969.
[274] B. P. Dailey, *J. Chem. Phys.* **41**, 2304 (1964).

Another criticism of the usage of the neighbor anisotropy model is the following: McConnell's[8] point-dipole model is valid only for large distances, usually of the order of several bond lengths.[10] However, the nuclei for which the point-dipole assumption is most valid are usually those for which the neighbor-anisotropy effect is small, and for which other effects become more significant.

In spite of these theoretical shortcomings the ring-current or magnetic anisotropy effect has been reasonably effective in predicting and identifying relative proton chemical shifts in many large ring systems with relatively few empirical parameters. It appears likely that the local effects such as those of C–H and C–C bonds and the ones discussed by Dailey[274] are relatively invariant among benzene protons and protons in general polycyclic benzenoid molecules.[271] The reader is referred to the several reviews for more detailed descriptions.

2. *Bond Anisotropies of Groups Other than Benzene Rings*

The early experimental recognition of the ring-current effect and its surprisingly reasonable classical description are one indication of the special nature of the magnetic anisotropy of a conjugated ring. Another is provided by a comparison of the susceptibility anisotropies of the aromatic and nonaromatic molecules as shown in this review and elsewhere;[257,275] i.e. the $\Delta\chi$'s of aromatics are significantly higher than those of the latter. This phenomenon has been described in terms of a susceptibility exaltation effect,[276] which is defined as the difference between the observed bulk susceptibility of an aromatic molecule and that predicted using bond or atom susceptibility additivity rules.

It was suggested previously that because of the dominant effect of the ring current, and also because of the comparative similarity of environments of aromatic protons, the local and bond effects did not require explicit descriptions. It has been demonstrated, however, that for almost all functional groups other than benzene rings, in order to get a reasonable account of shielding one requires a formulation which gives a better description of the overall features of the molecule.

The procedure is to establish a set of anisotropies for each of the bonds of the molecule. One of the problems encountered in using this procedure is that it is normally not possible to study the effects of individual bonds independent of the other bonds in the molecule. Thus, in many cases, particularly when utilizing observed NMR shifts, the tendency has been to

[275] See Tables XVII through XX.

[276] H. J. Dauben, J. D. Wilson, and J. L. Laity, *J. Amer. Chem. Soc.* **90**, 811 (1968).

ascribe the major effect to the most important bond, e.g. C≡C, and in effect neglect other bonds, e.g. C–H. Because of this and other difficulties such as the inadequacy of the point-dipole approximation and lack of assessment of bond polarity and symmetry, a large number of conflicting values for bond and group anisotropies have been proposed. For example the C–C bond anisotropy has been estimated from 0 to 10×10^{-6} cm^3/mole. In addition there have been conflicting opinions about the reasonableness of paramagnetic (positive) contributions to bond susceptibility components.

It can be shown that the use of NMR-derived bond anisotropies often gives erroneous values for the molecular susceptibility anisotropy. Thus Zeil and Buchert[277] determined C≡C and C≡N bond anisotropies of $\sim -34 \times 10^{-6}$ cgs. These numbers are clearly incompatible with the observed anisotropies of C_2H_2 and HCN of -7.7 and -7, respectively, as they would require unreasonably large C–H bond anisotropies of $+14$ or $+28 \times 10^{-6}$ cm^3 mole^{-1}. Zurcher's[278] deduction of very small anisotropies for both the C–Cl and the C≡N bonds is in conflict with the observed anisotropy of Cl–C≡N of -10.8×10^{-6} cgs. Other estimations of bond anisotropies from NMR shifts are also unable to give correct molecular $\Delta\chi$'s.

One of the problems involved is that in an effort to eliminate effects other than anisotropy, many of these studies use fairly homogeneous series which do not provide sufficient variations in the parameter of interest to permit an accurate analysis. The great disparity found in the bond anisotropies of different groups attests to the uncertainty in the method. Other reviewers, including Bothner-By and Pople,[181] Ditchfield,[17] Raynes,[210] and Davies[206] have come to similar conclusions.

Recently attempts have been made to correct for the inadequacies of the above model. Probably the most general and extensive attempt to give a full description of the susceptibility anisotropy and other contributions to shielding has been that of ApSimon *et al.*[279–281] They modify the McConnell equation to include cases in which the distances between shielding groups and protons are small compared to the lengths of induced dipoles, an approach which has been criticized by Raynes.[210] Taking into

[277] W. Zeil and H. Buchert, *Z. Phys. Chem.* (*Frankfurt am Main*) **38**, 47 (1963).

[278] R. F. Zurcher, *Helv. Chim. Acta* **44**, 1755 (1961).

[279] J. W. ApSimon, W. G. Craig, P. V. Demarco, D. W. Mathieson, L. Saunders, and W. B. Whalley, *Tetrahedron* **23**, 2339, 2357, 2375 (1967); *Chem. Commun.* p. 359 (1966).

[280] J. W. ApSimon, W. G. Craig, P. V. Demarco, D. W. Mathieson, A. K. G. Nasser, L. Saunders, and W. B. Whalley, *Chem. Commun.* p. 754 (1966).

[281] J. W. ApSimon, P. V. Demarco, D. W. Mathieson, W. G. Craig, A. Harim, L. Saunders, and W. B. Whalley, *Tetrahedron* **26**, 119 (1970).

account bonds, angles, and distances, ApSimon *et al.* calculated bond anisotropies along with principal bond susceptibilities for C–C, C–H, C═C, and C═O bonds, each in a variety of environments. The change in shielding is then computed by utilizing all the bonds which must be broken and created to form the new molecule.

Another attempted rigorous approach to anisotropy is that of Raynes.[210] He divides the "nonlocal" proton shielding term into effects due to "through-bond," e.g. inductive, conjugative, and "through-space," e.g. magnetic anisotropy, electric field effects. He shows that the total neighbor-anisotropy effect is not sufficient to account for the observed relative shifts among protons in methyl-substituted acetylenes. In this study he used a set of standard bond anisotropies obtained from other sources. The results support the contention that magnetic anisotropies derived from NMR shielding constants are not reliable. A more definitive evaluation of the merits of the ApSimon[279–281] and Raynes[210] approaches has been hampered by the lack of sufficient reliable data.

There are several important unanswered questions to be resolved. Probably the first step is to determine whether the anisotropies of bonds are sufficiently constant and transferable among molecules to provide new, useful, and reliable information about unknown systems. Ditchfield[17] has questioned this assumption and notes that theoretical studies of the electronic structure of ground states indicate that substituents can have significant effects upon adjacent bonds, particularly multiple bonds. However, note that in spite of these factors, the early theoretical work of Guy and Tillieu[282] was remarkably successful in predicting total susceptibilities using a variational wavefunction to calculate bond susceptibilities, although the anisotropies were significantly in error.

The ultimate determination of the feasibility of bond anisotropies can be made by establishing absolute principal bond susceptibilities and anisotropies for certain prototype molecules such as HC≡CH, $H_2C{=}CH_2$, H_3CCH_3, and H_2CO. These quantities could be compared directly with the semiempirical ones to provide an assessment of the models. The extension of the procedure to large systems is straightforward.

Very recently, Flygare and co-workers[283] have been able to devise sets of empirical bond and atom susceptibilities. The success of these schemes in reproducing experimental principal molecular susceptibilities apparently indicates that the concept of localized susceptibility contributions is valid for the tensorial components of a wide range of non-strained, non-aromatic compounds.

[282] J. Guy and J. Tillieu, *J. Chem. Phys.* **24**, 1117 (1956).
[283] T. G. Schmalz, C. L. Norris, and W. H. Flygare (personal communication) (1973).

D. Conclusions

There is now available a good variety of methods for the determination of magnetic shielding and susceptibility tensors and the extra information they provide. The *ab initio* methods of theoretical calculations have been increasingly successful for the complete shielding tensors of small molecules. In conjunction with shielding anisotropies of linear molecules, they have also helped to establish reference points for absolute shielding scales, which in turn have provided the absolute isotropic shifts for other less symmetric molecules. The availability of better *ab initio* values for diamagnetic shielding tensors has helped make possible the development of reliable empirical methods for predicting these quantities for a wider range of molecules. The well-known method for deducing paramagnetic shielding tensors from spin-rotation constants has aided and been aided by this latter development of the availability of σ^{d}. Spin-rotation constants have become a significant source of absolute shielding constants and anisotropies for the gas phase. This and other experimental methods for determining shielding anisotropies have provided checks upon one another, and in some cases have helped to establish spin-lattice relaxation mechanisms. An important new source for the complete principal shielding tensor of some normally hard-to-study nuclei is the recently developed cross-polarization technique.

The numerous experimental methods for determining susceptibility anisotropies have been compared, and in most cases the agreement of results is quite satisfactory. The highly productive method of microwave spectroscopy has yielded very precise and detailed descriptions of the susceptibility tensors. It has promoted much discussion about similarities of bonds and structures and has also been a primary source for quadrupole and second moments. The more easily obtained molecular g-values have been combined with accurate theoretical and empirical second moments to provide another source of susceptibility tensors. The magnetic birefringence technique applied to gases as well as liquids, and the molecular-beam methods are seen to be important complements to the high resolution microwave spectroscopic method which is limited by the requirement that a molecule possess a permanent dipole moment.

The relationship between the magnetic shielding and susceptibility tensors was considered in some detail. It was shown that for many small systems there is some correlation between the magnitudes of their respective anisotropies. However, it was concluded that the occurrence of a $1/r^3$ factor in the shielding results in this property being a predominantly local quantity and only slightly influenced by distant electrons and bonds. The susceptibility, on the other hand, is a property of the entire molecule and increases in magnitude roughly with the size of the molecule. This major difference was assumed to be responsible for the relatively greater success of *ab initio* methods for shielding and also for the remarkable success of the empirical additivity schemes for molecular susceptibilities.

The use of sets of bond susceptibilities and anisotropies, which has been criticized on theoretical grounds, has recently had some experimental corroboration. Proton NMR shifts have been shown to be an unsuitable source of data for determining bond anisotropies.

Acknowledgments

The authors thank Professor Robert Ditchfield for making available unpublished calculations and Professors Ditchfield and C. W. Kern for several helpful discussions. The National Institute of Health, Institutes of General Medicine is acknowledged for its partial support of this work.

Subject Index

A 4
B 5
C 6
D 7
E 8
F 9
G 0
H 1
I 2
J 3